IET CONTROL, ROBOTICS AND SENSORS SERIES 06

Motion-Induced Eddy Current Techniques for Non-Destructive Testing and Evaluation

The IET International Book Series on Sensors

IET International Book Series on Sensors—Call for Authors

The use of sensors has increased dramatically in all industries. They are fundamental in a wide range of applications from communication to monitoring, remote operation, process control, precision and safety, and robotics and automation. These developments have brought new challenges such as demands for robustness and reliability in networks, security in the communications interface, and close management of energy consumption. This Book Series covers the research and applications of sensor technologies in the fields of ICTs, security, tracking, detection, monitoring, control and automation, robotics, machine learning, smart technologies, production and manufacturing, photonics, environment, energy, and transport.

Proposals for coherently integrated international multi-authored edited or co-authored handbooks and research monographs will be considered for this Book Series. Each proposal will be reviewed by the IET Book Series Editorial Board members with additional external reviews from independent reviewers. Please email your book proposal to: vmoliere@theiet.org or author_support@theiet.org.

Motion-Induced Eddy Current Techniques for Non-Destructive Testing and Evaluation

Hartmut Brauer, Marek Ziolkowski, Konstantin Weise,
Matthias Carlstedt, Robert P. Uhlig, Mladen Zec

The Institution of Engineering and Technology

Published by The Institution of Engineering and Technology, London, United Kingdom

The Institution of Engineering and Technology is registered as a Charity in England & Wales (no. 211014) and Scotland (no. SC038698).

First published 2018

The Institution of Engineering and Technology
Michael Faraday House
Six Hills Way, Stevenage
Herts SG1 2AY, United Kingdom

www.theiet.org

British Library Cataloguing in Publication Data
A catalogue record for this product is available from the British Library

ISBN 978-1-78561-215-2 (hardback)
ISBN 978-1-78561-216-9 (PDF)

Typeset in India by MPS Limited

Contents

Author Biographies

Dr. Hartmut Brauer was born near Berlin, Germany, in 1953. He received the Diploma and Dr.-Ing. degrees in 1975 and 1982, respectively, both in electrical engineering from the Technische Universität Ilmenau, Germany. In June 2014, he was awarded with the honorary doctorate (Dr. h.c.) from the University of Niš, Serbia.

He has been with the Advanced Electromagnetics Group at the Technische Universität Ilmenau since 1975, where he is currently Senior Lecturer and Senior Researcher. He has authored or co-authored over 90 papers in reviewed journals and books. His current research interests include the theory and computation of electromagnetic fields and inverse field problems/optimization in electrical engineering, with applications to electromagnetic nondestructive evaluation and bioelectromagnetism. He is member of the International Compumag Society since 1994, and of the International Steering Committee of the International Workshops on Optimization and Inverse Problems in Electromagnetism since 2008.

Dr.-Ing. habil. Marek Ziolkowski is currently a scientific co-worker at the Advanced Electromagnetics Group, Technische Universität Ilmenau, Germany and assistant professor at the Applied Informatics Group, West Pomeranian University of Technology, Szczecin, Poland. He teaches electromagnetics and computational methods and algorithms. His current research interests include numerical and analytical modeling of electromagnetic fields, especially in relation to sensors, eddy currents, theoretical issues in computation, nondestructive testing of materials at low frequencies as well as in moving objects. He is the author or co-author of three monographs and over 200 scientific papers.

Konstantin Weise was born in Leipzig, Germany, in 1986. He received the B.Eng. degree in electrical engineering from the University of Applied Science Leipzig, Germany, in 2009, and the M.Sc. degree in electrical engineering from the Technische Universität Ilmenau, Germany, in 2012 where he also received his Dr.-Ing. degree in 2016. In his dissertation, he developed new numerical techniques to tackle advanced problems in Lorentz force eddy current testing. He is currently working toward his habilitation degree at the Max Planck Institute for Human Cognitive and Brain Sciences Leipzig, Germany, in the field of noninvasive brain stimulation using eddy currents.

Matthias Carlstedt was born in Nordhausen, Germany, in 1984. He received B.Sc. and M.Sc. degrees in mechanical engineering from the Technische Universität Ilmenau, Ilmenau, Germany, in 2010 and 2012, respectively. In 2017, he received the Ph.D. degree from the Faculty of Mechanical Engineering of the Technische Universität Ilmenau for his work at the Advanced Electromagnetics Group. Since 2016, he is the chief engineer at the Kompass GmbH in Ilmenau. His main research interests include embedded systems for industrial and environmental monitoring systems, time series analysis, model building and evaluation. In addition to his work at Kompass GmbH, he supports senior high school students from several Ilmenau grammar schools in their own research projects.

Robert Peter Uhlig was born in Berlin, Germany, in 1983. He obtained the academic degrees graduate engineer (Dipl.-Ing.) in mechatronics and Doctor of Engineering (Dr.-Ing.) in mechanical engineering from the Ilmenau University of Technology, Germany, in 2009 and 2013, respectively. From 2009 to 2013, he had been with the Advanced Electromagnetics Group, Ilmenau University of Technology, Germany. From 2013 to 2016, he had worked as a development engineer of new technologies in nondestructive testing at Institut Dr. Foerster GmbH & Co. KG in Reutlingen, Germany. Since 2016, he has been working as a sensor system architect at Robert Bosch GmbH in Reutlingen, Germany. His main interests include experimental validation, control technology, signal analysis, and model-based systems engineering (MBSE).

Mladen Zec was born in Bihac, Bosnia and Herzegovina, in 1983. He obtained the academic degree graduate engineer (Dipl.-Ing.) in electrical engineering at the Faculty of Electrical Engineering Banja Luka, Bosnia and Herzegovina in 2008. In 2013, he received his Doctor of Engineering (Dr.-Ing.) in electrical engineering from Ilmenau University of Technology, Germany. From 2008 to 2010, he was a teaching assistant at the Power Engineering Department at the Faculty of Electrical Engineering Banja Luka, Bosnia and Herzegovina. From 2010 until 2013, he was a part of Advanced Electromagnetics Group at Ilmenau University of Technology, Germany. Currently, he works as a product technology expert for electrical drives at Bühler Motor GmbH in Nürnberg, Germany. His main interests include finite element modeling, simulation and control of electrical machines. Furthermore, he is interested in programming, signal and numerical analysis.

Preface

Nondestructive testing (NDT) of materials and products is of great interest in a variety of modern engineering applications. Nowadays, the expectancy in performance of new devices is increasing. NDT in general enables the initial inspection of test samples to confirm the structural integrity of safety-relevant components without causing damage. In this sense, it provides quality control while being cost effective in the same way. The presence of NDT is hard to perceive in everyday life. However, NDT provides ground to identify and prevent failure of socially relevant parts of our life such as airplanes, railroads, and power plants. It is therefore essential to maintain a uniform quality level to avoid accidents and to ensure safety of human life. Besides that, it allows in service monitoring of test pieces before assembling. It also plays a major role in the framework of process control to prevent undesirable and dangerous operation of systems. Combining all this, manufacturers and other users are interested to apply methods, which are reliable and accurate while being cost effective.

The present book focuses on electromagnetic NDT methods and more specifically on motion-induced eddy current testing and evaluation (MIECTE). Traditional eddy current testing (ECT) methods make use of time-dependent magnetic fields to induce eddy currents in the object under test. Those are altered in the presence of physical irregularities such as flaws, cracks, or inclusions. In ECT, the variations are detected by measuring the magnetic flux through a pickup coil produced by the disturbed eddy currents in the specimen. In contrast to traditional eddy current methods, MIECTE makes use of relative motion between the object under test and a permanent magnet to induce eddy currents. The induced eddy currents interact with the applied magnetic field and result in a Lorentz force. Considering Newton's third law, this force acts on both, the specimen and the magnet itself, where it is measured. This quantity is used to evaluate the integrity of the structure under test.

It is the intent of this book to introduce the technology of MIECTE to those who are interested in this rather new approach of the conventionally applied ECT methods. There are several excellent reference books on various methods that can provide additionally more in-depth information, if desired. Thus, MIECTE differs from traditional ECT methods in the way how the eddy currents are induced and how signals are evaluated.

One may wonder why the title of this book contains (additionally) the word "evaluation". There can be found several definitions in the literature but most of them do not really apply. "Evaluate", has a definition that seems to be more fitting for the intent of this book: in our opinion, "evaluation" is indicating much more than "testing" because the aim is not only the detection of a defect or a conductivity anomaly. Finally, the goal is the identification of the defect/anomaly including the position and the depth, the shape, its characteristics, and perhaps the reconstruction of the defect from the measured signals. Thus, it seems to be reasonable why this book has been entitled with "MIECTE".

Beside the increasing use of plastics, most of materials being used in modern lightweight constructions are electrically conductive. Whether carbon-fiber reinforced structural components or fiber-metal laminates, like in modern aircrafts, these materials are often investigated using electromagnetic testing methods. In many applications, manufacturing errors and material aging occur at the surface of components which are important for operation. Therefore, the use of surface testing methods, such as conventional ECT, is an indispensable tool for the evaluation of structural integrity. However, in many cases, the volumetric examination is indispensable to identify material failure, often derived from an existing defect, at an early stage.

The Lorentz force eddy current testing (LET) method, a rather new MIECTE method, was invented to complement established methods by overcoming the well-known detection limitations for subsurface defects. In fact, the MIECTE method was originally proposed by Hartmut Brauer and Marek Ziolkowski in 2008. Decent advantages are lying in the application of stationary magnetic fields, which potentially allow the detection of defects lying deep inside the object under test. The working principle of LET permits the inspection of moving parts as it can frequently be found in industrial settings. The fabrication process of aluminium for example takes place at velocities of up to 20–30 m/s. Even higher velocities can be observed in wire drawing processes where speeds in the range of 40–60 m/s are applicable. This necessitates a NDT method like LET which is capable to test moving objects.

In the past more than 10–12 years, the authors have been conducting basic research, by means of a considerable funding from the German Research Foundation. This support ensures the continuous improvement of the related technologies in theory and academic practice, i.e. it allowed to perform basic research and experimental studies in university labs only. Consequently, much of the content of this book comes from the doctoral and master's theses guided by the authors. These former doctoral students include Robert P. Uhlig, Mladen Zec, Bojana Petkovic, Matthias Carlstedt, Konstantin Weise, Judith Mengelkamp, Jan Marc Otterbach, Reinhard Schmidt, and others. The authors thank all those people sincerely.

Furthermore, the authors particularly acknowledge for the support by the German Research Foundation (DFG) with grants in the Research Training Group GRK 1567 "Lorentz Force," by the Federal Ministry of Economics (BMWi) of Germany with grant ZIM 16KN020332 "FiVe-Net - LETRA," and by the German Federal Ministry of Education and Research (BMBF) with grant VIP 03V0590 "LOFOTEST".

This book also relates the research results to the relevant counterparts in the field of ECT. With the growing demands for NDT, research in electromagnetic NDT is drawing again more attention to both theoretical aspects and industrial applications more intensively. The few practical applications, presented and briefly illustrated in Chapter 6, should indicate that there is a high potential for industrial applications.

The authors hope that this book will not only provide references for the research and development staff or for students and teachers at universities, but can also be of interest for engineers or technicians, which would like to solve problems of NDT and evaluation in industrial environment. We highly appreciate your feedback.

Hartmut Brauer
Ilmenau, February 2018

Chapter 1
Introduction

Hartmut Brauer[1]

What is Nondestructive Testing (NDT)?

NDT is a wide group of analysis techniques used in science and industry to evaluate the properties of a material, component or system without causing damage. Because NDT does not permanently alter the article being inspected, it is a highly valuable technique that can save both money and time in product evaluation, troubleshooting, and research.

The field of NDT is a very broad, interdisciplinary field that plays a critical role in assuring that structural components and systems perform their function in a reliable and cost-effective fashion. NDT technicians and engineers define and implement tests that locate and characterize material conditions and flaws that might otherwise cause planes to crash, reactors to fail, trains to derail, pipelines to burst, and a variety of less visible, but equally troubling events. These tests are performed in a manner that does not affect the future usefulness of the object or material. In other words, NDT allows parts and material to be inspected and measured without damaging them. Because it allows inspection without interfering with a product's final use, NDT provides an excellent balance between quality control and cost-effectiveness. In general, NDT includes their application to the industrial inspection. The technologies that are used in NDT, are very similar to those used in the medical industry, but nonliving objects are the subjects of the inspections.

NDT methods are all those evaluation methods by which the integrity of different components or assembled pieces of equipment is being examined nondestructively. The examination can be performed directly after manufacturing, during acceptance testing or on-line as a tool for preventive maintenance as well as for the location of damages, the analysis and the aftercare of damages. The diagnostic methods utilize physical phenomena to monitor the health of materials or devices and to make prognosis of the future use. This is more-and-more often done online without interrupting the industrial process.

The subject of NDT has no clearly defined boundaries; it ranges from simple techniques such as visual inspection of surfaces, through the well-established methods

[1] Advanced Electromagnetics Group,Technische Universität Ilmenau, Germany

of radiography, ultrasonic testing (UT), magnetic particle crack detection, to new and very specialized methods such as the measurement of Barkhausen noise. NDT methods can be adapted to automated production processes as well as to the inspection of localized problem areas.

- NDT is the process of inspecting, testing, or evaluating materials; components or assemblies for discontinuities; or differences in characteristics without destroying the serviceability of the part or system. In other words, when the inspection or test is completed the part can still be used.
- NDT is the application of measurement techniques in order to identify damage and irregularities in materials. NDT often provides the only method of obtaining information about the current "health" of the examined object.
- NDT is a measurement of a physical property or effect from which the presence of damage or irregularity can be inferred. It is not a measurement of an absolute parameter such as temperature or pressure. The distinction between what would be considered changes in material properties and what would be considered a defect is not distinct. This can lead to NDT missing defects and also producing false calls, i.e. a defect is reported when in fact the signal is not produced by a defect. Also, NDT is applied to a greater or lesser extent by human operators who introduce human error and subjectivity into the process.
- NDT is the branch of engineering concerned with all methods of detecting and evaluating flaws in materials. Flaws can affect the serviceability of the material or structure, so NDT is important in guaranteeing safe operation as well as in quality control and assessing plant life. The flaws may be cracks or inclusions in welds and castings, or variations in structural properties which can lead to loss of strength or failure in service.
- NDT is commonly used in forensic engineering, mechanical engineering, petroleum engineering, electrical engineering, civil engineering, systems engineering, aeronautical engineering, medicine, and art. Innovations in the field of NDT have had a profound impact on medical imaging, including on echocardiography, medical ultrasonography, and digital radiography.
- NDT is never 100% effective at detecting defects of concern. Like all measurements, defect positioning and sizing measurements with NDT techniques are subject to errors. As these techniques are often a combination of separate measurements, these errors can be significant.

If done well, NDT can provide useful information to assist in the management of plant safety. If inappropriate NDT is applied or NDT is not applied correctly, then the results are likely to give a false impression of the integrity and safety of the plant. In contrast to NDT, other tests are destructive in nature and hence they are applied to a limited number of samples, rather than on the materials, components, or assemblies actually being put into service. These destructive tests are often used to determine the physical properties of materials such as impact resistance, ductility, yield and ultimate tensile strength, fracture toughness, and fatigue strength, but discontinuities and differences in material characteristics are more effectively found by NDT.

NDT is a little-known, yet crucial industry. NDT quite simply is the backbone of the industrial society. NDT keeps the technology running smoothly, and the transportation running safely. Today modern nondestructive tests are used in manufacturing, fabrication and in-service inspections to ensure product integrity and reliability, to control manufacturing processes, lower production costs, and to maintain a uniform quality level. During construction, NDT is used to ensure the quality of materials and joining processes during the fabrication and erection phases, and in-service NDT inspections are used to ensure that the products in use continue to have the integrity necessary to ensure their usefulness and the safety of the public. It is also used for measurement of components and spacing and for the measurement of physical properties such as hardness and internal stress. The essential feature of NDT is that the test process itself produces no deleterious effects on the material or structure under test.

Today the increased competition in industry and the expectation of shorter return-of-invest periods for complex and expensive machinery, as well as occupational safety, health and environmental requirements, presuppose a high availability of the production machinery and a high and stable quality of the products. These goals are met only if the machinery is kept in proper working condition by utilizing a functioning maintenance philosophy and the right machine diagnostic methods for preventing machinery breakdowns and loss of profit.

What is Nondestructive Evaluation (NDE)?

A wide variety of test schemes exist, some destructive and some nondestructive. Strictly speaking NDT has no clearly defined boundaries. Usually, NDT is the development and application of technical methods to examine material of components in ways that do not impair future usefulness and serviceability in order to detect, locate, measure and evaluate discontinuities, and other imperfections; to assess integrity, properties, and composition; and to measure geometrical and physical characteristics. The terms NDT and nondestructive inspection (NDI) are taken to be interchangeable.

Along with further development and sophistication of NDT methods together with larger diversity of the techniques a rather new term came into use, the nondestructive evaluation (NDE). NDE comprises many terms used to describe various activities within the field. Some of these terms are NDT, NDI, and nondestructive examination (which is often called NDE as well, but should probably be called NDEx). These activities include testing, inspection, and examination, which are similar in that they primarily involve looking at (or through) or measuring something about an object to determine some characteristic of the object or to determine whether the object contains irregularities, discontinuities, or flaws. Consequently, nondestructive evaluation is a term that is often used interchangeably with NDT. However, technically, NDE is used to describe measurements that are more quantitative in nature. For example, an NDE method would not only locate a defect, but it would also be used to measure something about that defect such as its size, shape, and orientation. NDE may be used to determine material properties, such as fracture toughness, formability, and other physical characteristics [1,2].

The terms irregularity, discontinuity, and flaw can be used interchangeably to mean something that is questionable in the part or assembly, but specifications, codes, and local usage can result in different definitions for these terms. Because these terms all describe what is being sought through testing, inspection, or examination, the term NDE has come to include all the activities of NDT, NDI, and NDEx used to find, locate, size, or determine something about the object or flaws and allow the investigator to decide whether-or-not the object or flaws are acceptable. Thus, a flaw that has been evaluated as rejectable is usually termed a defect.

In NDT, in flaw detection applications, the end-product is taken to be a description of the flaws, which have been detected in terms of their nature, size, and location. From this, either in conjunction with a standard for acceptable/rejectable flaws, or a knowledge of, for example, fracture mechanics, a decision is made by the designer, but in practice may be left to the NDT personnel, or the NDT inspector.

In NDE, it is assumed that this acceptance/rejection of flaws is part of the NDT process. Thus, NDE includes much more than NDT. While during the NDT process the defects/anomalies are detected and localized, the goal of the NDE process is the defect identification, where shape and spatial extension of the defect are estimated. The final step of the NDE process will be the reconstruction of whole defects. On the other hand, these differences seem to be important only from a rather academic point of view. Usually, under the harsh conditions in an industrial testing environment, it is more-or-less impossible to realize this final step and determine the real size, shape, extension, and depth of the defect.

Brief history of NDT

Although history does not provide a precise starting date for NDT, its use dates back many years. It is said that flour and oil were used during Roman times to find cracks in marble slabs. For centuries, blacksmiths used sonic NDT when listening to the ring of different metals as they were being hammered into shape; a technique also used by early bell makers.

Before a historical review of the NDT methods is considered, one has to remember the definition of an NDT method, which is the utilization of a physical phenomenon for the noninvasive testing of a product or a material.

With this rough definition of NDT in mind, the oldest NDT method by far is visual testing (VT) which is as old as mankind starting most likely from the visual checking of knives for cutting meat and spears for hunting [3].

Acoustics would be the second oldest method because it has been used for testing since ancient times when man started to make the first pottery vessels. The earliest known pottery vessels may be those made by the people in China about 20,000 years ago [4] and acoustics was surely used much later on for the testing of glassware. The same technique was used in the Middle Ages when testing for instance brass castings such as a huge church bell. This was, however, testing with audible sound.

The third oldest classical method would be magnetic flux testing of gun barrels. According to Aristotle magnetism was first discovered by Thales of Miletus and was utilized for compasses during the Middle Ages. One of the first recorded uses of NDT was in 1868, when Englishman S.H. Saxby relied on the magnetic characteristics of a compass to find cracks in gun barrels. Here he utilized the remanence of the steel giving detectable leakage fields.

NDT/NDE has been practiced for decades. Over the years, technological advances spurred rapid developments in techniques and instrumentation. It is impossible to identify exactly when this science began; however, we know that it has been evolving for centuries. For example, blacksmiths used a sonic technique (listening to a ring of different metals) to shape them as desired.

The roots of modern NDT/NDE began prior to the 1920s, but awareness of different methods truly came in the 1920s. During this time, there was an awareness of some of the magnetic particle tests (MT) and radiography testing (RT), especially in the medical field.

MT was patented in the United States in 1922 when W.E. Hooke, working with precision gage blocks at the American Bureau of Standards, devised a method for magnetizing an object producing a leakage field and using iron powder to delineate cracks invisible to the naked eye. The real breakthrough for MT came, however, about a decade later after F.B. Duane and A.V. de Forest had started a partnership in 1929 that later on in 1934 became the Magnaflux Corporation.

Before the breakthrough of MT, the first classical NDT method was RT. NDT, as it is considered today, started with Professor Wilhelm Conrad Roentgen in 1896. After his discovery of the X-rays, he took a radiograph of four soldered pieces of zinc and one of his own hunting rifle. The radiograph of the rifle showed some cast defects in the material and was thus the start of industrial radiography. Professor Roentgen disguised the publicity around him and never made any attempts to claim a patent on his discovery.

In the early days of railroad, a technique referred to as the "oil and whiting test" was used and staged the ground for the present days penetrant test (PT). This was the second method to be patented. F.B. Duane was awarded a patent for the fluorescent penetrant method in 1948. PT was already used before MT. The method was used for testing the heavy cast parts of huge locomotives in the beginning of the 20th century. They applied used oil that had a dark pigment, i.e. contained dirt, and whiting was simply a water-based chalk-slurry that dried out to white film and worked as the developer [3].

The basis for ultrasonic testing (UT) was established in 1940 when F.A. Firestone achieved a patent for his invention concerning a flaw detection device. Then, in 1942, Firestone was the first to use his method for the sonar. In Germany, two physicists and brothers Herbert and Josef Krautkrämer, who had studied works by Firestone, made a bet of being able to tell if a cannonball, too thick to be radiographed with existing equipment, would have a casting flaw inside or not. They used ultrasound transmission for the bet and finally won it. They founded a company that was to become the biggest UT equipment manufacturer ever. These two German brothers did a lot of research on the method and greatly contributed to the development of the UT method [5]. Since their time the method has gone through several phases of development and made enormous achievements in many countries and still has great potential for further applications.

The sixth classical method is eddy current testing (ECT). This method was being developed primarily in the United States at the beginning of the 20th century [6]. There were some useful applications like the equipment for sorting materials, but a working theory of the method was lacking. Then, it was the German F. Foerster who in the 1950s clarified the theory for ECT and devised the necessary formulae [7,8]. Today production testing of austenitic tubes and in-service testing of heat-exchanger tubes are well-known applications of ECT.

The breakthrough for the use of NDT methods took place during the Second World War starting from the testing of submarines and airplanes. During the last 50 years, the use has then incorporated the inspection of nuclear power plant components, pressure vessels, bridges, elevators, and car parts, which if measured in numbers are the biggest user today.

Methods and techniques

NDT is usually classified into various methods, each based on a particular scientific principle. These methods may be further subdivided into various techniques. The various methods and techniques, due to their particular natures, may lend themselves especially well to certain applications and be of less or no value at all in other applications. Therefore, choosing the right method and technique is an important part of the performance of NDT [1,6,9]. Test method names often refer to the type of penetrating medium or the equipment used to perform that test. Current NDT methods are acoustic emission testing, electromagnetic testing (ET), guided wave testing (GWT), ground penetrating radar, laser testing methods, leak testing (LT), magnetic flux leakage (MFL), microwave testing, PT, MT, neutron radiographic testing, RT, infrared testing, UT, vibration analysis, and VT. Figure 1.1 shows an overview of available methods and their classification with respect to the corresponding frequency range for appropriate applications.

The six most frequently used NDT methods are VT, MT, PT, RT, UT, and ECT:

VT: The most commonly applied NDT method is quite often enhanced by the use of magnification, borescopes, cameras, or other optical arrangements for direct or remote viewing.

MT: Another commonly used NDT method used particularly on ferrous materials involves the application of fine iron particles (either suspended in liquid or dry powder–fluorescent or colored) that are applied to a part while it is magnetized, either continually or residually. The particles will be attracted to leakage fields of magnetism on or in the test object, and form indications (particle collection) on the objects surface, which are evaluated visually. This is a great technique for use on welds, or other areas cracking may be of concern. MT uses high contrast paint, in conjunction with magnetic particles in order to look for very fine cracking and other defects.

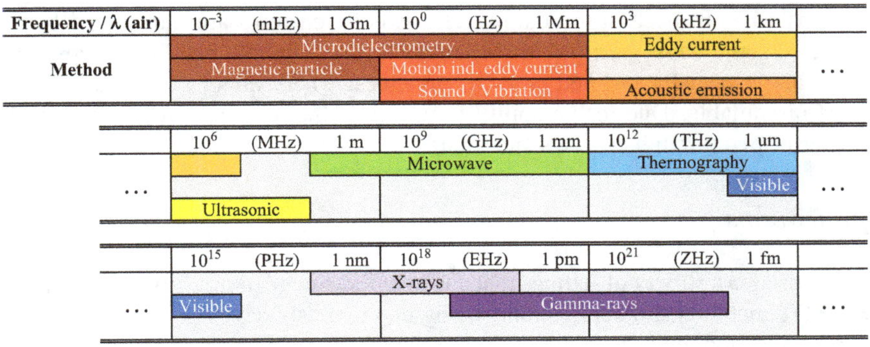

Figure 1.1 Overview of NDT methods

PT: Penetrant testing is a technique to look for cracking and other surface defects. Contrast and probability of detection for a visual examination by the unaided eye is often enhanced by using liquids to penetrate the test article surface, allowing for visualization of flaws or other surface conditions. This method (PT) involves using dyes, fluorescent or colored (typically red), suspended in fluids and is used for nonmagnetic materials, usually metals. Allowing dye to seep into cracks and other surface breaking defects, and then washing and highlighting the area using a separate paint, cracks and other defects will hold the dye in, allowing them to be plainly seen. Ultraviolet dyes can be used as well for extra visibility.

UT: In the case of ultrasonic testing, another volumetric NDT method, sound waves are utilized, where the mechanical signal (sound) being reflected by conditions in the test article and evaluated for amplitude and distance from the search unit (transducer). UT is a common technique employed, providing reliable information, with great precision and accuracy. Ultrasound can be used to check the thickness of an asset with ease, and giving remaining wall thickness. This is a very common technique, often utilized because of its ability to determine corrosion rates, provide accurate data and the speed of inspection.

RT: The internal structure of a sample can be examined for a volumetric inspection with penetrating radiation, such as X-rays, neutrons, or gamma radiation. Radiography is used in a wide range of applications including medicine, engineering, forensics, security, etc. In NDT, radiography is one of the most important and widely used methods. RT offers a number of advantages over other NDT methods. However, one of its major disadvantages is the health risk associated with the radiation. In general, RT is method of inspecting materials for hidden flaws by using the ability of short wavelength electromagnetic radiation to penetrate various materials. The intensity of the radiation that penetrates and passes through the material is either captured by a radiation sensitive film or by a planer array of radiation sensitive sensors. The film radiography is the oldest approach, and it is still most widely used in NDT.

ECT: It is an electromagnetic technique that uses electromagnetic coils to generate eddy currents and a secondary magnetic field within that asset. This allows subsurface inspection to take place. The eddy currents are interrupted by the defects and cause permeability changes in the original coil. ECT is great for welds, especially in lined tanks or pipelines, where direct access to the weld is unavailable. Due to the ability to penetrate the material, inspection can be performed through coatings and linings.

Applications

NDT is used in a variety of settings that covers a wide range of industrial activity, with new NDT methods and applications, being continuously developed. NDT methods are routinely applied in industries where a failure of a component would cause significant hazard or economic loss, such as in transportation, pressure vessels, building structures, piping, and hoisting equipment.

NDT methods rely upon use of electromagnetic radiation, sound and other signal conversions to examine a wide variety of articles (metallic and nonmetallic, food product, artifacts and antiquities, infrastructure) for integrity, composition, or condition with no alteration of the article undergoing examination.

The introduction and application of NDT in industry is grossly misrepresented and misunderstood. It is often said that introduction of this expensive technology does not give any tangible returns or at least does not give returns proportional to the investment made. The facts, however, are exactly opposite to this notion and thinking. When appropriately applied, NDT gives tremendous returns by way of savings in scrap, by lowering the ultimate rates of rejection, saving valuable manufacturing time, increasing the overall quality and reliability of manufactured goods, providing an extension of plant life through preventive maintenance, saving unnecessary shutdowns, particularly through in-service inspection, and enhancement of a particular industry's reputation and consequent increased sales and profits. Therefore, even from a purely commercial viewpoint, NDT is of utmost importance for an industrial concern. The additional considerations of NDTs role in safety, failure, and consequent accident prevention leave no doubt at all about the value and need of NDT.

Summary

As new materials have been developed and products manufactured that demanded higher quality levels, along with the imposition of stricter standards, the challenges to NDT are great. The NDT equipment available today should send the message that "we've come a long way". Those involved with NDT today should appreciate the significant improvements to this technology made possible through the innovative equipment at our disposal. Compare today's equipment with that of the early days and you cannot help but realize how much better off we are today.

Those early practitioners had to deal with significant challenges, including a general absence of standards, archaic equipment with limited reliability, little or no formalized training programs, and an overall lack of acceptance of NDT as a technology. To many in those early days, NDT was considered a necessary evil. While NDT has grown to become a recognized and sophisticated technology, some concerns remain. NDT is still generally unknown or misunderstood by many. The many different personnel qualification and certification programs can be confusing and complicated. It will be interesting to see if the next period in the history of NDT will be considered the dark ages or a period of enlightenment.

Today there can be found many new challenges in engineering to be overcome by NDT/NDE. The development of new materials, such as fiber reinforced composites, laminated structures, soft magnetic composites (SMC), meta-materials, or micro- and nanostructures, cause a lot of problems concerning NDT. As a reasonable consequence of this development, the established NDT methods and techniques have to be enhanced and refined further. On the other hand, faced with the new materials there is also a need to develop new NDT methods. This will lead to a large market where in particular NDE strategies can play an important role, maybe much more important than today.

1.1 Electromagnetic testing

ET technologies involve test methods that use magnetism and electricity to detect or measure fractures, faults, corrosion, or other damage in conductive materials. There is a number of electrical methods, which can be used for NDT, such as resistance measurement, electrical conductivity measurement, the use of triboelectric, thermo electric, or exoelectron effects. Eddy current, penetrating radar, and other electromagnetic techniques are used to detect or measure flaws, bond or weld integrity, thickness, electrical conductivity, detect the presence of rebar or metals, whereas the major group is ECT.

All electrical methods are working indirectly, i.e. a material property is measured as an electrical property variation. The methods are applicable to all electrically conducting materials. Furthermore, analyses of specific magnetic properties and component geometries are used to identify the optimum ET method. Electromagnetic effects resulting from the interaction of electricity and magnetism, form the basis of a number of NDT methods, including ECT, magnetic particle inspection, magnetic flux leakage testing, alternating current field measurement, or others.

Although there are numerous ET methods, most often used is the ECT. It is a no contact method for the inspection of metallic parts. Eddy currents are the result of alternating electromagnetic fields that are created when an alternating electric current is passed through one or more coils in a probe assembly. When the probe is linked with the part under inspection, the alternating magnetic field induces eddy currents in the test part. Material defects cause interruptions in the flow of the eddy currents which alert the inspector to the presence of a defect or other change in the material. Discontinuities or property variations in the test part change the flow of the eddy current and are detected by the probe.

Eddy currents do not depend on the frequency only, they are also affected by the electrical conductivity and magnetic permeability of materials. Therefore, eddy current measurements can be used to sort materials and to tell if a material has seen high temperatures or been heat treated, which changes the conductivity of some materials. Thus, ECT allows material thickness measurements or searching for defects such as cracks and corrosion.

Electromagnetic induction tests are applied to all stages of metal and alloy fabrication and processing. Over the years, probe technology and data processing have advanced to the point where ECT is recognized as being fast, simple, and accurate. The technology is now widely used in the aerospace, automotive, petrochemical, and power generation industries for the detection of surface or near-surface defects in materials such as aluminum, stainless steel, copper, titanium, brass, Inconel, and even carbon steel (surface defects only).

1.1.1 Brief historical review

The term eddy current comes from analogous currents seen in water in fluid dynamics, causing localized areas of turbulence known as eddies giving rise to persistent vortices. In fluid dynamics, an eddy is the swirling of a fluid and the reverse current created when the fluid is in a turbulent flow regime. Fluid behind an obstacle flows into the void creating a swirl of fluid on each edge of the obstacle, followed by a short reverse flow of fluid behind the obstacle flowing upstream, toward the back of the obstacle.

The first person to observe eddy currents was Francois Arago (1786–1853), a French politician, who was also a mathematician, physicist, and astronomer. In 1824, he observed what has been called rotatory magnetism, and that most conductive bodies could be magnetized; these discoveries were later completed and explained by Michael Faraday (1791–1867).

In 1834, Heinrich Lenz (1804–1865) formulated the principle that the properties of the test object react on the test system. The Lenz law describes that the current flow in the test object is directed in a way that the magnetic field caused by this current is counteracting the primary magnetic field. Thus, eddy currents cause a secondary magnetic flux in the test coil which is compensating that part of the flux in the coil that is equivalent to magnitude and phase of the flux caused by the eddy currents.

French physicist Léon Foucault (1819–1868) is credited with having discovered eddy currents, and for this reason eddy currents are sometimes called Foucault currents. In 1855, he discovered that the force required for the rotation of a copper disc becomes greater when it is made to rotate with its rim between the poles of a magnet, the disc at the same time becoming heated by the eddy current induced in the metal.

ECT technique has its origin with the English scientist Michael Faraday and his discovery of electromagnetic induction in the beginning of 19th century. Faraday was a chemist in England during the early 1800s and is credited with the discovery of electromagnetic induction, electromagnetic rotations, the magneto-optical effect, diamagnetism, and other phenomena.

Michael Faraday invented in 1831 that the penetration of a time-variant magnetic field into a conductor as well as the movement of a conductor in a magnetic field cause a current that is distributed over the entire conductor (electromagnetic induction law).

This finding has been the prerequisite for the development of a today widespread ET technique, the ECT. Faraday continued the work of Léon Foucault and found that due to the relative movement of a conductor and a magnetic field, a voltage is induced in the conductor, causing a current that nowadays is called "eddy current". This means, that exerting of an alternating magnetic field of a coil leads to an induced voltage in a conducting specimen, driving a current flow in the test object.

Thus, the electromagnetic induction is the working principle of the ECT, which can be applied to NDT of conducting materials only. Additionally, it has to be noticed that the eddy currents have been invented once more but this time rather theoretically. This is due to James Clerk Maxwell (1831–1879) who has formulated in 1864 the equations defining the theory of electromagnetic fields.

In 1879, another breakthrough was reached when the English scientist, David Hughes (1831–1900), demonstrated how the properties of a coil change when placed in contact with metals of different conductivity and permeability. Thus, the first use of eddy current for NDT has been done by David Hughes when he used the principles to conduct metallurgical tests for ore sorting.

However, it was not until the Second World War that these developments in the transmitting and receiving of electromagnetic waves were put to practical use for materials testing.

The development of ECT was growing up significantly many decades later. Beginning of 1933, the German physicist Friedrich Foerster (1908–1999) adapted ECT to industrial use, developing instruments for measuring conductivity and for sorting mixed-up ferrous components. In 1948, Foerster founded his own company in Reutlingen, Germany, a business based on ECT that continues to this day. In the late 1960s, after Dr. Friedrich Foerster has founded the Institute Dr. Foerster in 1948, he developed several ECT devices for industrial applications. The next important milestone was the introduction of the multi-frequency technique by a French manufacturer in 1974. Later several special techniques have been developed (e.g. magnetic flux leakage, remote ECT, and modulation analysis inspection) leading to a remarkable extension of the spectrum of practical applications. Other companies soon followed, especially in the aircraft and nuclear industries.

1.1.2 Electromagnetic NDT methods

ET is a general test category that includes ECT, alternating current field measurement (ACFM), and remote field testing. Although MT is also an electromagnetic testing method but, due to its widespread use it is treated more as an independent test method than as an ET technique. All these techniques use the induction of an electric current or magnetic field into a conductive part, then the resulting effects are recorded and evaluated. The group of ET methods included the following:

1.1.2.1 Eddy current testing

Eddy current is the most widely applied electromagnetic NDT technique. Eddy current testing uses an electromagnet to induce an eddy current in a conductive sample. The response of the material to the induced current is sensed. Discontinuities appear as variations in the eddy currents produced, which are indicated by a signal in the testing device.

The ECT method is frequently used to detect flaws in pipes and tubings by inducing a strong external magnetic field around the subject material. ECT is further beneficial in detecting discrepancies in nonferrous materials such as heat exchanger tubes, condensers, boilers, and tubings. Since the probe does not have to contact the work surface, ECT is useful on rough surfaces or surfaces with wet films or coatings. The eddy current method is also useful in sorting alloys and verifying heat treatment. It is used on tubing, wire, bearings, rails, nonmetallic coatings, aircraft components, turbine blades and disks, automatic transmission shafts, and many others. Due to the low penetration depths (typically <5 mm), only surface and slightly subsurface flaws can be detected. Although it is a noncontacting technique, ECT requires customized probes and closed proximity of probe and specimen.

1.1.2.2 Pulsed eddy current testing

As one kind of ECT technology, the pulsed eddy current testing (PECT) technology is based on the principle of electromagnetic induction and is used to detect the defects in conductive materials. The principle of PECT is basically the same as that of traditional ECT, and the differences are the means of excitation and the signal analysis method.

Unlike ultrasonic thickness measurement it measures average wall loss over an area (footprint). A transmitter coil produces a magnetic pulse which induces eddy currents within the component wall. The eddy currents in turn produce a second magnetic pulse, which is detected by the receiving coil. The system monitors the rate of decay of the eddy current pulse within the steel wall. The average thickness is derived from the comparison of the transient time of certain signal features with signals from known calibration pieces. It is important that the operator is given information regarding the component to allow the NDT equipment to be set up correctly and the results to be accurately interpreted. This technique is quick to apply, can test through nonconductive and nonmagnetic material (passive fire protection, concrete) up to 100 mm thick. It is only suitable for low alloy steels and is unable to differentiate defects on the top and bottom surfaces. Thus, PECT is a technique preferable for detecting corrosion and erosion and measuring average remaining wall.

1.1.2.3 Remote field eddy current testing

This technique provides an alternative to eddy current NDT for ferromagnetic tube inspection due to the presence of a strong skin effect found in such tubes. The technique monitors the magnetic field produced by induced eddy currents at some distance from the exciting coil. Compared to standard eddy current techniques, remote field testing provides better results throughout the thickness of the tube, having approximately equal sensitivity at both the inside and outside surfaces of the tube. The system gives poorer resolution and has a lower test speed than a high frequency eddy current test. The technique is highly sensitive to gradual wall thinning but detection of localized thinning requires special probes and electronic control. For nonferromagnetic tubes, eddy current tends to provide more sensitivity.

1.1.2.4 Magnetic particle testing

Defects on the inspection surface interrupt the lines of magnetic flux. Magnetic particles sprayed onto the surface are attracted to these defects identifying their position. MT is used to discover surface and subsurface discontinuities, such as cracks and seams in ferromagnetic materials, where the position, size, shape, and scope of imperfections is estimated. This method only detects abrupt changes in the magnetic field and therefore only supplies capability for defects that break the inspection surface. However, because the sensitivity of MT decreases significantly a short depth under the surface being examined, it is principally used to catalogue surface discontinuities. A magnetic field is induced into the examined component and iron particles are applied to the surface as a dry powder or in medium such as water, solvent, or light oil. A variety of particle colors are applied to get the best contrast with the material being examined. Fluorescent magnetic inks are used to increase the contrast of indications making them more visible to the operator and hence increasing the sensitivity of the technique. Discrepancies are highlighted as the particles collect and form around the magnetic flux leakage that surrounds such defects. Welds, castings, forgings, valves, machined parts, pressure vessels, and structural steel are typical candidates for MT inspection. Care needs to be taken to avoid false calls which may arise due to changes in geometry or the presence of residual magnetic fields. Magnetic particle inspection is generally the preferred NDT method for the detection of surface or slightly subsurface flaws in ferritic material. The detection of flaws is limited by the field strength and the direction of the magnetization field with respect to the flaw extension. It needs clean and relatively smooth surfaces and the depth of the flaws can usually not be indicated. Furthermore, the test piece which can be difficult for some shapes and magnetizations.

1.1.2.5 Liquid or dye penetrant testing

PT is used to detect discontinuities that are open to the surface in both ferrous and nonferrous test materials. Dye is drawn into any surface breaking defects which are then highlighted by the application of a developer which draws the dye back out of the defect. This NDT method can only detect defects which are open to the inspection surface. Dye penetrant is the preferred surface technique for nonmagnetic materials. PT can distinguish surface discontinuities such as cracks, seams, laps, cold shuts, laminations, and porosity. Dye penetrant is better suited to the detection of volumetric defects like pits but is more susceptible to the surface condition than magnetic particle inspection. Detection of tight cracks will require the dye to be left on the surface for a long time. Red dyes visible in ordinary light and fluorescent dyes visible under ultraviolet light are used along with a developer. The developer is a thin powder coating, normally applied by an aerosol, which draws the penetrant out of any discontinuities while providing a contrasting background to increase the discernibility of discontinuities. The component surface needs to be cleaned prior to the application of dye penetrant inspection. Mechanical cleaning methods can lead to crack openings being closed, subsequently preventing detection. Fluorescent dyes are used to increase the contrast of indications making them more visible to the operator and hence increasing the sensitivity of the technique. Typical candidates for PT include welds and weld overlays, castings, forgings, valves, machined parts, pressure vessels, vessel linings, and structural steel, but it can be applied to tubing, brazing, billets, turbine blades and disks, gears, and in particular to aluminum parts as well. It is often used on nonferromagnetic materials for which techniques, such as magnetic particle inspection, are not appropriate. But it needs access to the test surface, the defects must be surface breaking and decontamination or precleaning of the test surface may be needed. Furthermore, PT requires nonporous material surfaces and very tight and shallow defects are difficult to find, the flaw depth is not indicated.

1.1.2.6 Magnetic flux leakage

This technique relies on the detection of the magnetic flux, which is "squeezed" out of the metal wall under test by any decrease in the wall thickness. The amplitude of the signal obtained from any wall loss is proportional to the volume that is missing from the region interrogated. This means that the amplitude does not necessarily correspond to the decrease in thickness of the wall. The technique is not able to discriminate between material loss on the near surface and material loss on the far surface. Surface roughness, surface corrosion, distortion, build-up of debris on the magnets, and any physical disturbance of the scanning system as it moves across the component will adversely affect the results. MFL is a qualitative technique and is unable to give an accurate assessment of the remaining wall. It has found wide use in the NDT of tank floors because it is quick to apply and can detect material loss on both surfaces of the floor. The requirement for the sensor to be placed between the poles of a magnet mean that the technique is unable to give 100% coverage of a floor up to vertical obstructions and side walls. The wall thickness that can be inspected by magnetic flux leakage is limited by the requirement to achieve magnetic saturation. The high level of set-up effort makes the technique susceptible to human error. Procedures need to be clear and sufficiently detailed and operators need to be qualified and experienced in the application of the technique.

1.1.3 Capabilities of electromagnetic techniques

1.1.3.1 Thickness measurement

The commonest damage found on process plant is corrosion and so techniques which allow remaining wall thickness to be measured are widely applied. Ultrasound (high-frequency sound) provides an accurate point measurement of wall thickness. The surface on which the transducer is placed needs to be clean and, as it provides a point measurement, the measurement positions need to be selected with consideration of the type of corrosion damage so that the minimum wall thickness can be detected. When using a grid to survey a large surface area, the pitch of the grid needs to be selected so that it will detect the damage of concern. Care needs to be taken when taking measurements on plant which is painted or coated to ensure that the measurement is just that of the remaining wall. Newer instruments have facilities to assist the operator in this task but older equipment require more care on the part of the operator. Other thickness techniques are radiography, MFL, or Pulsed eddy currents, some of them are discussed later. These techniques are more limited in their application by material type, accuracy of measurement, wall thickness, or geometry than ultrasound but offer other advantages such as speed of application or the ability to inspect under insulation.

1.1.3.2 Defect detection

Defect detection techniques fall into two categories:

- Surface techniques – can only detect defects on or near to the surface of a component;
- Volumetric techniques – can detect both surface and embedded/internal defects.

There are several electromagnetic surface techniques, such as PT, ECT, or MT, which can be used to detect surface defects. But there is in fact no electromagnetic volumetric technique available or even used in industrial applications. Such testing can only be done if methods like radiography or ultrasound are applied. They enable the fully investigation of the specimen's volume in the testing procedure, i.e. even defects in large depth can be detected.

On the other hand, several ET techniques are used in industry, which cannot clearly be classified into one of these two categories. These methods are utilized at the borderline between surface and volumetric studies, depending on the application and current measurement conditions. Methods such as ACFM, PECT, remote field eddy current testing (RFECT), or MFL, which are already well-established on the market, can be counted to this group.

1.1.4 Present state of eddy current inspection

1.1.4.1 Perspectives of NDT

The NDT market is expected to be worth USD 12.06 billion by 2023, growing at a compound annual growing rate (CAGR) of 7.83% between 2017 and 2023 [10]. NDT has its applications in various industries such as manufacturing, aerospace, automotive, oil and gas, infrastructure, and power generation.

The UT segment is expected to hold the largest market share between 2017 and 2023. UT uses high-frequency sound waves to detect flaws or variations in properties of the materials. The UT is used to determine the thickness and detect the depth of internal flaws of metallic and nonmetallic materials. Ultrasonic rays have a high penetrating power, sensitivity, and accuracy; also, they are nonhazardous. Other techniques such as terahertz imaging and near-infrared spectroscopy have niche applications, and the market for the same is expected to grow at the highest CAGR between 2017 and 2023.

Inspection services accounted for the largest market share in 2016. The need for regular inspection and maintenance of the equipment in the oil and gas industry vertical leads to the growing demand for NDT inspection services. The booming automotive and manufacturing verticals in the Asia-Pacific (APAC) region are the major drivers for the growing NDT market. The market for training services is expected to grow at highest CAGR between 2017 and 2023. The lack of skilled and qualified technicians and the need to upgrade the skills of existing technicians is one of the drivers for the growth of the NDT market for training services.

The oil and gas vertical accounted for the largest market share in 2016. The demand for testing in this sector is largely driven by the mandates pertaining to the safety and environmental regulations by various governments to extend the asset's life and productivity, minimize repair cost, manage risks, and avoid catastrophic disasters, and other accidents. The market for the manufacturing vertical is expected to grow at the highest rate between 2017 and 2023. High investment in various projects, such as "Make in India" by the Indian Government or the high-tech strategy of the German Government Industries 4.0, is expected to uplift the manufacturing vertical during the forecast period.

North America held the largest share of the NDT market in 2016, and it is expected to grow at a moderate CAGR between 2017 and 2023. The market in the APAC region, which includes India, China, Japan, and Korea, is estimated to grow at the highest rate during the forecast period. The demand for NDT in the APAC region is expected to be driven by increasing infrastructure projects and power plants.

There are few restraints for this market such as the high cost of automated NDT equipment and the lack of skilled and qualified personnel. The setup cost is high for automated equipment; as a result of which, many small-scale organizations use simple inspection systems that might not be as reliable as the automated ones. The lack of awareness about new technologies and less number of training centers are hampering the NDT market.

1.1.4.2 Perspectives of ECT

According to the report released by MarketsandMarkets in August 2016 [11], the global ECT market is expected to grow from 866 million USD (2015) to 1.68 billion USD by 2022, at a CAGR of 9.82% between 2016 and 2022. The base year considered for the study is 2015, and the forecast period is between 2016 and 2022.

The ECT report provides a detailed analysis of ECT market based on type, service, vertical, and geography. This market report gives detailed information regarding the market dynamics influencing the growth of the market. The market within this study has been classified on the basis of ECT types into conventional eddy current, remote field testing, eddy current array (ECA), alternating current field measurement, PECT, near-field testing, near-field array, and partial saturation eddy current. The application in the oil and gas vertical is expected to hold the largest market share between 2016 and 2022. The demand for testing in this sector is largely driven by various governments' mandates for ensuring the safety of the environment by avoiding pipeline leakages, oil spills, and other accidents. The demand for ECT is expected to be driven by increasing number of power plants, since ECT is used to inspect heat exchangers, tanks, and other equipment, which are key parts of power plants. The demand is increasing all over the world due to government regulations for the overall safety of industrial assets, workforce, and the surrounding environment.

1.1.4.3 Summary

There have been many recent developments in ECT, leading to improved performance and the development of new applications. ECT is now a widely used and well understood inspection technique for flaw detection as well as for thickness and conductivity measurements.

Eddy current inspection is used in a variety of industries to find defects. One of the primary uses of ECT is for defect detection when the nature of the defect is well understood. In general, the technique is used to inspect a relatively small area and the probe design and test parameters must be established with a good understanding of the flaw that is to be detected. Since eddy currents tend to concentrate at the surface of a material, they can only be used to detect surface and near surface defects.

In thin materials such as tubings and sheet stocks, eddy currents can be used to measure the thickness of the material. This makes ECT a useful tool for detecting corrosion damage and other damage that causes a thinning of the material. The technique is used to make corrosion thinning measurements on aircraft skins and in the walls of tubing used in assemblies such as heat exchangers. ECT is also used to measure the thickness of paints and other coatings.

Computer-based systems are also available that provide easy data processing features for scientific purposes. Signal processing software has been developed for trend removal, background subtraction, and noise reduction. Sometimes impedance analyzers are used to allow improved quantitative eddy current measurements. Some laboratories have multidimensional scanning capabilities that are used to produce images of the scanned regions.

More details of these frequently applied techniques used since many years can be found in textbooks, handbooks, or review papers like [3,6,7,12]. But there are only very few references considering more recent developments [13,14] or presenting new techniques applied to new fields of research [15,16]. Electromagnetic NDT technologies have developed rapidly in recent years, and there have been some new methods or new applications. In recent years, the theories and applications of electromagnetic ultrasonic guided wave testing (UGWT), RFECT, defect quantification in MFL, and PECT have achieved rapid development, and these techniques are widely used in the online defect detection of oil and gas pipeline, rail track, pressure vessel, etc.

The new book, published recently by Huang and Wang [15], introduces new methods and technologies in the electromagnetic NDT field, as UGWT or metal magnetic memory testing, as well as some new developments of PECT, low-frequency eddy current testing (LFECT), RFECT, and MFL.

1.1.4.4 Electromagnetic UGWT

As an important branch of the NDT field, UT is widely used in the steel, electric power, petroleum, transportation, medical, industry. In the process of UT, the ultrasonic transducer is the core component of excitation and reception of ultrasonic waves, mainly including the piezoelectric ultrasonic transducer and the electromagnetic acoustic transducer (EMAT). Compared with the piezoelectric ultrasonic transducer, EMAT has many advantages, such as being contact free, without the need for the coupling medium, and easy to produce shear horizontal waves. Furthermore, it can be applied in a high-temperature environment.

1.1.4.5 Metal magnetic memory testing

The mechanical stress is directly related to the spontaneous magnetization and the residual magnetic field of the ferromagnetic material. Residual stresses and stress concentration in the component of structures impinge on the mechanical properties, erosion-resistance, dimensional precision, and cause fatigue failure. They also have an impact on the magnetic characteristics of ferromagnetic materials.

1.2 Eddy current testing

1.2.1 Eddy current and ECT

1.2.1.1 Eddy currents

Eddy currents (also called Foucault currents) are loops of electrical current induced within conductors by a changing magnetic field in the conductor, due to Faraday's law of induction. Eddy currents flow in closed loops within conductors, in planes perpendicular to the magnetic field. They can be induced within nearby stationary conductors by a time-varying magnetic field created by an AC electromagnet or transformer, for example, or by relative motion between a magnet and a nearby conductor.

The magnitude of the current in a given loop is proportional to the strength of the magnetic field, the area of the loop, and the rate of change of flux, and inversely proportional to the resistivity of the material. By Lenz's law, an eddy current creates a magnetic field that opposes the magnetic field that created it, and thus eddy currents react back on the source of the magnetic field. For example, a nearby conductive surface will exert a drag force on a moving magnet that opposes its motion, due to eddy currents induced in the surface by the moving magnetic field. This effect is employed in eddy current brakes, which are used to stop rotating power tools quickly when they are turned off. The current flowing through the resistance of the conductor also dissipates energy as heat in the material. Thus, eddy currents are a source of energy loss in alternating current (AC) inductors, transformers, electric motors and generators, and other AC machinery, requiring special construction such as laminated magnetic cores to minimize them. Eddy currents are also used to heat objects in induction heating furnaces and equipment, and to detect cracks and flaws in metal parts using ECT instruments. The study of traditional ECT in materials with no ferromagnetic properties, such as aluminum, provides the knowledge base needed to implement the ECT method to ferromagnetic materials as well.

This chapter includes first an explanation of the characteristics, capabilities, and drawbacks of the conventional ECT method. Then the different options for ECT techniques are discussed, starting with the different strategies to implement such techniques. In particular, an introduction to motion-induced eddy current testing

(MIECT) is given, explaining the potential benefits of researching this topic to inspect materials where motion is involved.

1.2.1.2 Eddy current testing

Traditional ECT is an NDT method that is used to inspect conductive materials to detect and characterize defects caused by corrosion, impact or material fatigue. In the traditional method, a time-varying magnetic field is applied to the conductive material which induces eddy currents. These currents generate a secondary magnetic field which opposes the excitation field. In the presence of a discontinuity in the conductivity of the sample material (such as a crack), the eddy currents are disturbed (from the normal path without a crack) and in turn so is the opposing magnetic field. By measuring the total magnetic flux (resulting from the excitation field plus secondary field), the changes in the paths of the induced eddy currents can be detected, and the presence of a defect is detected. A magnetic sensor can be used to measure the total magnetic flux and detect the anomalies caused by defects. This method allows the detection, location, and characterization of defects in metals without changing its mechanical properties, and without requiring direct contact with the metal. It is also a very sensitive method, allowing the detection of defects at a very early stage. The current induction is affected by several factors that occur due to the physical nature of the induction process. Such factors are the electromagnetic characteristics of the sample material like conductivity, permittivity, and permeability (distance between the probe and the surface of the material), and frequency of excitation. For instance, the magnetic permeability increases the skin-effect which restricts the defect detection capability to superficial defects. Also, the shape of the coil and the shape and orientation of the defects directly affect the perturbation of the measured magnetic flux.

In short, one could claim that primarily the idea of using NDT methods is to find discontinuities in the material, either originating from the manufacturing process or from overstraining in use. The sought discontinuities are mostly cracks stemming from false manufacturing techniques or from fatigue or thinning caused either by corrosion or erosion. In other words, the discontinuities have to be located first. Thereafter, the dimensions and directions of the discontinuities have to be evaluated from the measurements, and then the flaws have to be categorized in terms of conformance to stipulated acceptance criteria. These criteria evolve from fracture mechanical calculations based on critical flaw size and the speed of the extension of the flaw.

1.2.2 ECT principles

ECT is an extensively used method for the inspection of electrically conducting objects. The method allows the contactless detection of defects in ferromagnetic and nonferromagnetic materials. The general principle of ECT is shown in Figure 1.2. It is based on the induction of eddy currents inside the object under test. This is achieved by a coil, driven by an alternating current, which generates a time-dependent primary magnetic field $\mathbf{B}^{(p)}$. Conductivity anomalies are revealed by measuring the variations in the magnetic field resulting from a perturbed eddy current distribution $\mathbf{J}$.

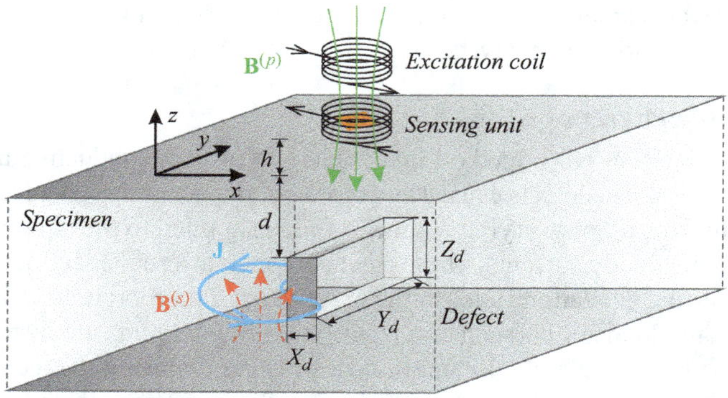

Figure 1.2 *Basic principle of ECT. An excitation coil, including a sensing unit, is located at a lift-off distance h above the specimen. A defect of length X_d, width Y_d, and height Z_d is located at a depth d*

The origin of the method can be traced back to the work done by Dr Friedrich Foerster in the 1960s, whereas early work on the analytical analysis of the field problem in ECT has been done by Dodd and Deeds [17]. They evaluated the impedance variations in secondary pick-up coils as a direct consequence of a perturbed eddy current profile in case of defective conductors.

Typical areas of application include the evaluation of safety-sensitive parts in nuclear power plants, aircraft structures [18,19] as well as in the petroleum or automotive industry, respectively, or for the inspection of printed circuit boards [20]. Very recently, its application is extended to inspect carbon-fiber-reinforced polymers [21,22], which shows that the continued development of eddy current techniques is still of great interest. The challenging task in ECT is to detect deep-lying defects. The measurement task lies in the detection of weak changes in the magnetic field in close vicinity of the excitation coil. The induced eddy currents generate a secondary magnetic field **B(s)** for itself. This counteracts the exciting primary field **B(p)**. The total magnetic field $\mathbf{B} = \mathbf{B}^{(p)} + \mathbf{B}^{(s)}$ is expelled out of the conductor with electrical conductivity σ and magnetic permeability μ in case of high frequencies $\omega = 2\pi f$. Thus, the eddy currents are concentrated near the surface of the conductor. This phenomenon is called skin-effect and can be approximated by the skin-depth δ

$$\delta = \sqrt{\frac{2}{\omega\sigma\mu}}. \tag{1.1}$$

It approximates the depth below the surface of the conductor at which the current density is decreased to *1/e* of its surface value. This definition is derived from the case of a sinusoidal current in a homogeneous conducting half-space. It represents a physical limit, which cannot be overcome. Therefore, in order to inspect deep-lying defects, the excitation frequency has to be reduced. However, the detection of subsurface defects requires sensors having a high sensitivity and low intrinsic magnetic noise to improve the signal-to-noise ratio. At the early stage of ECT, the most common sensors were secondary pick-up coils. In that case, the defect signal consists of the relative change in impedance with respect to the sensor position. When decreasing the frequency, the rate of change of the magnetic flux density perturbation due to impurities is reduced and pick-up coil-type sensors become ineffective. In the following, some alternative magnetic field sensors, applied in the framework of ECT, are presented. It is emphasized that those are not restricted to ECT and are also applied to some extend in the framework of MIECT.

To overcome the disadvantages of pick-up coils, Hall probes [2] or fluxgate sensors [23] are often used. However, the currently most prevalent magnetic field sensors in ECT are highly sensitive giant-magnetoresistive (GMR) sensors [24]. Some disadvantages are lying in their hysteretic nature and the need to bias them with a distinct external magnetic field in order to reach the linear operating point. A promising alternative to the previously mentioned sensors are spin-dependent tunneling (SDT) devices [25]. This type of sensor makes use of the principle of the electron spin-dependent quantum mechanical tunneling through a thin insulating layer (e.g. 1–2 nm, Al_2O_3) located in between two magnetic layers (e.g. FeCo/CrPtMn and NiFe). This is in contrast to GMR sensors, which make use of a conducting layer. The relative magnetization direction between the two magnetic layers determines the resistance of the device, which is in turn proportional to the external magnetic field to be sensed. Their application in the framework of ECT is reported by Wincheski *et al.* [25]. They showed that the use of SDT sensors allows the detection of defects in close vicinity to ferromagnetic fasteners located in a depth of around 5 mm considering an excitation frequency of 500 Hz. The application of superconducting quantum interference devices (SQUIDs) seems inherent when it comes to most challenging measuring tasks considering magnetic fields [26]. This technology offers an unrivaled sensitivity which enables the detection of very deep faults. However, these systems are disadvantageous in terms of increased cost and the requirement of cooling. Due to the presence of the cooler, the lift-off distance between the sensor and the specimen is considerably higher compared with room temperature systems. The increased distance between source and sensor influences the effective sensitivity. Typical lift-off values of SQUID ECT systems are in the range of about 7–20 mm [27] compared with 0.5–2 mm of traditional setups. Initial work on this topic can be dated back to the late 1970s [28]. Following this, SQUID-based applications in NDT were published in the early 1980s [26]. In general, there are two kinds of systems. These are either shielded systems, which are based on standard SQUIDs or unshielded systems which make use of SQUID gradiometers. One challenge in such systems is the cancellation of the excitation field at the location of the SQUID. This is done either by a double D-shaped excitation coil [29] or by a circular primary coil in combination with a local compensation coil [27]. It is reported that with the latter it was possible to achieve a considerably better compensation. In 1995, Tavrin *et al.* demonstrated a gradiometric-based SQUID ECT system which worked in a magnetically unshielded laboratory environment [30]. This study confirmed that this kind of system could find practical application. The group around Tavrin measured very deep-lying slot like flaws covered by 34.5 mm of aluminum. An overview about potential applications and developed SQUID-based ECT systems is given in [31,32].

Another rather new approach uses multi frequency excitation and the spectrogram eddy current method, e.g. for the detection of surface cracks and dissections in clad materials [33–35]. The system consists of small differential eddy current transducers and subsystems for scanning, excitation, and data acquisition. The signal measured during transducer movement can be presented in form of a spectrogram. Properties of the spectrograms can be used for defect's detection and identification.

Recently, the ECT method became greatly enhanced in terms of testing time by the development of array-based systems (ECA testing). Studies by Mook *et al.* [36], Postolache *et al.* [37], and Jun *et al.* [38] proposed arrays made of secondary coils, GMRs, and Hall sensors, respectively. Industrial applications can be found preferably in the oil and gas industry. Over the years, a variety of modifications of the ECT method came into existence as for example PECT [39] or RFECT [40]. Further readings about the ECT method, its extensions, and the application of different sensor technologies can be found in [2,13,24,41].

The main disadvantage of ECT is the frequency-dependent field attenuation. In order to provide an overview about the current detection limits in ECT, a summary of selected publications distinguished by the applied sensor technology is given in Table 1.1. It focuses on the detection of subsurface defects located deep inside the specimen. The defect depth d is defined by the distance from the surface of the specimen up to the upper surface of the defect such that it represents the amount of flawless material covering the defect (see Figure 1.2). The size of the investigated defect $[X_d, Y_d, Z_d]$ plays a major role during the investigations of the detection limit. Most of the studies listed in Table 1.1 do assume slit-like cracks such that the length of the defect is much larger than the characteristic diameter of the sensor system itself. This considerably increases the depth limit compared with isolated inclusions of finite size which are surrounded by conductive material. That circumstance has to be considered when comparing the results of the studies to each other. The advancements of the eddy current technique by applying alternative magnetic field sensors can be seen clearly from Table 1.1. As expected, SQUID sensors outperform all other sensors which operate at room temperature. However, the requirement of cooling and eventually shielding leads to increased maintenance and cost.

1.2.2.1 Current-induced ECT

Eddy currents are defined in the general theory of electromagnetism and they can be explained using Faraday's law of induction. This law states that when a time-varying magnetic field is applied to a conductor, an electromotive force (EMF) will be induced in it. This EMF creates currents that generate a magnetic field that opposes the initially applied field. This law is represented in (1.2) in its integral form and states that the line integral of the electric field around a closed loop L that bounds the surface S is equal to the negative rate of change (along time) of the magnetic flux through the same surface.

$$\oint_L \mathbf{E} \cdot \mathrm{d}\mathbf{l} = -\iint_S \frac{\partial \mathbf{B}}{\partial t} \cdot \mathrm{d}\mathbf{S}. \tag{1.2}$$

Table 1.1 Overview of ECT studies using different magnetic field sensors to detect deep-lying subsurface defects. The geometrical dimensions $[X_d, Y_d, Z_d]$ and d are defined in Figure 1.2

Author	**Year**	**Ref.**	**Frequency**	X_d **(mm)**	Y_d **(mm)**	Z_d **(mm)**	d **(mm)**
			Secondary pick-up coils				
Mook *et al.*	2006	[42]	350 Hz	3	100	3	8.5
			100 Hz	<0.1[a]	>100[b]	25	22.5
			50 Hz	<0.1[a]	>100[b]	25	28.8
Almeida *et al.*	2013	[43]	100 kHz	2-3[c]	2-3[c]	7	3
Carlstedt *et al.*	2014	[44]	100 Hz	12	2	2	6
			Fluxgate sensors				
Gasparics *et al.*	1998	[45]	20 kHz	10	<0.1[a]	1	4
Kreutzbruck *et al.*	2000	[23]	180 Hz	<0.1[a]	40	0.6	12.4
			GMR sensors				
Dogaru *et al.*	2001	[46]	1.5 kHz	15	0.5	2	1.5
Sikora *et al.*	2003	[47]	20–120 Hz	0.5	>50[c]	4	16
Tsukada *et al.*	2006	[48]	50 Hz	1	25	1	6
Yamada *et al.*	2008	[49]	50 Hz	1	25	1	8–14[d]
Wincheski *et al.*	2010	[19]	185 Hz	0.13	14	1	9
Hamia *et al.*	2010	[50]	325 Hz	0.5	50	2	8
Cacciola *et al.*	2010	[51]	60 kHz	2	2	4	4
			SQUID sensors				
Tavrin *et al.*	1996	[30]	10 Hz	<0.1[a]	200	1.5	34.5
Krause *et al.*	2002	[26]	90 Hz	0.15	40	1.2	12.7
Horng *et al.*	2002	[52]	400 Hz	1	50	1.5	7.2
Jeng *et al.*	2002	[53]	2–20 kHz	1	50	1.5	7.2
Allweins *et al.*	2003	[54]	15 Hz	20	<0.1[a]	15	31
Fardmanesh *et al.*	2009	[55]	20 Hz	0.05	>100[b]	5	24

[a] The defect was represented by a thin cut whose explicit width is not provided.
[b] The defect is assumed as infinitely long and extends along the whole specimen; the explicit value is not provided.
[c] Approximated; explicit value not provided.
[d] Maximum detectable defect depth for solid and layered specimens, respectively.

This phenomenon is illustrated in Figure 1.3 where the coils, a conductive specimen with a defect, the eddy currents and magnetic fields involved are present. The excitation coil is carrying a sinusoidal current imposed by the current source. This current generates the primary magnetic field $\mathbf{B}^{(p)}$, which is proportional to the intensity of the excitation current at any moment in time. Therefore, the primary magnetic field is also sinusoidal, and as it varies in time, EMF is induced in the conductive plate. The EMF in turn induces eddy currents $\mathbf{J}^{(i)}$ in the specimen. As the EMF is proportional to the derivative of the magnetic flux, the induced currents are delayed (in

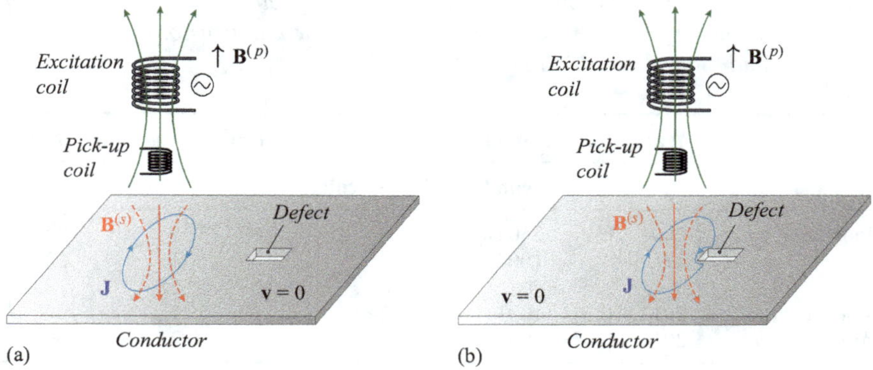

Figure 1.3 Conventional eddy current induction method using an excitation coil with sinusoidal current. (a) Far from the defect. (b) Close to the defect

time) compared with the excitation current applied. The secondary magnetic field $\mathbf{B}^{(s)}$ produced by the eddy currents opposes the primary field and affects the resulting magnetic flux.

The induced eddy current distribution and intensity depend on parameters, such as the intensity and frequency of the primary field or the conductivity and magnetic permeability of the sample material. An alternating current of a given frequency is generated in the primary or exciting coil. An alternating magnetic flux is consequently produced. This induces an alternating current of the same frequency in the secondary coil. With the introduction of the specimen, the alternating flux of the primary magnetic field induces in the specimen an eddy current flow which gives rise to an alternating magnetic flux in the opposite direction. The current in the secondary coil as well as resultant field, which will change the coil impedance, are consequently reduced. For given conditions, the reduction in current should be equal for all identical specimens placed in the same position relative to the coils. Any observed inequality in the value of the reduced current could indicate the presence of a defect, a change in dimensions, or a variation in the electrical conductivity or in the magnetic permeability of the test specimen due to a change in its physical or chemical structure. Thus, the change in the eddy current density can be detected and used to characterize the discontinuity causing that change.

The coil impedance, which is usually measured in practice instead of the current or flux, is a vector quantity having resistive and inductive components. These are phase shifted by 90° to each other. The other quantity that may be measured in practice is the voltage across the coil. The coil impedance as well as the voltage are related to the effective permeability of the test specimen, the test frequency of the coil, the limiting or boundary frequency of the test specimen, and the fill factor of the coil. By varying the type of the coils, the test method can be applied to flat surfaces or tubular products. This technique works best on smooth surfaces and has limited penetration, usually less than 6-7 mm. The skin depth, which is a function of the permeability and conductivity of the material as well as the frequency, determines the depth of penetration of the eddy currents. In ferromagnetic materials, the skin depth is very small and the technique will only detect surface breaking defects. In nonmagnetic material, it provides some subsurface capabilities and may give some indication of the depth of a defect. Encircling coils are used to test tubular and bar-shaped products. The tube or bar can be fed through the coil at a relatively high speed, allowing the full cross-section of the test object to be interrogated. However, due to the direction of the flux lines, circumferentially oriented discontinuities may not be detected with this application. The inspection frequencies used in eddy current inspection range from 200 Hz to about 6 MHz. The choice of the frequency depends on the thickness of material, the desired depth of penetration, the degree of sensitivity or resolution, and the purpose of inspection. Selection of inspection frequency is normally based on a compromise between the depth of interest and sensitivity to flaws. Increasing the frequency lowers the depth of penetration but increases the resolution and vice versa. Normally the highest inspection frequency compatible with the penetration depth required is selected. For surface flaws, frequencies up to several MHz may be used. For the inspection of ferromagnetic materials, due to the limiting skin-effect, relatively low frequencies are normally used. The inspection probe will give a certain indication on the instrument when placed in air. This indication will begin to change as the probe is brought close to the test piece and will continue to change until the probe is directly on the piece. This change in indication with change in spacing between the probe and the material to be tested is termed lift-off. The lift-off has a drawback as well as an advantage. The drawback is that many indications resulting from conditions of primary interest are masked by small changes in spacing. The advantage is that by utilizing the lift-off effect, the ECT instrument can be employed much easier for measuring nonconductive coating such as paint and anodized coating in metals. When an eddy current inspection probe approaches the edge of a part, the eddy currents are distorted because they are unable to flow beyond it. The indication obtained from it is called *edge-effect* and is very dominant, thereby limiting inspection near edges. It is not advisable to inspect any closer than about the penetration depth from the edge of a part. The distribution of eddy currents in the part being inspected is densest at the surface (due to skin-effect) closest to the probe and progressively become less dense with increasing distance from the surface.

1.2.2.1.1 Equipment and measurements

The main component of eddy current equipment is the probe of which there are several different types. The probe could be of encircling type, of internal type, or of external type. The main coil arrangements which may be present in these probes can be divided mainly into three categories depending upon the methods of measurement. In the absolute method, the primary and secondary coils are matched so that in the absence of any test specimen the voltages across them are equal and opposite. Introduction of the test piece results in a change in impedance and a voltage change appears which is measured. The comparison method consists of the use of two identical coil assemblies. A standard defect-free specimen is placed in one coil and the test specimen in the other. Changes arising from the differences in the two samples are measured.

In the auto-comparison method, two different parts of the same sample are compared with each other. Such equipment can be used for testing ferromagnetic as well as nonferromagnetic materials, provided a DC magnetic saturation unit is used. There is equipment available which is used for testing tubes, rods, and bars that are passed through an encircling coil assembly.

The eddy current equipment for measuring conductivity of materials employs a single probe coil acting simultaneously as an exciter and pick-up. The probe is moved by hand over the surface of the test material. The impedance of the coil is initially balanced with that of a similar coil inside the main body of the apparatus. Changes in the impedances of the probe coil due to eddy currents in the material under test give rise to an out-of-balance voltage which is indicated by a meter directly in units of conductivity. The frequency chosen for operation depends on the range of values of conductivity to be measured and the thickness of the material. If this frequency has been chosen too high, the measurements are performed, due to the skin-effect, close to the surface of the specimen. This procedure can cause problems for those cases where a volumetric measurement is preferred. Applications of this type of equipment include sorting of mixed materials, hardness testing, control of homogeneity, measurement of porosity, and investigating degrees of heat treatment for nonferromagnetic materials.

The thickness of nonconducting coatings on nonferromagnetic metal surfaces with the help of eddy current equipment is determined by measuring the lift-off effect for a probe coil. The probe coil is coupled by a transformer to a tuned circuit which is connected to a highly sensitive and stable frequency oscillator. When the probe is placed in contact with the surface of the coating, the oscillations decrease in amplitude by an amount depending on the coating thickness.

Ferromagnetic materials can be tested by subjecting them to magnetic hysteresis. The equipment for this includes two identical coil assemblies of either the encircling or probe type which are located at right angles to one another in order that the flux passing through one set of coils does not pass through the other.

The two signals are superimposed on one another, and in the absence of a test sample, the phases cancel out, and a horizontal straight line is observed. When a test specimen is introduced in one of the coils, the material undergoes magnetic hysteresis the loop of which is modified by the action of induced eddy currents. The straight line becomes disturbed and the trace assumes a shape that is characteristic of the electrical conductivity, the magnetic permeability, and the dimensions of the material. On applying an identical specimen to the second coil in exactly the same relative position, the trace again becomes a straight line. If, however, the permeability, conductivity, or dimensions of two specimens differ in any way, the trace assumes a shape which is characteristic of this difference. The equipment can be used to test ferromagnetic components of various shapes and sizes for such properties as hardening, the existence of internal stresses, machinability, etc. Manufacturers usually supply along with the equipment standard shapes of traces characteristic of some of these properties.

1.2.2.1.2 Limitations of ECT

ECT can be carried out on all materials which conduct electricity. Both ferromagnetic and nonferromagnetic materials can be tested. The method has the advantage that contact with the test specimen is not necessary. No couplant is therefore needed. The probe coils can be made with very small diameters and thus can detect the presence of very small flaws.

The sensitivity of the coils can be increased by the insertion of high-permeability cores such as ferrite rods which produce very sensitive focused coils. Long wires, tubes, rods, etc. can be tested by feeding them through the coils at a constant speed. The relative cost of inspection is therefore low. Under certain circumstances, the indications produced are proportional to the actual size of the defect. Thus, the tests can be useful for grading and classifying.

Due to the skin effect, the depth of penetration into the test specimen is limited and therefore the application of the technique is limited to the detection of surface and close-to-surface defects. Because of this phenomenon, the measurement of wall thicknesses is limited to thin wall tubing and to smaller thicknesses of materials. The lift-off effect is undesirable in most testing cases. The technique is limited to inspecting materials that are good conductors of electricity. It presents some difficulties considering absolute measurements. In the case of manual testing, it requires the presence of properly trained, qualified, and experienced operators.

In contrast to classical ECT, alternative methods exist which make use of relative motion instead of alternating currents to induce eddy currents in the object under test. These methods are presented in the next sections since they obey decent advantageous regarding the penetration of the electromagnetic fields when considering moderate velocities in the range of a few m/s.

1.2.2.2 Motion-induced ECT

The induction of eddy currents in the object under test is not restricted to the use of alternating magnetic field sources. If an electrical conductor and a magnetic field source experience relative movement, eddy currents are induced inside the conductor. Besides in the field of NDT, the calculation of the involved electromagnetic fields and retarding forces is of great theoretical interest in electromagnetism. But it is a difficult problem of evaluating motion-induced eddy currents. The results available in the literature are restricted to a few canonical cases, like the movement (with constant velocity) of a conducting loop in the field of a magnetic dipole [56] or the constant movement of a nonmagnetic conducting sphere in the field of a static magnetic dipole [57]. Only in such canonical problems closed form solutions for this kind of "Foucault currents" can be found. But these solutions are valuable for a better understanding of the physical phenomenon involved. Some years later, Saslow [58] provided a comprehensive review about the theory of motion-induced eddy currents and Maxwell's receding image theory. In the past, this topic was of special practical interest in the framework of magnetic levitation and transportation which was initially proposed in 1912 by Bachelet [59]. However, as late as in the 1970s, this topic became popular and Reitz [60,61], Richards [62], Borcherts and Davis [63–65], Lee and Menendez [66], and many others [67–69] studied the behavior of the electromagnetic fields in the vicinity of moving conductors extensively.

In recent years, an increase of a variety of methods can be observed which make use of relative motion between a magnetic field source producing a stationary magnetic field and the object under test. Techniques based on this principle can be classified as MIECT methods. A few examples of MIECT type methods, which were developed in the recent years, are given in the following.

The group of Chady *et al.* recently has been realized a prototype equipment consisting of an eddy current transducer and rotating permanent magnets [70]. The eddy current transducer is designated for testing of planar conducting plates. A rotating head with permanent magnets is used to induce eddy currents in the specimen. The two Hall-effect devices connected in a differential manner are utilized to measure an eddy current reaction. This inspection system is effective especially in cases of thick metallic elements, when it is necessary to utilize low excitation frequency or systems without power supply. Thus, the system is suitable for applications where it is important to achieve high penetration depth. Furthermore, advantages are the freedom of designing the shape of the exciting magnetic field and the availability of measuring two quantities at the same time. The changes of the magnetic field produced by disturbed eddy currents, and the changes of the torques, caused by the different forces between a rotating head with magnets and the sample in case of element with and without flaws.

The general principle of the MIECT method is shown in Figure 1.4(a). The group of Brauer *et al.* presented almost ten years ago a technique, which was called Lorentz force eddy current testing (LET) because Lorentz forces have been measured [71,72], whereas the group of Ribeiro and Ramos proposed in 2013 with velocity-induced eddy current testing a slightly different approach, where the magnetic fields are measured. Ramos *et al.* [24,73–75] investigated the applicability of moving stationary magnetic field sources using DC coils in the framework of NDT. They measured the disturbances of the magnetic field resulting from a defect directly by means of GMR sensors. Following this approach, these studies are extended in [76,77] to the use of single or differential pick-up coils, respectively. Moreover, they also exchanged the magnetic field source with a permanent magnet in order to achieve higher flux densities and an increased induced eddy current density inside the moving specimen. The sensor orientation has to be chosen carefully when using GMR sensors in order to avoid saturation effects. This can be overcome by applying differential coils as magnetic field sensors as it is also done in the framework of ECT. In this way, only the temporal change of the magnetic flux resulting from a passing defect is measured. Rocha *et al.* extended the analysis to the application of Hall sensors instead of GMRs and pick-up coils in [78]. They also investigated the defect response signals for different permanent magnet configurations and proposed the use of sensor arrays to expedite the assessment of larger areas. In a subsequent study, the application of GMRs, differential coils, and Hall sensors is compared in the framework of MIECT [76]. As a result, it turned out that GMRs were able to detect defects when crossing the edges of the defect. In contrast, pick-up coils and Hall sensors also provided signals when the probe passed the defect in its centerline (see Figure 1.4(a)).

In 2015, another MIECT technique is proposed by Tan *et al.* [79]. In contrast to previous studies, which made use of translational motion, they proposed a system using rotational motion of the magnetic field source to induce eddy currents inside the object under test. The basic principle of the method is shown in Figure 1.4(b).

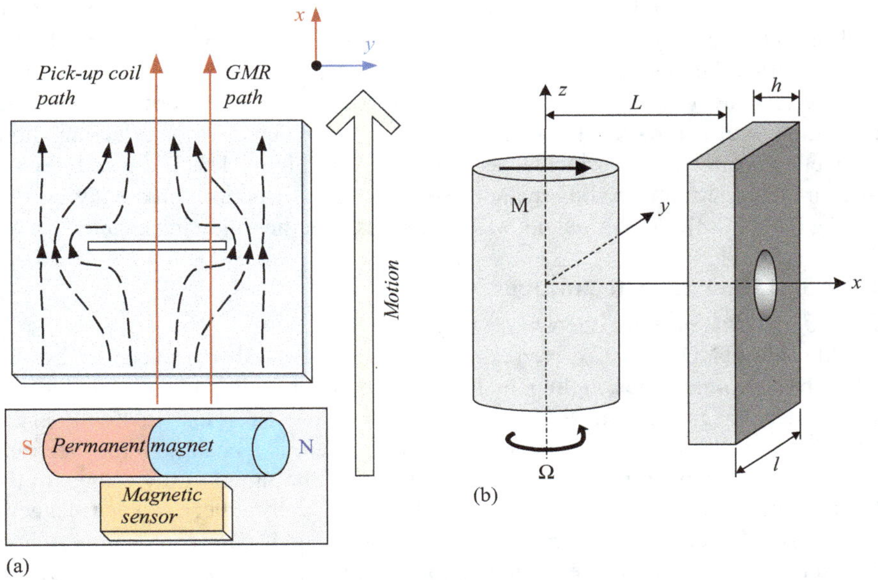

Figure 1.4 *Two different MIECT methods reported in the literature. (a) Linear motion using a lying axially magnetized permanent magnet [77] and (b) rotary motion of a diametral magnetized permanent magnet in close vicinity of a defective conductor [79]*

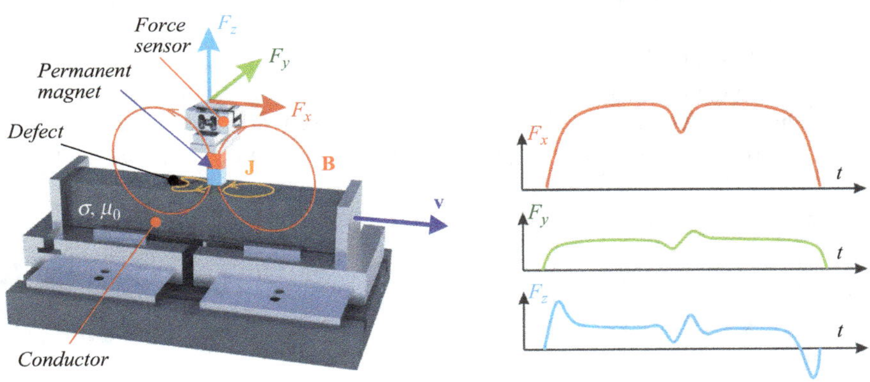

Figure 1.5 *Lorentz force eddy current testing for contactless evaluation of electrically conducting material*

A diametral magnetized, cylindrical permanent magnet rotates in close vicinity of a conductive object and anomalies are analyzed by means of the variations in the electromagnetic torque. The use of rotational motion provides the opportunity to design portable MIECT systems according to ECT devices nowadays available.

The presented studies are all limited to the analysis of surface touching defects indicating the early state of MIECT systems. However, it is emphasized that MIECT is not restricted to detect flaws on the surface. As Brauer and Ziolkowski [71] have already shown in 2008, their MIECT system consisting of a permanent magnet combined with a force sensor can be used to detect defects in conducting, non-magnetic specimen. The principal setup is shown in Figure 1.5. Because there are measured Lorentz forces, this method has been called Lorentz force eddy current testing (LET) [80]. Further details will be discussed in the following chapters.

1.2.2.3 Other ECT techniques

1.2.2.3.1 Pulsed eddy current technique

In eddy current NDT, an AC-driven excitation coil induces eddy currents in the sample through electromagnetic coupling. In turn, the circulation of the eddy currents induces a secondary magnetic field $\mathbf{B}^{(s)}$ as illustrated in Figure 1.6. This field will vary if flaw that impedes the eddy currents is present or there is a change in the electrical conductivity, magnetic permeability, or thickness of the sample. Any change in the field will be picked up by a sensing device, which is typically either a coil or a magnetic field sensor. In contrast to the conventional sinusoidal eddy current technique, where the excitation is limited to one frequency component, PECT excites the induction coil with a pulse waveform.

The first and main advantage is that, compared with single frequency ECT, PECT inherently has a broadband of frequencies, which is advantageous for any eddy-current-based NDT&E techniques due to the frequency-dependent skin effect. Another benefit is that PEC signals are relatively easier to interpret, while it requires a special skill of the operators for interpreting conventional ECT signals, which are presented in the impedance plane trajectory.

Conventional ECT only applies a single frequency for excitation, which makes it unable to detect both surface and sub surface defects reliably. The improved technique is the multi frequency ECT, which applies different excitation frequencies, one after another. Compared with multi frequency ECT, PECT can potentially be applied in

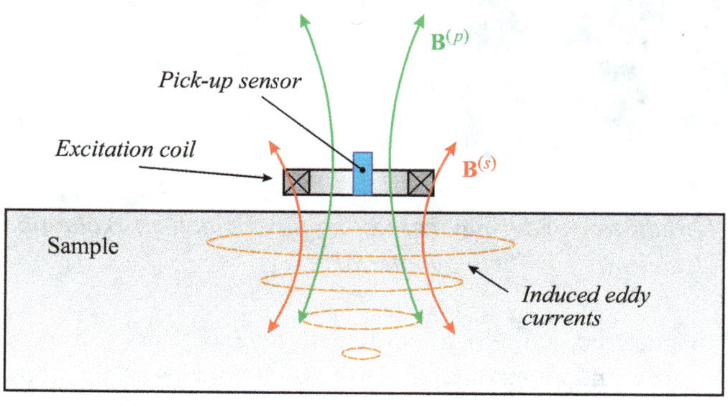

Figure 1.6 Illustration of the working principle of PECT

shorter time for inspection of different depths as PECT applies a wideband of frequencies in a single pulse. This allows to reduce the measurement time depending on the sample characteristics.

Similar to other ECT techniques, PECT, in general, requires no surface preparation, which leads to reduction of inspection time and costs efficiency is improved. The inspection can also be done without interrupting the operation or service of the structure being tested, unlike for example X-ray testing. In many applications where the sample is coated, no removal of the coating is required when ECT is used. Any eddy current systems are relatively cost-effective and reliable.

Thanks to its versatility, PECT has been used in numerous different NDT applications, both in material characterization and structural integrity inspection. In material characterization, PECT has been suggested to be used for measurement of electrical conductivity and magnetic permeability of materials. While in the structural integrity testing, PECT has been applied for defect detection and characterization, evaluation of corrosion, measurement of insulation thickness, plate thickness, and wall thickness of pipes. This covers both insulated and noninsulated, coated and noncoated materials. Furthermore, still within the area of integrity testing, detection of cracks under fasteners and between fasteners in aircraft structures using PECT has also been explored and implemented [14].

1.2.2.3.2 Remote field eddy current testing

RFECT is an electromagnetic method of NDT whose main application is finding defects in ferromagnetic steel pipes and tubes since conventional eddy current techniques have difficulty inspecting the full thickness of the tube wall due to the strong skin effect in ferromagnetic materials.

Both ECT and RFECT, use the principles of electromagnetic induction to detect defects in condenser and heat exchanger tubes. The basic RFECT probe consists of an exciter coil (or sending coil) which sends a signal to the detector (or receive coil). The exciter coil is feeded with an AC current and emits a magnetic field. The field travels outwards from the exciter coil, through the pipe wall, and along the pipe. The detector is placed inside the pipe, two to three pipe diameters away from the exciter and detects the magnetic field that has traveled back in from the outside of the pipe wall (for a total of two through-wall transits). In areas of metal loss, the field arrives at the detector with a faster travel time (greater phase) and greater signal strength (amplitude) due to the reduced path through the steel. Hence, the dominant mechanism of RFECT is through transmission. The main difference between RFECT and conventional ECT is in the coil-to-coil spacing. The RFECT probe has widely spaced coils to pick up the through-transmission field, whereas the typical ECT probe has coils or coil sets that create a field and measure the response within a small area, close to the object being tested.

Although both eddy current and remote field techniques rely on electromagnetic induction as a function of the inspection process, they are very different in operation and application. ECT relies on direct coupling between the inspection coil and the test material and works very well for nonferromagnetic materials. Magnetic materials have a major impact on the penetration of the eddy current field. Also, the permeability varies throughout the material and causes erratic signals and increased noise. RFECT has been designed to overcome this permeability effects in ferromagnetic tubing such as carbon steel and ferritic stainless steels. As the name implies, remote field testing does not work in the direct coupled zone. The remote field zone is the region in which direct coupling between the exciter coil and the receiver coil(s) is negligible. Coupling takes place indirectly through the eddy currents and their resulting magnetic field. The remote field zone starts to occur at approximately two tube diameters away from the exciter coil. RFECT does theoretically work on nonpermeable materials but it is not as accurate or effective as conventional ECT. Defects in the tube wall, such as pitting or cracking, and changes in wall thickness will interrupt or alter the amplitude and pattern of the eddy currents, changing its magnetic field. This change in the magnetic field then affects the coil by varying its electrical impedance, which is monitored by the test instrument.

Unfortunately, remote field testing does not easily lend itself to the variety of frequencies and signal mixing that ECT does. Due to the characteristic low-frequency operation, one or two test frequencies are typical for an RFECT inspection. Adding too many low frequencies has an impact on production by reducing the sample rate and in turn forces slower scanning speed.

The RFECT method has the advantage of allowing nearly equal sensitivities of detection at both the inner and outer surfaces of a ferromagnetic tube. The method is highly sensitive to variations in wall thickness and tends to be less sensitive to fill-factor changes between the coil and tube. RFECT can be used to inspect any conducting tubular product, but it is generally less sensitive than conventional eddy current techniques when inspecting nonferromagnetic materials [81,82]. Although RFECT works in nonferromagnetic materials such as copper and brass, ECT is preferred for such cases.

1.2.2.3.3 Low frequency electromagnetic testing

The low frequency electromagnetic testing (LFET or LFECT) is a special kind of conventional ECT technique and can be used to inspect, detect, and measure changes in the structure of materials such as changes in thickness caused by corrosion or others factors in storage tanks or other convex/concave ferrous surfaces, as well as nonferrous metal tubing/piping surfaces. A low-frequency electromagnetic field is induced into the plate, piping, or tubing to be inspected using a horseshoe-shaped electromagnet. By using low frequencies, the penetration of the magnetic field is more uniform throughout the plate wall thickness and defects on both the top and bottom sides of the plate can readily be seen. Any flaw in the plate, piping, or tubing will distort the returning field, which is picked up by a magnetic field sensor. The sensor registers the changes in the electromagnetic field as the scanner pass a flaw in the metal. The data are analyzed to determine the condition of the test material. The waveform will show a signal increase from the material baseline to indicate, where the wall loss has been detected by the sensor. Since the probe does not have to contact the work surface, LFET is useful and reliable technique on rough surfaces or surfaces with wet films where coatings are on pipes or plates. Thus, LFET is widely used to detect pits, holes, material loss caused by corrosion for storage tanks, tank shells, pipes, heat exchangers, and many other equipment.

1.2.2.3.4 Alternating current field measurement

The ACFM technique or electromagnetic field imaging, is very similar to the LFET. It is also a noncontact electromagnetic technique capable of both detecting and sizing (length and depth) defects in metals. The basis of the technique is an alternating current flow in a thin skin near the surface of any conductor. By introducing a remote uniform current into an area of the component under test, when there are no defects present, the electrical current will be undisturbed. If a crack is present, the current flows around the ends and down the faces of the crack. The current flowing in the surface has an associated magnetic field above the surface and this magnetic field will be disturbed as well, if the current is disturbed by a defect. The ACFM method involves the measurement of this magnetic field.

As the technique requires no electrical contact with the surface, it can be used to inspect through paint and coatings. The technique is widely used for weld and thread inspection and for subsea inspection of offshore platforms. It can also be used on both magnetic and nonmagnetic components. This technology is ideal for inspection applications in many industries, such as oil and gas, where the focus is on the detection and measurement of pipeline defects, and stress corrosion cracking. Thus, it is often applied to the petrochemical, power generation, aerospace, infrastructure, and manufacturing industries [83,84]. Because electromagnetic field shapes are created and measured, ACFM allows fast detection of surface breaking defects through dirt, paint, and nonconductive coatings, usually up to 15 mm thick. High-resolution digital data can be obtained at inspection speeds of about 300 mm per second.

Array probes containing large numbers of sensors can be deployed typically in situations where larger areas need to be inspected or where *pick and place* deployment is preferable to probe scanning. Thus, ACFM array systems can be incorporated into automated inspection systems to give simple PASS/FAIL reporting, avoiding the need for skilled operators. The ACFM method should only be applied to surface-breaking defects when used on carbon steels but is suitable for subsurface flaws in some nonmagnetic materials.

1.2.2.3.5 Eddy current array (ECA) testing

Eddy current array (ECA) and conventional ECT share the same basic principle and physics, the magnetic coupling of a probe sensor (coil) close to a test specimen (conductive material, ferromagnetic, or nonferromagnetic), generating eddy currents inside the test specimen, and displaying signals on the instrument's impedance plane.

ECA technology uses several individual coils grouped together in one assembly. An ECA, in its simplest form, is a series of single elements arranged in a row, allowing users to cover a larger area in a single pass than conventional, single-coil probes (i.e. pencil probes using ECT). However, this could lead to disturbed measurement signals. This is the reason why ECA probes use multiplexing. Multiplexing involves activating and deactivating coils in specific sequences to leverage the probe's width. Multiplexing also minimizes the interference between coils in close proximity (mutual inductance) and maximizes the resolution of the probe. Thus, ECA probes effectively eliminate the raster scanning necessary when using ECT pencil probes.

Most conventional eddy current flaw detection techniques can be reproduced with an ECA inspection. Compared with single-channel ECT, ECA technology provides the following benefits:

- Inspection of complex shapes using probes customized to the profile of the part
- Reduction of the inspection time
- Covers a large area in one single pass while maintaining a high resolution
- Reduces the complexity of mechanical and robotic scanning systems
- Automated ECA probes yield more consistent results compared with manual raster scans
- Simple manual scan is often enough improving flaw detection and sizing with C-scan imaging
- Easier analysis because of simpler scan patterns
- Probes can easily be designed to be flexible or shaped to specifications, making hard-to-reach areas easier to inspect
- Improves reliability and probability of detection

This method is widely used for a number of industrial applications. It can be used for both measuring the thickness of steels and detecting corrosion. ECA can be used on materials as diverse as vessels, columns, storage tanks and spheres, piping systems, and even structural applications.

1.2.3 Applications

Some of the applications of eddy current testing have already been mentioned while describing the basic principles, equipments, and procedures in the previous sections. In the following, a summary of these applications is being given.

All these techniques working on the same principle, where a coil driven by an alternating current induce eddy currents in the conducting specimen, whose distribution in the material enables some estimation of its properties. The frequency-dependent penetration depth of the electromagnetic field in the conductor as well as the low-spatial resolution for low frequencies are limiting, for example, the identification of deep internal defects in the test object. Consequently, ECT is considered as a surface-oriented method which enables preferably the detection of flaws at the surface or close to the surface. Additionally, metallic alloys or wall and coating thicknesses can be estimated with ECT as well.

Furthermore, it is remarkable that ECT was used rather lately in industrial practice. Whereas portable ultrasonic devices are available since about 1960, ECT devices first came up in the 1980s. A reason for this is that the ECT theory was much better well-understood and introduced into professional training. Today, it is well-known that ECT methods can be used to verify different material parameters, such as electrical conductivity, magnetic permeability, detection of discontinuities, material thickness or coating thickness of metallic objects, the effect of the distance between test coil and test object (lift-off distance), or the distances of conductors in laminated materials (e.g. composites). The result is a wide spectrum of applications for ECT methods, from pipe inspection in power plants, in the chemical or petrochemical industry, in nuclear submarines or air conditioning devices via the inspection in aircraft and automotive industries through the manufacturing of pipes, wires, rods, and bars.

ECT is employed for the detection and measurement of defects such as cracks, porosity, blowholes, inclusions, overlaps, shrinkages, and soft spots in a wide variety of test specimens in solid cylindrical, hollow cylindrical, or other complex shapes. Corrosion and cracking due to stress corrosion can also be detected. Changes in electrical conductivity and permeability can be measured which in turn have a bearing upon the material properties such as hardness, homogeneity, degree of heat treatment, existence of internal stresses, decarburization, diffusion, alloy composition, presence of impurities, etc. Thickness measurements can be made on metallic plates, foils, sheets, strips, tubes, and cylinders. Typically, it is possible to determine the thickness of nonmetallic coatings on metals such as for example the insulating layers on cables, nonconducting paints on some aircraft castings and anodic coating on aluminum alloy surfaces. Dimensions such as diameters of cylindrical specimens can also be determined. The materials can be automatically sorted in a production process. Since the method is adaptable to high-speed inspection, evaluation of small diameter tubes, such as those used in steam generators or heat exchangers is possible. It is also possible to inspect welded small-bore-piping. By using encircling type probes, large diameter pipes can be inspected. Similarly, long bars and wires can be inspected in short time. In tube testing, the eddy current method also allows high-speed detection of inter-granular corrosion on the inside surface.

1.3 Motion-induced ECT

1.3.1 Introduction

Material moving in a magnetic field experiences an EMF acting in a direction perpendicular both to the motion and to the magnetic field. This discovery was one of the foundations of electromagnetism. That it should occur even when the material was fluid did not escape the attention of early investigators such as Faraday, who reported to the Royal Society of London in 1832 how he had tried vainly to measure the voltage induced across the river Thames by the motion of the water in the vertical component of the earth's magnetic field [85]. The measurement was made between large electrodes, lowered into the river from Waterloo Bridge. Such signals were spurious one due to electrochemical and thermoelectric effects, two factors which can still trouble us when we try to apply the principle of electromagnetic induction to measuring a fluid velocity or bulk flow rate. Faraday's experiments failed chiefly because the river bed would short-circuit much of the genuine signal. However, he lived to hear of Wollaston's measurements of voltages induced tidally in the English channel in 1851. Faraday's method which consists of exposing a flow to a magnetic field and measuring the induced voltage using two electrodes has evolved into a successful commercial application known as the inductive flowmeter. The theory of such devices has been developed and comprehensively summarized by [86]. While inductive flowmeters are widely used for flow measurement in fluids at low temperatures such as beverages, chemicals, and wastewater, they are not suited for flow measurement in metallurgy. Since they require electrodes to be inserted into the fluid, their use is limited to applications at temperatures far below the melting points of practically relevant metals. Consequently, there have been several attempts to develop flow measurement methods which do not require any mechanical contact with the fluid. Among them is the eddy current flowmeter [87] which measures flow-induced changes in the electric impedance of coils interacting with the flow. More recently, a noncontact method was proposed [88] in which a magnetic field is applied to the flow and the velocity is determined from measurements of flow-induced deformations of the applied field. Today Faraday's invention, the electromagnetic flowmeter, enjoys broad success in the chemical and food industries. But it has fallen short of solving the grand challenge of flow measurement in high temperature melts such as steel, aluminum, or glass. Thess *et al.* [89,90] describe a technique which has been termed "Lorentz force velocimetry" (LFV), based on measuring the drag force on magnetic field lines which cross the melt flow. This noncontact technique is suited for high-temperature applications as well because it is free from the unavoidable electrode corrosion problem that has plagued Faraday's classical method.

1.3.2 Lorentz force eddy current testing

LET belongs to the group of MIECT type methods. It is a technique for nondestructive and contactless evaluation of electrically conducting specimens. The basic principle, shown in Figure 1.7, is based on the interaction between a permanent magnet and a moving specimen. As a consequence of this motion, eddy currents are induced inside the object under test, which in turn react with the magnetic field, producing a Lorentz force acting on both, the specimen and the permanent magnet. The novelty of the method lies in the determination of the measurement signal. In contrast to ECT and other MIECT techniques, the force acting on the magnet is measured using force sensors. In the presence of a defect, the eddy current profile and hence the resulting Lorentz force are perturbed. The physical principle of LET is an analogy to LFV [89]. In LFV, the main goal is to determine the flow rate of a conducting liquid by means of the Lorentz force which is proportional to the velocity of the liquid [90].

LET was initially demonstrated as an alternative NDT method by Brauer *et al.* [71]. Ziolkowski *et al.* [72] tackled the numerical analysis of the reported experimental setup and proposed techniques to analyze the electromagnetic field problem

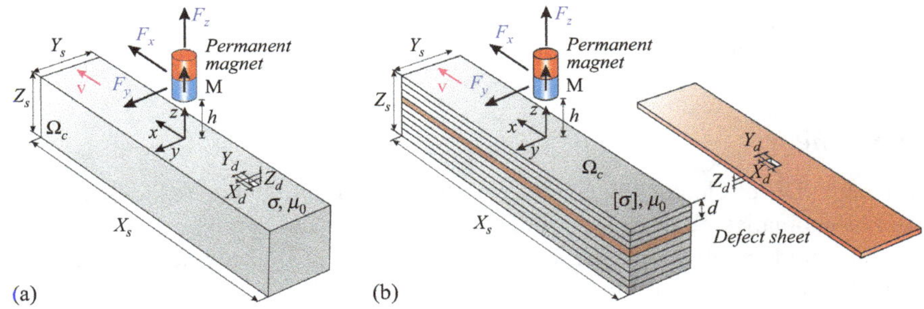

Figure 1.7 General principle of Lorentz force eddy current testing for contactless evaluation of electrically conducting material. The specimens and the geometrical parameters of the LET problem under investigation are shown in (a) for solids and (b) for layered structures

with increased computational efficiency. The work on LET in an experimental and numerical framework was continued by Uhlig [91] and Zec [92].

A very important study is related to the investigation of the effect of defect depth on the Lorentz force signals exerting on the magnet. To study the impact of the defect depth, a layered specimen containing a number of aluminum sheets of same thickness (usually 2 mm) has been used. The defect depth can be changed easily, if the position of the layer containing the defect is modified (Figure 1.7).

A conceptional model of LET is proposed and investigated in [93]. It consists of a modification to the well-known creeping magnet experiment, where a permanent magnet is slowly falling down a copper pipe [94]. The modification in this study consists of adding defects into the pipe wall such that the eddy current distribution and Lorentz force profile is disturbed. The LET method is extended to the determination of the electrical conductivity of the specimen assuming that the object under test is free of defects [95]. This technique is called Lorentz force sigmometry. It is shown that the lift-to-drag ratio of the Lorentz force components is proportional to the conductivity of the specimen such that $\sigma = \alpha F_z/F_x$ with a calibration factor α, which is determined experimentally (see Chapter 6.1). Besides the mentioned investigations, fundamental studies exist on the influence of the Lorentz force on geometrical parameters such as the lift-off distance, the size of the magnet, and the size and depth of the defect [96,97]. These studies are accomplished with the analysis how the velocity or conductivity affects the resulting Lorentz force profile.

The state-of-the-art of LET has been summarized by Brauer *et al.* [80]. It includes a summary about the experimental setup, the numerical modeling techniques and currently applied defect reconstruction methods. The investigations on the forward models were supported by Petkovic *et al.* [98], addressing the inverse problem, i.e. the identification of the defect called Lorentz force evaluation (LFE). She proposed reconstruction algorithms to determine the shape and the location of the defects solely out of the Lorentz force profiles. There were following up several studies of the LFE problem, i.e. considering the defect identification as an extension and improvement of this first approach [98–101].

1.3.3 Theory

NDT and NDE of electrically conductive objects require reliable methods to detect material anomalies or deep lying defects. Besides of radiographic, ultrasonic, or optical techniques, electromagnetic methods such as ECT find a wide range of application due to low cost, easy to use equipment and low demands to the measurement environment [6,12]. However, one of the most limiting factors in ECT is the frequency-dependent skin depth [71]. This restricts the capability to detect deep lying defects. With LET a novel electromagnetic NDT technique is presented [89,92,93,95]. The aim is to overcome this limitation. Lorentz force eddy current testing is based on setting an electrically conductive specimen into relative motion to a constant magnetic field. Due to Ohm's law for moving conductors, eddy currents are induced in the conductor under test

$$\mathbf{J} = \sigma\left(-\frac{\partial \mathbf{A}}{\partial t} - \nabla\varphi + \mathbf{v} \times \mathbf{B}\right), \tag{1.3}$$

where $\mathbf{J}$ denotes the induced current density, φ the scalar electric potential, $\mathbf{A}$ the magnetic vector potential ($\mathbf{B} = \nabla \times \mathbf{A}$,$\nabla \cdot \mathbf{A} = 0$), $\mathbf{v}$ the conductor velocity, and $\mathbf{B}$ the total magnetic flux density. $\mathbf{B}$ can be divided into a primary magnetic field (caused by a permanent magnet) and a secondary magnetic field generated by the eddy currents. The interaction of the constant magnetic field and the induced eddy currents results in a Lorentz force $\mathbf{F}^{(L)}$ acting on the specimen. Due to Newton's third law, an equal force $\mathbf{F}^{(PM)}$ exerts on the permanent magnet in the opposite direction

$$\mathbf{F}^{(PM)} = -\mathbf{F}^{(L)} = \iiint_{V_c} \mathbf{J} \times \mathbf{B}\,\mathrm{d}V \tag{1.4}$$

with V_c describing the volume of the specimen. If a defect is present in the conductive material, perturbations in the measured Lorentz force occur. Based on these perturbations the defect can be detected and reconstructed.

In contrast to LET, common eddy current testing uses a time changing current in a primary coil which generates a time changing primary magnetic field $\mathbf{B}^{(p)}$. Usually, the signal used to evaluate the material, is the change in impedance of the secondary coil.

Both principles are based on the induction of eddy currents, whereas major differences arise in shape and magnitude of the induced current densities as well as in the method of signal evaluation. Figure 1.8 shows the comparison of both methods and illustrates the perturbation of eddy currents due to defects. In both methods, a secondary magnetic field $\mathbf{B}^{(s)}$ is generated which interacts with the primary magnetic field $\mathbf{B}^{(p)}$. The total magnetic field is given by the sum of both fields $\mathbf{B} = \mathbf{B}^{(p)} + \mathbf{B}^{(s)}$. The formalism to describe the LET and ECT problem in theory is given

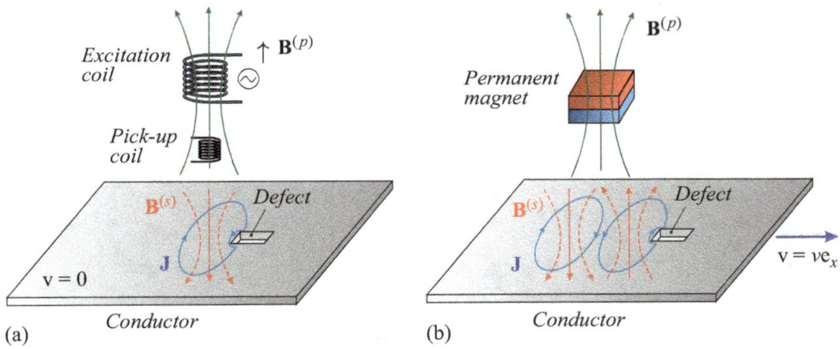

Figure 1.8 Comparison of characteristic eddy current profiles observed in ECT and LET. (a) In ECT, the eddy currents show a circular profile, which are similar to an imprint of the primary excitation coil. (b) In LET, the eddy currents follow a figure-of-eight and do cross below the magnet

by the magnetic convection diffusion equation [92,102], which can be written in its potential form as

$$\nabla \times \left(\frac{1}{\mu_0} \nabla \times \mathbf{A} - \mathbf{M} \right) = -\sigma \left(\frac{\partial \mathbf{A}}{\partial t} + \nabla \varphi - \mathbf{v} \times \nabla \times \mathbf{A} \right) + \mathbf{J}^{(e)}, \tag{1.5}$$

where a linear and nonferromagnetic material was assumed. In (1.5), $\mathbf{M}$ denotes the magnetization vector, $\mathbf{J}^{(e)}$ is the external current density, and $\mathbf{v}$ is velocity of the object under test. The limiting factor of ECT is the skin depth δ (1.1), which results in a fast decay of the information signal for subsurface defects.

A similar factor, namely, the magnetic Reynolds number R_m, can be defined for moving conductors. By transforming the magnetic convection diffusion equation into its nondimensional form, it can be derived [93]:

$$R_m = \mu \sigma |\mathbf{v}| L. \tag{1.6}$$

The parameter L is the typical length-scale of the problem. In general, for $R_m \ll 1$ diffusion of the magnetic field dominates and the resulting field is primarily determined by the boundary conditions and the primary magnetic field $\mathbf{B}^{(p)}$. For $R_m \gg 1$, the magnetic field lines are deformed in the moving direction, which results in a similar phenomenon as skin effect.

1.3.4 Experiments

Throughout this work, two different kinds of specimens are investigated. These are either solid or layered specimen. Both types are shown in Figure 1.9 together with the corresponding geometrical parameters given in Table 1.2. The conductivity of solid specimen, shown in Figure 1.9(a), can be assumed as isotropic. Meanwhile, the layered specimens are advantageous when varying the depth and size of defects which is demanding and expensive in case of solid bars.

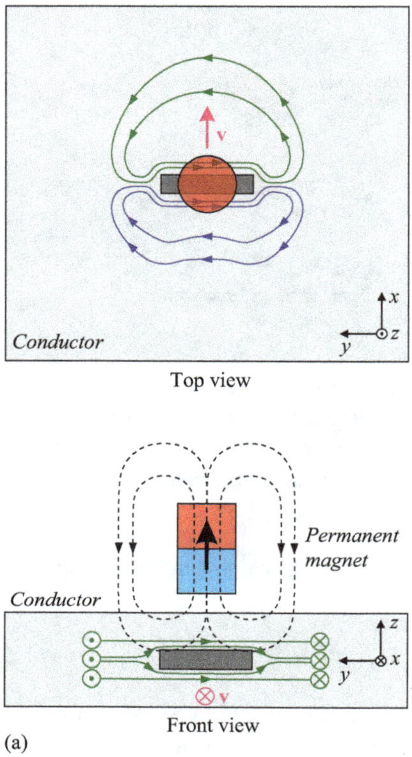

Figure 1.9 Sketch of the perturbed eddy currents induced in a moving plate located below a permanent magnet. Differences between (a) solid and (b) layered specimen. The layered structure restricts the eddy currents to flow in the respective sheet

Table 1.2 Parameters of the LET setup and characteristic values

Parameter	**Value**	**Description**
B_r	~1 T ... 1.4 T	Remanence of the magnet (NdFeB)
d	~1 mm ... 10 mm	Depth of the defect
h	1 mm	Lift-off distance
v	~0.1 m/s ... 2 m/s	Velocity of the specimen
X_d	~1 mm ... 10 mm	Length of the defect
Y_d	~1 mm ... 10 mm	Width of the defect
Z_d	~1 mm ... 10 mm	Height of the defect
X_s	250 mm	Length of the specimen
Y_s	50 mm	Width of the specimen
Z_s	50 mm	Height of the specimen
$\sigma_\square$	(19.88±0.5) MS/m	Electrical conductivity of solid specimen
$\sigma_\parallel$	(30.61±0.20) MS/m	Electrical conductivity of layered specimen

As a direct consequence, their conductivity profile has to be treated differently compared with solid bodies. Due to their stratified structure, an oxidation layer on the surface of each conducting sheet is present. This prevents the current to flow from one layer to the next. As an admissible approximation, which was tested experimentally in [80], the anisotropic conductivity profile is homogenized assuming a vanishing vertical conductivity ($\sigma_{zz} = 0$) throughout the whole conductor. The characteristic eddy current profiles are illustrated in Figure 1.9 for both types of specimen. Thus, in the following the defect is assumed to be nonconductive and nonmagnetic. The eddy current profiles would considerably change when one of both assumptions is violated. A permanent magnet, which is magnetized perpendicular to the surface of the specimen, generates a characteristic eddy current profile in the xy-plane with a shape of an eight. The induced eddy current density is highest right under the permanent magnet. The conductivity anisotropy in the z-direction does not affect this general behavior. In the case of a defect, the induced eddy currents circumvent the defect. The major difference between both conductivity profiles is that in case of anisotropic specimen, the induced eddy currents are restricted to flow in the respective sheet omitting any z-component.

The laboratory LET setup, shown in Figure 1.10, has been developed by Uhlig [91] and Carlstedt [103]. The 3D force sensor K3D40 [104] (ME-Messsysteme GmbH) based on the strain gauge technology is used to determine the dynamic forces acting on the permanent magnet. The data are acquired using the commercial PXI system NI PXI-1036 (National Instruments Corporation) together with the signal acquisition module NI PXI-4472. The output voltage of the force sensor amplifier is sampled with a frequency of $f_s = 10\,\text{kHz}$. The specimen is moved by a customized linear belt-driven drive (Jenaer Antriebstechnik GmbH), which realizes the required relative movement between magnet and specimen with a velocity of up to 3.75 m/s and a maximum acceleration of 20 m/s^2. The permanent magnet provides the

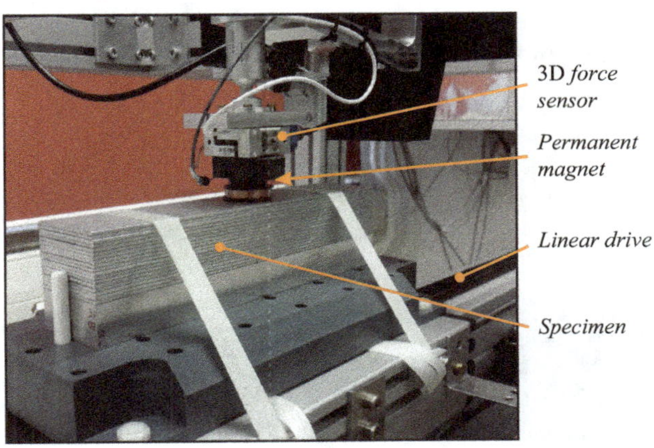

Figure 1.10 Laboratory LET setup developed by Uhlig [91] and Carlstedt [103]

stationary primary magnetic field in the LET system. In this case, a cylindrical NdFeB magnet is mounted on a 3D force sensor that is based on strain gauge technology. A 2D positioning stage enables the alignment of the permanent magnet in yz-plane. All devices and the data acquisition of the force sensor are controlled by a PXI-system by National Instruments. The specimen under test is a stack of 25 aluminum sheets each 250 mm × 50 mm × 2 mm, where well-defined defects at different positions and depths can be realized easily. The LET system investigated numerically is strongly related to the laboratory setup in order to compare the simulation results with experimental data. The problem geometry utilized in the numerical analysis is shown in Figure 1.9 and an overview of the involved parameters is given in Table 1.2. More information about the experimental setup can be found in [91,103].

An example of 3D FEM simulations is shown in Figure 1.11. Using the $\mathbf{A} - \varphi$ potential formulation, the LET field problem can be described by (1.5), but without the external current density on the right-hand side. This formulation separates the two induction phenomena into the moving part $\mathbf{v} \times \mathbf{B}$ and the time changing part on the right-hand side. Depending on the definition of the frame of reference, two equivalent types of the general magnetic field induction equation can be distinguished [92,102]. In the so-called moving frame of reference, the global coordinate system is associated with the moving permanent magnet, i.e. the conducting object moves in the direction along the x-axis with velocity $\mathbf{v}$. If the conducting object moves with a constant velocity and has a constant cross-section normal to the direction of motion, e.g. the object is free of defects, the time derivative $\partial\mathbf{A}/\partial t$ vanishes and (1.5) is reduced to a quasi-static approach.

In many experiments it has shown that the detection of subsurface defects in stacked aluminum sheets is possible for both testing techniques using the described experimental setup [105]. In the ECT method, the detection of a subsurface defect is mainly limited by the frequency-dependent penetration depth, i.e. if deep internal defects should be detected the testing frequency has to be as low as possible. On the other hand, if low frequencies are used, the performance of the electronic amplifier becomes more important due to the weak signals. Furthermore, the testing speed is strongly restricted depending on the properties of defects, e.g. characteristic length and shape. In LET, a relative movement between the permanent magnet

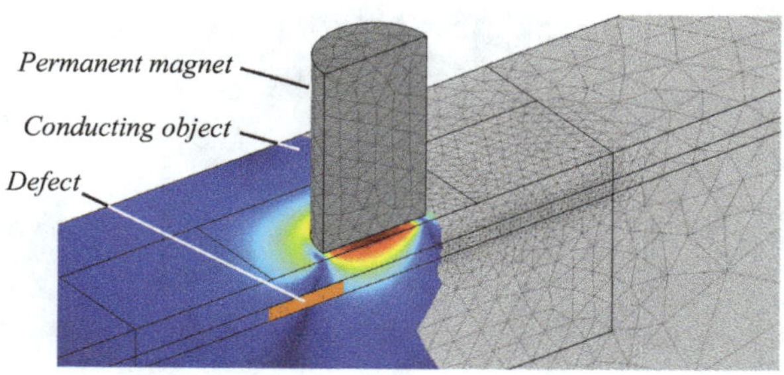

Figure 1.11 3D FEM model used for LET simulations

and the specimen is required to induce eddy currents. To create a sufficiently large Lorentz force, the relative velocity has to be high enough to detect small perturbations induced by subsurface defects. With increasing speed, the absolute force and the force perturbations increase linearly at magnetic Reynolds number $R_m < 1$. Therefore, the magnitude of the desired force signal is theoretically adjustable with the velocity for optimal utilization of the applied force sensor. In practice, the force sensor is sensitive to unwanted vibration of the environment and the system itself. To summarize, both testing techniques are highly dependent on the used sensors and measurement electronics as well as on the available testing speed. Consequently, many areas of application of LET and ECT, respectively, will be different.

1.3.5 Comparison of ECT and LET

A comparison between the ECT technique and LET is reported in [44,105]. To compare both methods from the numerical point of view, a detailed model of the applied ECT sensor is necessary. For that reason, it turns out that it was necessary to perform numerical simulations to get precise information of the internal probe structures. There have been no data (internal geometrical and material properties) available for the commercial ECT probe which should be used for the comparison of ECT and LET. The probe under investigation was a differential type probe PKA-48 (Rohmann GmbH), including secondary pick-up coils. It was used with the ECT device Elotest N300 (Rohmann GmbH). X-ray images were taken in order to get information about the internal structure of the probe. The result is given, together with the corresponding finite-element model, in Figure 1.12. First this analysis enabled the possibility to compare the defect response signals obtained experimentally with numerical simulations [106].

The major differences between ECT and LET are given by the shape and the magnitude of the induced eddy current profile as well as by the evaluated signal. The impedance variations of the imaginary part ΔZ_i and the back-induced voltage U_2 in the secondary pickup coil from ECT have been compared with the force perturbations in case of LET [44].

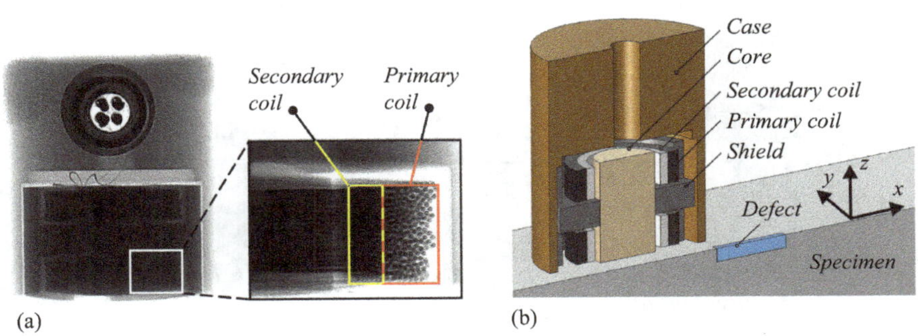

Figure 1.12 X-ray images and model of the ECT probe PKA-48 [106]. (a) X-ray images. (b) Reverse engineering model

The normalized force perturbations (LET) and impedance perturbations (ECT) representing the normalized defect response signals one will get from both methods are shown in Figure 1.13. The graph shows normalized signals of the drag-force F_x together with the imaginary part of the secondary coil impedance at comparable source dimensions.

Usually, the ECT method is applied in stationary applications. However, when the object under test is moving relative to the ECT probe, the induced voltage in the pick-up coil is modulated in the defect region. This effect is shown in Figure 1.14. If the velocity-to-frequency ratio v/f increases, the amount of sinusoidal periods in the defect region decreases. It is shown in [44] that the use of the Hilbert transform of the secondary induced voltage $\mathscr{H}[U_2(t)]$ is suitable to post process the modulated defect response signals to determine the envelope of the modulated signal. In practice, this requires additional adjustments of currently available ECT devices. A direct comparison between both methods in terms of defect depth and velocity showed that with ECT it was possible to detect defects of size $[X_d, Y_d, Z_d] = [12\,mm, 2\,mm, 2\,mm]$ up to a depth of 6 mm at a velocity of $v = 0.25$ m/s considering a frequency of $f = 100$ Hz. In contrast, the LET method was able to resolve the defect up to a depth of 8 mm at $v = 0.5$ m/s.

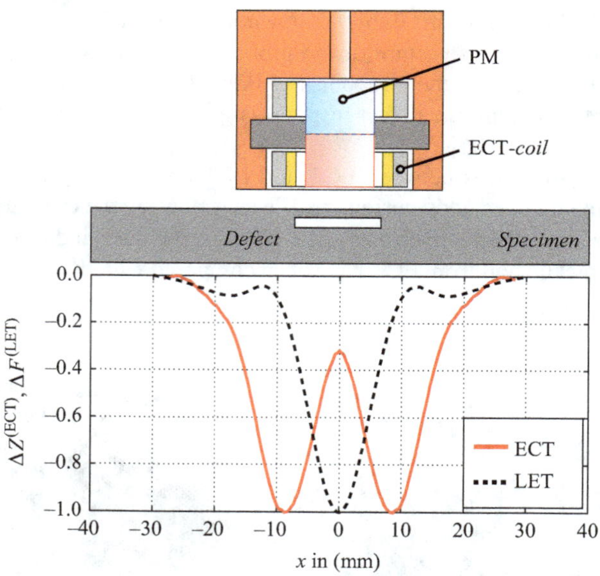

Figure 1.13 Normalized defect response signals in case of ECT and LET assuming equivalent dimensions [105]

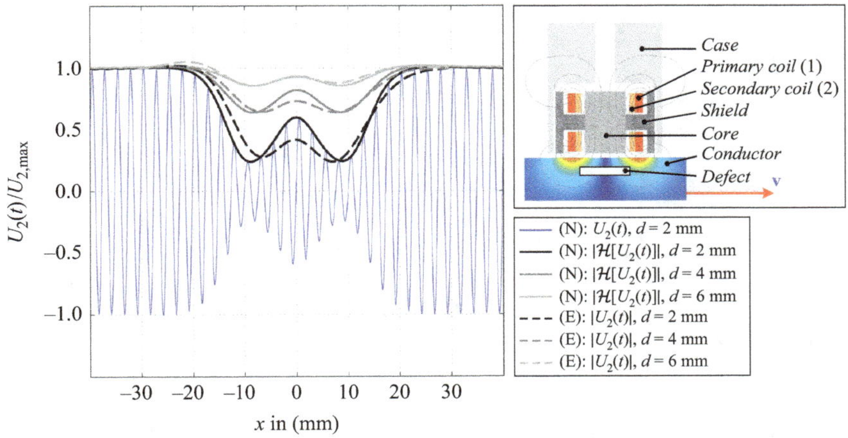

Figure 1.14 Modulated secondary-induced voltage $U_2(t)$ in the pick-up coils in case of moving objects under test ($v = 0.25$ m/s, $\sigma_0 = 30.61$ MS/m) [44]

It can be concluded that both methods obey individual advantages. The classical ECT method is suitable to inspect stationary objects which is not possible with LET or any other MIECT type method. However, if the object is in motion, the use of alternating currents can be omitted. In this way, it is possible to apply permanent magnets which produce considerably higher magnetic flux densities compared with current carrying coils. Comparative studies [44,105] showed that LET is a promising and competitive alternative to traditional ECT methods considering the contactless evaluation of moving electrical conductors.

Chapter 2
Forward simulation methods

Marek Ziolkowski,[1,2] *Mladen Zec*[1] *and Konstantin Weise*[1]

This chapter presents methods for simulation of electromagnetic problems related to MIECT. The main purpose of the presented methods is their application to development of the general LET systems. Furthermore, the main objective is the calculation of the Lorentz forces which result from the interaction between permanent magnets and moving, nonmagnetic and electrically conductive objects. In general, numerical simulations of LET problems can be relatively time-consuming. Therefore, an additional emphasis was put on the development of fast semi-analytical and simplified numerical methods that allow solving general LET problems with satisfactory accuracy.

Initially, the electromagnetic equations which model general LET systems are given in two equivalent frames of reference, namely, a stationary (laboratory) frame and a frame moving at a constant speed (Section 2.1). This is followed by several semi-analytical methods that can be applied for calculating Lorentz force and their perturbations resulting from typical LET systems. Most of the presented methods assume weak reaction of the secondary magnetic field which results from motion induced eddy currents.

- Section 2.2.1 presents solutions for 2D LET models in which the system dimension perpendicular to the direction of motion is much larger than in the motion direction. The systems discussed include models using equivalent simple and modified magnetic linear dipoles as well as analytical formulas for the description of 2D permanent magnets (PMs). The methodology of determining forces acting on a PM as well as reaction signals from one or more defects in a moving 2D object is also described in detail.
- Section 2.2.2 presents general semi-analytical solutions for calculating forces acting on a simple three-dimensional PMs (rectangular, cylindrical) placed above moving nonmagnetic and electrically conducting large plate without defects.
- Section 2.2.3 describes the general methodology for calculating Lorentz forces and defect response signals (DRS) for LET systems with test objects containing defects.
- Section 2.2.4 addresses the problems of oscillatory motion between the magnet system and the object under test.

[1] Advanced Electromagnetics Group, Technische Universität Ilmenau, Germany
[2] Applied Informatics Group, West Pomeranian University of Technology, Poland

- Section 2.2.5 presents the simplest method for calculating the force reaction signals from defects of different shapes.
- Sections 2.2.6 and 2.2.7 give a description of the approach based on the so-called extended defect area implemented for anisotropic and isotropic conductors.
- Section 2.3 describes the basics of the surface charge simulation method (SCSM) adapted to LET problems with anisotropic and isotropic conductors. This section also gives the practical method for determining the magnetic field and the scalar electrical potential in the rectangular area using a 2D fast Fourier transform.
- Section 2.4 presents robust and accurate finite element methods (FEM) used to analyze complex LET problems. The main emphasis is on techniques which simplify the motion modeling between stationary and moving parts without reduction of the calculation accuracy, namely, the moving magnet (MMA) and the moving defect (MDA) approaches. Additional FEM based techniques, such as quasi-static approach (QSA) and weak-reaction approach (WRA) which can considerably reduce the simulation time are introduced as well.

2.1 Moving coordinate systems—transformations

The principal postulates of special theory of relativity are as follows [107]:

- *postulate of relativity*, the laws of physics are the same in all inertial systems,
- *postulate of a universal light speed*, the same speed of light is measured in all inertial systems.

Two inertial systems are considered, namely, stationary system $\mathcal{O}$ (reference frame) with coordinates (x, y, z, t) and the moving system $\mathcal{O}'$ (moving frame) with coordinates (x', y', z', t'). The system $\mathcal{O}'$ moves with a constant velocity $\mathbf{v}$ relatively to $\mathcal{O}$ (Figure 2.1).

According to the first postulate, for problems where the displacement current can be omitted ($\partial\mathbf{D}/\partial t = 0$), Maxwell's equations take in both systems the following form:

$$\begin{array}{ll} \text{Reference frame } \mathcal{O} & \text{Moving frame } \mathcal{O}' \\ \nabla \times \mathbf{H} = \mathbf{J} & \nabla' \times \mathbf{H}' = \mathbf{J}' \\ \nabla \cdot \mathbf{B} = 0 & \nabla' \cdot \mathbf{B}' = 0 \\ \nabla \cdot \mathbf{J} = 0 & \nabla' \cdot \mathbf{J}' = 0 \\ \nabla \times \mathbf{E} = -\dfrac{\partial \mathbf{B}}{\partial t} & \nabla' \times \mathbf{E}' = -\dfrac{\partial \mathbf{B}'}{\partial t'} \\ \mathbf{B} = \mu_0(\mathbf{H} + \mathbf{M}) & \mathbf{B}' = \mu_0(\mathbf{H}' + \mathbf{M}') \end{array} \tag{2.1}$$

where

$$\nabla = \frac{\partial}{\partial x}\mathbf{1}_x + \frac{\partial}{\partial y}\mathbf{1}_y + \frac{\partial}{\partial z}\mathbf{1}_z, \quad \nabla' = \frac{\partial}{\partial x'}\mathbf{1}_{x'} + \frac{\partial}{\partial y'}\mathbf{1}_{y'} + \frac{\partial}{\partial z'}\mathbf{1}_{z'}.$$

As a consequence of the second postulate, Lorentz transformation of time and space coordinates can be formulated as:

$$t' = \gamma\left(t - \frac{\mathbf{v} \cdot \mathbf{r}}{c}\right), \tag{2.2}$$

$$\mathbf{r}' = \gamma(\mathbf{r} - \mathbf{v}t). \tag{2.3}$$

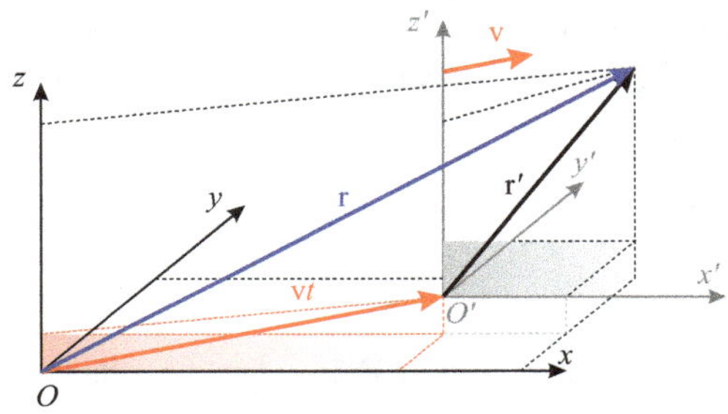

Figure 2.1 Relative motion of two inertial systems

where

$$\gamma = \frac{1}{\sqrt{1+\beta^2}}, \qquad \beta = \frac{v}{c}, \tag{2.4}$$

and c is the speed of light in a vacuum.

Considering only the cases for which $v \ll c$, i.e., $\beta \simeq 0$, $\gamma \simeq 1$, Lorentz transformation is reduced to Galilean transformation

$$t' = t, \tag{2.5}$$

$$\mathbf{r}' = \mathbf{r} - \mathbf{v}t. \tag{2.6}$$

To find the relationship between vectors describing the electromagnetic field in both inertial systems (2.1), the basic operations on scalar and vector functions in these systems will be defined using the Galilean transformation [108–110]. Considering the scalar function $f'(x', y'z', t')$ defined in the system $\mathcal{O}'$, spatial partial derivatives in the system $\mathcal{O}$ can be calculated as follows

$$\frac{\partial f'}{\partial x} = \frac{\partial f'}{\partial x'}\frac{\partial x'}{\partial x} = \frac{\partial f'}{\partial x'}. \tag{2.7}$$

Using (2.7), the following relation between nabla operators is received

$$\nabla' f' = \nabla f'. \tag{2.8}$$

The relation between time derivatives of f' in both system is calculated using a chain rule as:

$$\begin{aligned}\frac{\partial f'}{\partial t} &= \frac{\partial f'}{\partial t'}\frac{\partial t'}{\partial t} + \frac{\partial f'}{\partial x'}\frac{\partial x'}{\partial t} + \frac{\partial f'}{\partial y'}\frac{\partial y'}{\partial t} + \frac{\partial f'}{\partial z'}\frac{\partial z'}{\partial t}\\ &= \frac{\partial f'}{\partial t'} - v_x\frac{\partial f'}{\partial x'} - v_y\frac{\partial f'}{\partial y'} - v_z\frac{\partial f'}{\partial z'}\\ &= \frac{\partial f'}{\partial t'} - \mathbf{v}\cdot\nabla' f' = \frac{\partial f'}{\partial t'} - \mathbf{v}\cdot\nabla f'\end{aligned} \tag{2.9}$$

and finally

$$\frac{\partial f'}{\partial t'} = \frac{\partial f'}{\partial t} + \mathbf{v}\cdot\nabla f'. \tag{2.10}$$

Similar expressions can be derived for vector functions. Let $\mathbf{F}(x', y', z', t')$ be a vector function in the $\mathcal{O}'$ system. It is easy to show that

$$\nabla' \cdot \mathbf{F}' = \nabla \cdot \mathbf{F}' \tag{2.11}$$

and

$$\nabla' \times \mathbf{F}' = \nabla \times \mathbf{F}'. \tag{2.12}$$

Relationship between time derivatives of $\mathbf{F}'$ is defined by the following expression:

$$\frac{\partial \mathbf{F}'}{\partial t'} = \frac{\partial \mathbf{F}'}{\partial t} + (\mathbf{v} \cdot \nabla)\mathbf{F}'. \tag{2.13}$$

Using (2.11), (2.12), and the vector identity

$$\nabla \times (\mathbf{A} \times \mathbf{B}) = (\mathbf{B} \cdot \nabla)\mathbf{A} - (\mathbf{A} \cdot \nabla)\mathbf{B} + \mathbf{A}(\nabla \cdot \mathbf{B}) - \mathbf{B}(\nabla \cdot \mathbf{A}) \tag{2.14}$$

Equation (2.13) can be rewritten as

$$\frac{\partial \mathbf{F}'}{\partial t'} = \frac{\partial \mathbf{F}'}{\partial t} + \mathbf{v}(\nabla \cdot \mathbf{F}') - \nabla \times (\mathbf{v} \times \mathbf{F}'). \tag{2.15}$$

Finally, using (2.11), (2.12), and (2.14) in (2.1) for the moving frame $\mathcal{O}'$, the following set of equations is received:

$$\begin{aligned} \nabla \times \mathbf{H}' &= \mathbf{J}', \\ \nabla \cdot \mathbf{B}' &= 0, \\ \nabla \cdot \mathbf{J}' &= 0, \\ \nabla \times (\mathbf{E}' - \mathbf{v} \times \mathbf{B}') &= -\frac{\partial \mathbf{B}'}{\partial t}. \end{aligned} \tag{2.16}$$

Comparing (2.1) to (2.16), the following Galilean transformations of electromagnetic field vectors can be found:

$$\begin{aligned} \mathbf{H}' &= \mathbf{H}, \\ \mathbf{B}' &= \mathbf{B}, \\ \mathbf{E}' &= \mathbf{E} + \mathbf{v} \times \mathbf{B}. \end{aligned} \tag{2.17}$$

The transformation for the magnetization density vector takes the form

$$\mathbf{M}' = \mathbf{M}. \tag{2.18}$$

The law of Ohm for moving conductors has to be written as

$$\mathbf{J}' = \sigma(\mathbf{E} + \mathbf{v} \times \mathbf{B}). \tag{2.19}$$

It can be concluded that this law applies to any frame of reference that moves across magnetic flux lines (or in which the body carrying the current $\mathbf{J}$ moves with respect to the magnetic field source) [109].

2.2 Semianalytical methods used in LET systems

The semianalytical methods presented in this section assume that the magnetic field produced by the permanent magnet is not affected by the secondary magnetic field from motion-induced eddy currents in the conducting object. This is further referenced as the so-called weak reaction approach (WRA). The main purpose of these methods is to quickly and accurately calculate profiles of Lorentz forces acting on permanent magnet in the LET system. The first part presents methods that are used for 2D models of LET systems. Next, the method of calculating Lorentz forces for configuration 3D permanent magnet over non-magnetic, conducting wide plate is described. The last part describes semianalytical methods and conditions of their application for calculating forces and DRSs in 3D LET systems.

2.2.1 Calculation of forces in 2D LET systems

In 2D LET models, it is assumed that the perpendicular dimension l of the LET system is much larger than the lateral dimensions of the system. In this section 2D model of the LET system is analyzed. A long permanent magnet (PM) of a rectangular cross-section ($w \times h$) is located at the lift-off distance h_0 above the moving, nonmagnetic ($\mu = \mu_0$), conducting plate with a $L \times D$ cross-section and homogeneous electrical conductivity σ_0 (Figure 2.2).

The permanent magnet is magnetized along $0Z$-axis. The magnetization is described using the magnetization vector $\mathbf{M} = M\mathbf{1}_z$. In the plate, an artificial, ideal defect ($\sigma_d = 0$) with dimensions $c_x \times c_z$ is drilled at the depth d. The center of the defect is located at $\mathbf{x}_0 = [x_n, -d - c_z/2]^{\mathrm{T}}$.

In LET measurement systems, the force $\mathbf{F}_{PM}$ exerted on the permanent magnet resulting from relative movement of the magnet and conductor is measured by a sensor directly attached to the magnet. However, in the calculations, this force is determined indirectly using the third Newton's axiom which says that the force exerted on the permanent magnet is of the same magnitude as the Lorentz force acting on the conductor $\mathbf{F}_{LF}$ but with the opposite direction ($\mathbf{F}_{PM} = -\mathbf{F}_{LF}$). The reason is greater

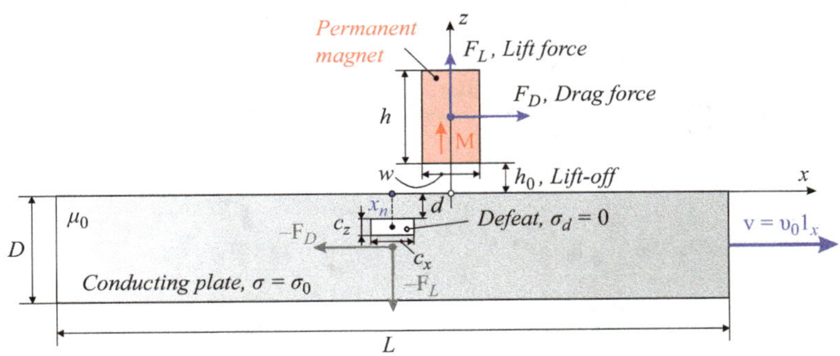

Figure 2.2 Permanent magnet above moving nonmagnetic conductive plate with rectangular defect

accuracy of Lorentz force calculations in the conductor than the accuracy of methods which can be used for a direct estimation of the force exerted on permanent magnet, e.g. based on Maxwell's tensor or the virtual work principle. In 2D LET systems the force exerted on permanent magnet has only two components $\mathbf{F}_{PM} = [F_x, F_z]^{\mathrm{T}}$. The force exerted on permanent magnet can be described in two equivalent coordinate systems:

1. Coordinate system fixed to the PM with the conducting object (CO) moving
2. Coordinate system fixed to the conductor with the PM moving.

In both coordinate systems, the component F_z exerted on the permanent magnet is always positive regardless of the direction of the velocity vector and it is called the lift force (F_L). The sign of the F_x component depends on the direction of the velocity vector as well as on the used coordinate system. In the coordinate system fixed to the permanent magnet (see 1 and 2 in Figure 2.3), the sign of F_x component follows the direction of the vector $\mathbf{v}$ and, in this case, F_x is called the drag force (F_D). In the coordinate system fixed to the conductor ($1'$ and $2'$ in Figure 2.3), the sign of F_x component is opposite to the direction of the vector $\mathbf{v}$ and the component is called the brake force (F_B) in this case.

To calculate analytically the force exerted on the permanent magnet above the moving conducting object with defect, some simplifications have to be introduced. Firstly, it is assumed that the magnetic field produced by the permanent magnet is not affected by induced eddy currents in the conducting object, i.e., so-called WRA can be applied [102,111]. A necessary condition for the applicability of the WRA can be formulated as $R_m = v_0\mu_0\sigma_0 a \ll 1$, where R_m is the magnetic Reynolds number, a is the length parameter (specific for the analyzed problem), and v_0, σ_0, μ are the velocity, the electrical conductivity, and the magnetic permeability of the conducting object, respectively. Assuming that the coordinate system is assigned to a permanent magnet,

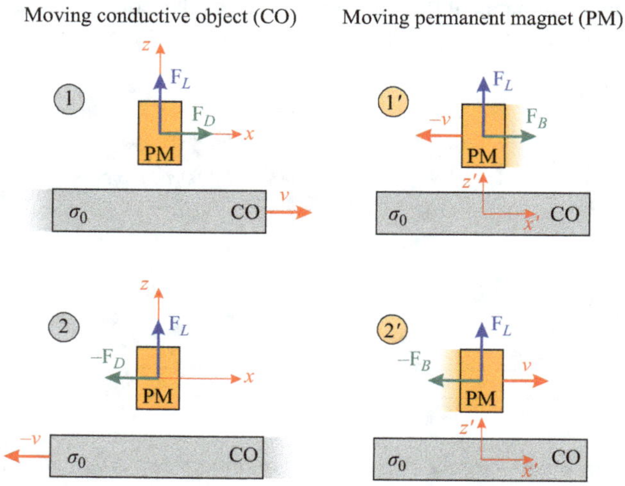

Figure 2.3 Force exerted on a permanent magnet in inertial systems

use of WRA allows to calculate the induced eddy currents **J** in a conducting object directly by Ohm's law (2.19). For 2D problems, the equation (2.19) can be reduced to

$$\mathbf{J} = \sigma_0(\mathbf{E} + \mathbf{v} \times \mathbf{B}_0) = \sigma_0(-\nabla\phi + \mathbf{v} \times \mathbf{B}_0) = \sigma_0 \mathbf{v} \times \mathbf{B}_0 = -\sigma_0 v_0 B_{z,0} \mathbf{1}_z, \tag{2.20}$$

because of $\nabla\phi = 0$. $\mathbf{B}_0$ denotes the magnetic flux density produced by the permanent magnet.

The force **F** exerted on the 2D permanent magnet above the moving conductor can be calculated as follows

$$\begin{aligned}\mathbf{F} &= -\mathbf{F}_{LF} = -l \int_{S_0} \mathbf{J} \times \mathbf{B}_0 \, \mathrm{d}S = F_x \mathbf{1}_x + F_z \mathbf{1}_z = F_D \mathbf{1}_x + F_L \mathbf{1}_z \\ &= l\sigma_0 v_0 \left[\left(\int_{S_0} B_{z,0}^2 \, \mathrm{d}S \right) \mathbf{1}_x - \left(\int_{S_0} B_{x,0} B_{z,0} \, \mathrm{d}S \right) \mathbf{1}_z \right],\end{aligned} \tag{2.21}$$

where $S_0 = L \times D$ is the cross-section area of the conductor, and l is the length of the system in the y-direction. It should be noted that directions of force components are independent from the direction of the magnetization vector **M**. They depend only on the direction of the velocity vector. For a conductor without defects, the second integral in (2.21) disappears due to anti-symmetry of $B_{x,0}$, i.e. the lift force F_L calculated by the WRA always equals 0 if the permanent magnet is far away from the front/back walls of the conductor.

Let $\mathbf{F}_0^{(n)}$ and denote profiles of forces exerted on the permanent magnet found for the conductor moving between x_1 and x_2 with velocity $\mathbf{v} = v_0 \mathbf{1}_x$ for the defect-free system and the system with a defect, respectively. The index n corresponds to the actual position of the conductor center $x_n \in < x_1, x_2 >$. The vector difference $\Delta\mathbf{F}^{(n)} = \mathbf{F}^{(n)} - \mathbf{F}_0^{(n)}$ describes the influence of the defect on Lorentz force component profiles and is called the DRS. In 2D LET problems eddy currents induced in the conductor have only y-component (see (2.20)). For a conductor of center located at x_n with an ideal defect ($\sigma_d = 0$) of a cross-section $S_D^{(n)}$, induced eddy currents $\mathbf{J}^{(n)}$ can be described by the following superposition

$$\mathbf{J}^{(n)} = \mathbf{J}_0^{(n)} - \mathbf{j}_D^{(n)},$$

where $\mathbf{J}_0^{(n)}$ denotes eddy currents density induced in the conductor without defect and $\mathbf{j}_D^{(n)} = \mathbf{J}_0^{(n)}|_{S_D^{(n)}}$ are eddy currents in the region of the defect filled with a material with electrical conductivity σ_0.

Using above and (2.21), the DRS $\Delta\mathbf{F}^{(n)} = [\Delta F_x^{(n)}, \Delta F_z^{(n)}]^\mathrm{T}$ can be calculated as

$$\begin{aligned}\Delta\mathbf{F}^{(n)} &= \mathbf{F}^{(n)} - \mathbf{F}_0^{(n)} = -l \int_{S_D^{(n)}} \mathbf{j}_D^{(n)} \times \mathbf{B}_0^{(n)} \, \mathrm{d}S = l \int_{S_D^{(n)}} \mathbf{J}_0^{(n)} \times \mathbf{B}_0^{(n)} \, \mathrm{d}S \\ &= l\sigma_0 v_0 \left\{ \left[-\int_{S_D^{(n)}} \left(B_{z,0}^{(n)}\right)^2 \mathrm{d}S \right] \mathbf{1}_x + \left(\int_{S_D^{(n)}} B_{x,0}^{(n)} B_{z,0}^{(n)} \, \mathrm{d}S \right) \mathbf{1}_z \right\}.\end{aligned} \tag{2.22}$$

In 2D LET problems, if the WRA can be applied, DRSs can be directly calculated also for a conductor containing more than one defect because induced eddy currents in the conductor flow only in the y-direction and are not disturbed by the defects.

The resultant DRS for conductors containing more defects is a simple superposition of single defect signals

$$\Delta\mathbf{F}^{(n)} = \sum_{k=1}^{K} \Delta\mathbf{F}_k^{(n)} = l\sum_{k=1}^{K} \int_{S_{D_k}^{(n)}} \mathbf{J}_0^{(n)} \times \mathbf{B}_0^{(n)} \, \mathrm{d}S, \tag{2.23}$$

where K is the number of defects and $S_{D_k}^{(n)}$ denotes the region covering the kth defect in the conductor with center located at x_n. The profile of the absolute force exerted on the permanent magnet above the moving conductor with K ideal defects can be obtained from

$$\begin{aligned}\mathbf{F}^{(n)} &= \mathbf{F}_0^{(n)} + \sum_{k=1}^{K} \Delta\mathbf{F}_k^{(n)} \\ &= -l\int_{S_0^{(n)}} \mathbf{J}_0^{(n)} \times \mathbf{B}_0^{(n)} \, \mathrm{d}S + l\sum_{k=1}^{K} \int_{S_{D_k}^{(n)}} \mathbf{J}_0^{(n)} \times \mathbf{B}_0^{(n)} \, \mathrm{d}S.\end{aligned} \tag{2.24}$$

The simplest model of a long permanent magnet of a rectangular cross-section $w \times h$ magnetized with the magnetization $\mathbf{M} = M\mathbf{1}_y$ consists of an equivalent 2D magnetic line dipole (l-dipole) of the moment $\mathbf{m}_l = m_l\mathbf{1}_z = Mwh\mathbf{1}_z$ located at the center of the magnet. The physical interpretation of the l-dipole is shown in Figure 2.4. The l-dipole can be understood as two infinitely thin line currents flowing in opposite y-directions located at a distance d from each other. The general formula for the magnetic field produced by the l-dipole can be easily derived [112]. The formula describing the magnetic flux density $\mathbf{B}_0$ from l-dipole has the following form

$$\mathbf{B}_0 = \frac{\mu_0}{2\pi(r')^2}\left[2\frac{\mathbf{m}_l \cdot \mathbf{r}'}{(r')^2}\mathbf{r}' - \mathbf{m}_l\right], \tag{2.25}$$

where $\mathbf{r}'$ denotes position of the calculation point in relation to the l-dipole.

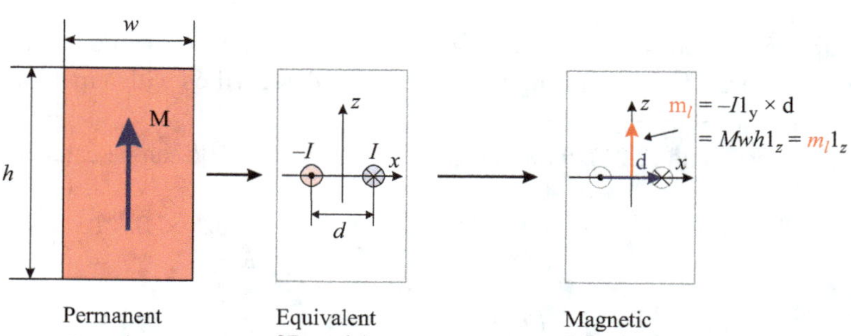

Figure 2.4 Construction of an equivalent magnetic line dipole for a permanent magnet of rectangular cross-section

In the Cartesian system, the magnetic flux density $\mathbf{B}_0^{(n)}$ at any point $\mathbf{r} = [x, z]^{\mathrm{T}}$ produced by the magnetic l-dipole $\mathbf{m}_l = m_l \mathbf{1}_y$ located at $\mathbf{r}_n = [x_n, z_n]^{\mathrm{T}}$ is given by

$$\begin{aligned}\mathbf{B}_0^{(n)} &= m_l \frac{\mu_0}{2\pi} \frac{2(x - x_n)(z - z_n)}{\left[(x - x_n)^2 + (z - z_n)^2\right]^2} \mathbf{1}_x \\ &\quad + m_l \frac{\mu_0}{2\pi} \frac{(z - z_n)^2 - (x - x_n)^2}{\left[(x - x_n)^2 + (z - z_n)^2\right]^2} \mathbf{1}_z = B_{x,0}^{(n)} \mathbf{1}_x + B_{z,0}^{(n)} \mathbf{1}_z. \end{aligned} \tag{2.26}$$

The DRS $\Delta\mathbf{F}^{(n)}$ of a single defect located in a conducting plate moving with the velocity $\mathbf{v} = v_0 \mathbf{1}_x$ along the x-axis below the magnetic l-dipole can be calculated using (2.22) and the setup shown in Figure 2.5.

Substituting (2.26) to (2.22), the DRS takes the following form

$$\begin{aligned}\Delta\mathbf{F}^{(n)} &= l\sigma_0 v_0 \left(m_l \frac{\mu_0}{2\pi}\right)^2 \left\{ -\int_{-\frac{c_x}{2}}^{\frac{c_x}{2}} \int_{-d-c_z}^{-d} \frac{\left[(z - z_n)^2 - (x - x_n)^2\right]^2}{\left[(x - x_n)^2 + (z - z_n)^2\right]^4} \mathrm{d}x \, \mathrm{d}z \, \mathbf{1}_x \right. \\ &\quad \left. + 2 \int_{-\frac{c_x}{2}}^{\frac{c_x}{2}} \int_{-d-c_z}^{-d} (x - x_n)(z - z_n) \frac{(z - z_n)^2 - (x - x_n)^2}{\left[(x - x_n)^2 + (z - z_n)^2\right]^4} \mathrm{d}x \, \mathrm{d}z \, \mathbf{1}_z \right\}. \end{aligned} \tag{2.27}$$

Integrals in (2.27) can be calculated analytically and expressed by the following functions

$$\begin{aligned} f_x(x, z, x_n, z_n) &= \frac{(x - x_n)(z - z_n)}{6\left[(x - x_n)^2 + (z - z_n)^2\right]^2} + \frac{1}{8(x - x_n)(z - z_n)} \\ &\quad + \frac{1}{8(x - x_n)^2} \arctan\left(\frac{z - z_n}{x - x_n}\right) + \frac{1}{8(z - z_n)^2} \arctan\left(\frac{x - x_n}{z - z_n}\right) \end{aligned} \tag{2.28}$$

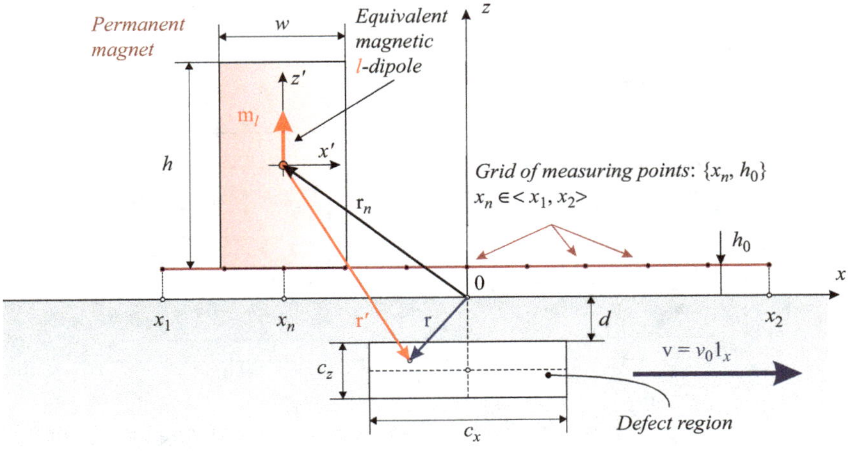

Figure 2.5 Setup of 2D LET system used for calculation of DRSs

and

$$f_z(x,z,x_n,z_n) = \frac{1}{12}\frac{(z-z_n)^2-(x-x_n)^2}{\left[(x-x_n)^2+(z-z_n)^2\right]^2}. \tag{2.29}$$

Using (2.28) and (2.29), the analytical forms of DRS components for the rectangular defect $\{S_d: c_x \times c_z\}$ located at the depth d can be written as

$$\begin{aligned}\Delta F_x^{(n)} = {} & l\sigma_0 v_0 \left(m_l \frac{\mu_0}{2\pi}\right)^2 \times \left[f_x\left(\frac{c_x}{2}, -d, x_n, z_n\right) - f_x\left(\frac{c_x}{2}, -d-c_z, x_n, z_n\right)\right. \\ & \left. - f_x\left(-\frac{c_x}{2}, -d, x_n, z_n\right) + f_x\left(-\frac{c_x}{2}, -d-c_z, x_n, z_n\right)\right]\end{aligned} \tag{2.30}$$

$$\begin{aligned}\Delta F_z^{(n)} = {} & l\sigma_0 v_0 \left(m_l \frac{\mu_0}{2\pi}\right)^2 \times \left[f_z\left(\frac{c_x}{2}, -d, x_n, z_n\right) - f_z\left(\frac{c_x}{2}, -d-c_z, x_n, z_n\right)\right. \\ & \left. - f_z\left(-\frac{c_x}{2}, -d, x_n, z_n\right) + f_z\left(-\frac{c_x}{2}, -d-c_z, x_n, z_n\right)\right]\end{aligned} \tag{2.31}$$

The Lorentz force exerted on the l-dipole located at $[x_n, h_0]$ above an infinitely wide plate without defects moving with the velocity v_0 can be expressed as

$$F_{x0} = l\sigma_0 v_0 \left(m_l \frac{\mu_0}{2\pi}\right)^2 \frac{\pi}{8}\left[\frac{1}{h_0^2} - \frac{1}{(h_0+D)^2}\right], \tag{2.32}$$

$$F_{z0} = 0. \tag{2.33}$$

In order to verify the introduced approach, the test problem shown in Figure 2.6 is solved using the FEM [72,102]. The test model consists of a long cylindrical permanent magnet located above a conducting plate moving with a constant velocity. It is easy to show that the magnetic flux density for $r \geqslant R$ produced by a uniformly,

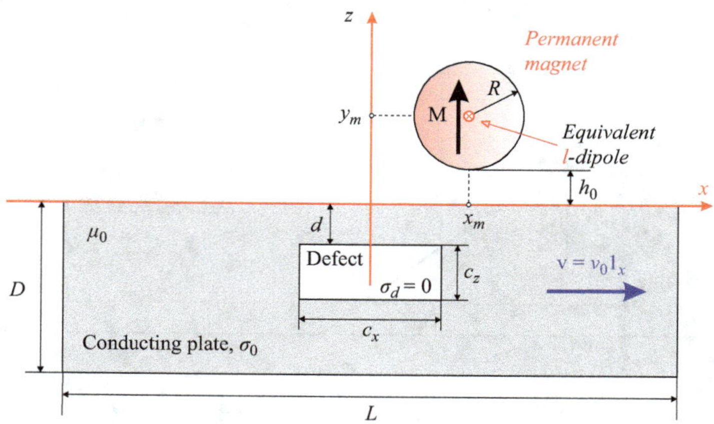

Figure 2.6 Uniformly magnetized long cylindrical permanent magnet above a conducting plate moving with the constant velocity $\mathbf{v} = v_0 \mathbf{1}_x$. The plate contains an ideal rectangular defect ($\sigma_d = 0$)

diametric magnetized infinitely long cylindrical permanent magnet of radius R is equal to the magnetic field (2.25) produced by the equivalent l-dipole of the magnetic moment $\mathbf{m}_l = \pi R^2 \mathbf{M}$ located at the center of the permanent magnet. Figure 2.7 shows results of the analytical and FEM simulations for the following LET configuration: (1) PM: $R = 2.5$ mm, $M = 931$ A/mm, $h_0 = 11$ mm, (2) conducting plate: $L \times D = 250\,\text{mm} \times 50\,\text{mm}$, $\sigma_0 = 30.61$ Ms/m, $v_0 = 1$ cm/s, (3) rectangular defect: $c_x \times c_z = 12\,\text{mm} \times 2\,\text{mm}$ located at the depth $d = 2$ mm.

To estimate the quality of l-dipole models quantitatively, a normalized root mean square error (NRMSE) is defined as

$$\text{NRMSE} = \sqrt{\epsilon_x^2 + \epsilon_z^2} \tag{2.34}$$

with

$$\epsilon_{x/z} = \frac{\sqrt{\frac{1}{N}\sum_{n=1}^{N}\left[S_{x/z}^{(n)} - S_{x/z,FEM}^{(n)}\right]^2}}{\max\limits_{n=1\ldots N}\left[S_{x/z,FEM}^{(n)}\right] - \min\limits_{n=1\ldots N}\left[S_{x/z,FEM}^{(n)}\right]} 100\%,$$

where N is a number of test points in the force/DRS profile S, and the subscript FEM denotes the reference solution calculated by the FEM.

Simulations show an excellent agreement between results received analytically using l-dipole and calculated by the FEM. The NRMSEs are 0.82% and 0.25% for Lorentz force and DRS profiles, respectively.

In the next example, an analysis of the model with a rectangular permanent magnet (cf. Figure 2.5) is presented. The equivalent l-dipole is located at the center of the long rectangular permanent magnet: $w \times h = 15\,\text{mm} \times 25\,\text{mm}$, $M = 931$ A/mm,

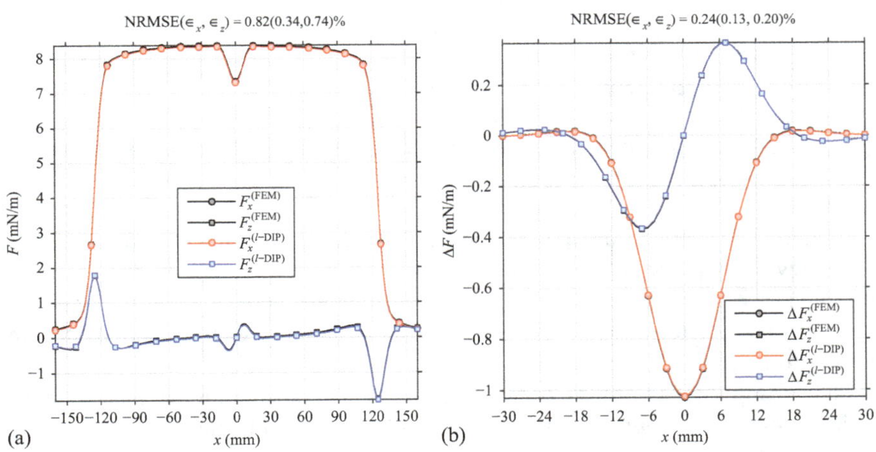

Figure 2.7 Long cylindrical permanent magnet—comparison of the equivalent l-dipole model with FEM (defect: $c_x \times c_z = 12\,\text{mm} \times 2\,\text{mm}$, $d = 2$ mm). (a) Lorentz force profiles. (b) Defect response signal profiles

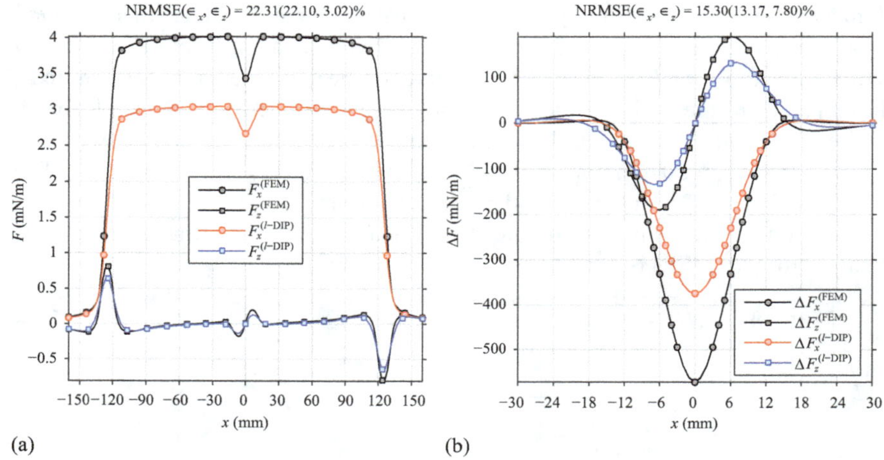

Figure 2.8 Long rectangular permanent magnet—Lorentz force and DRS profiles calculated analytically using l-dipole and simulated by FEM (defect: $c_x \times c_z = 12\,\text{mm} \times 2\,\text{mm}$, $d = 2\,\text{mm}$). (a) Lorentz force profiles. (b) DRS profiles

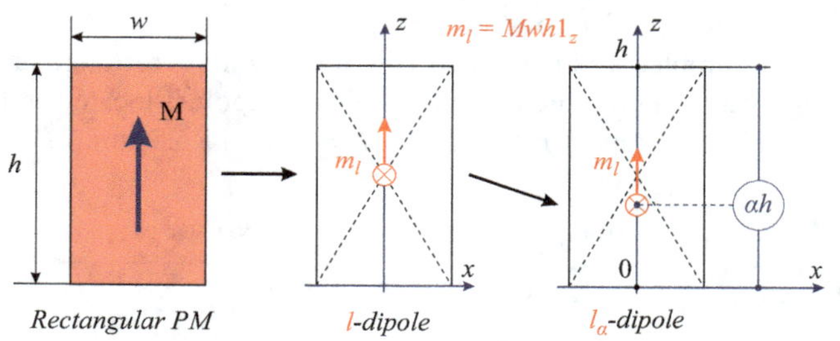

Figure 2.9 Long rectangular permanent magnet—equivalent l-dipole models

$h_0 = 1$ mm. The other parameters of the LET system remain unchanged in relation to the previously presented example. Results of analytical and FEM simulations are presented in Figure 2.8. In contrast to the previous example, the use of l-dipole is subject to large calculation errors compared to FEM, i.e., over 22% and 15% for Lorentz force and DRS, respectively.

The errors can be reduced after modifying the l-dipole and replacing it with the l_α-dipole shown in Figure 2.9. In the l_α-model, y-position of the equivalent dipole depends on the parameter α. The value of parameter $\alpha \in< 0, 1 >$ can be determined using a minimizing procedure that minimizes NRMSE between force profiles calculated by FEM and in a model using l_α-dipole. The force profiles that are used in the minimizing procedure are profiles determined for a conductive object without defects.

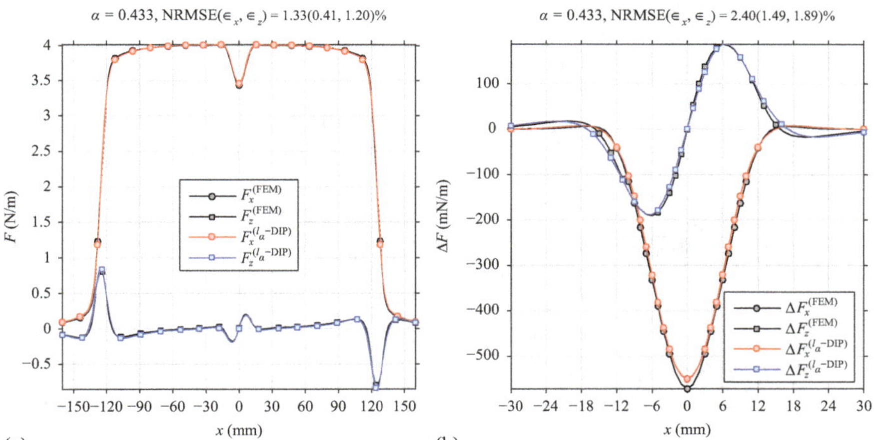

(a) (b)

Figure 2.10 Long rectangular permanent magnet—Lorentz force and DRS profiles calculated using the equivalent l_α-dipole model. (a) Lorentz force profiles. (b) DRS profiles

Although the α parameter found is not optimal for calculations with a defective conductor, the use of l_α-dipole guarantees much better results than for the l-dipole.

Figure 2.10 presents profiles calculated using l_α-dipole with $\alpha = 0.433$. The corresponding NRMSE is reduced to 1.33% and 2.40% for Lorentz force and DRS profiles, respectively.

The use of dipolar models (l-dipole, l_α-dipole) in modeling 2D LET systems enables to find analytical formulas for Lorentz force profiles exerted on the permanent magnet above moving plate of finite width and depth with and without defects (see (2.28) and (2.31)). In the further part of this section, a semianalytical method will be presented that allows the calculation of Lorentz force profiles based on analytical formulas of the magnetic field produced by a permanent magnet with a rectangular cross-section.

Let the infinitely long, rectangular permanent magnet ($w \times h$) be magnetized with a constant magnetization density $\mathbf{M} = M\mathbf{1}_z$. In this case, the magnetic field produced by the rectangular permanent magnet can be calculated using an equivalent 2D current sheet model shown in Figure 2.11. The surface sheet current densities $\mathbf{J}_S$ can be found from the cross product of $\mathbf{M}$ and the normal unit vector $\mathbf{n}$ emerging from the side edges of the permanent magnet.

The magnetic flux density $\mathbf{B}_0 = [B_{x,0}, B_{z,0}]^{\mathrm{T}}$ at any point $\mathbf{P}$ outside the permanent magnet can be calculated as a superposition of $\mathbf{B}_1$ and $\mathbf{B}_2$, the magnetic flux densities of the left and the right current sheet

$$\mathbf{B}_0 = \mathbf{B}_1 + \mathbf{B}_2 = \frac{\mu_0 M}{2\pi}\left(\int_{-h/2}^{h/2} \frac{\mathbf{1}_z \times \mathbf{r}_1}{r_1^2}\,\mathrm{d}z' - \int_{-h/2}^{h/2} \frac{\mathbf{1}_z \times \mathbf{r}_2}{r_2^2}\,\mathrm{d}z''\right), \tag{2.35}$$

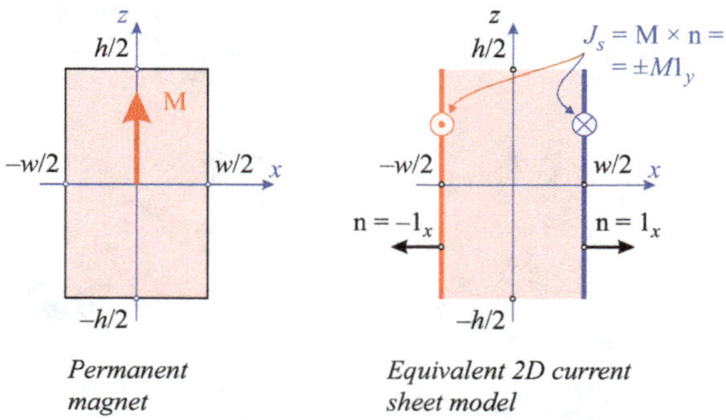

Figure 2.11 Long rectangular permanent magnet—equivalent current sheet model

where

$$r_1 = \sqrt{\left(x+\frac{w}{2}\right)^2 + (z-z')^2}, \qquad r_2 = \sqrt{\left(x-\frac{w}{2}\right)^2 + (z-z'')^2}.$$

The integrals in (2.35) can be calculated analytically and expressed as

$$B_x(x,z) = \frac{\mu_0 M}{4\pi}\ln\left[\frac{(x+\frac{w}{2})^2+(z-\frac{h}{2})^2}{(x+\frac{w}{2})^2+(z+\frac{h}{2})^2}\frac{(x-\frac{w}{2})^2+(z+\frac{h}{2})^2}{(x-\frac{w}{2})^2+(z-\frac{h}{2})^2}\right], \tag{2.36}$$

$$B_z(x,z) = \frac{\mu_0 M}{2\pi}\left[\arctan\left(\frac{z+\frac{h}{2}}{x+\frac{w}{2}}\right) - \arctan\left(\frac{z-\frac{h}{2}}{x+\frac{w}{2}}\right) + \arctan\left(\frac{z-\frac{h}{2}}{x-\frac{w}{2}}\right) - \arctan\left(\frac{z+\frac{h}{2}}{x-\frac{w}{2}}\right)\right]. \tag{2.37}$$

To calculate DRS and Lorentz force profiles, Eqs (2.36)–(2.37) are substituted into (2.22)–(2.24). Unfortunately in this case, it is not possible to determine analytical formulas as it was in the case of dipolar models. To overcome this problem, a concept of voxel grids is introduced (voxel = *vo*lume *x el*ement). It is assumed that any conducting region may be replaced by a uniform grid of conducting volumetric elements (voxels) $\Delta V = l\Delta S = l\Delta x\Delta z$ of conductivity σ_0 (l is the length of the system in the y-direction, and Δx, Δz is the size of element in x- and z-direction, respectively). In each voxel flow induced eddy currents of current density $\mathbf{J}_0$. If ΔS is sufficiently small, continuous distribution of eddy currents induced in the i-th voxel can be approximated by a constant current density vector $\mathbf{J}_{0,i}^{(n)}$ located at the center of voxel $\mathbf{r}_i$. The defect response profile $\Delta\mathbf{F}^{(n)}$ can be calculated using setup shown in Figure 2.12 and the equation (2.22) written as

$$\Delta\mathbf{F}^{(n)} = l\int_{S_d}\mathbf{J}_0\times\mathbf{B}_0\,\mathrm{d}S \simeq l\Delta S\sum_{i=1}^{N_V^D}\mathbf{J}_{0,i}^{(n)}\times\mathbf{B}_{0,i}^{(n)} = \Delta\widetilde{\mathbf{F}}^{(n)} \tag{2.38}$$

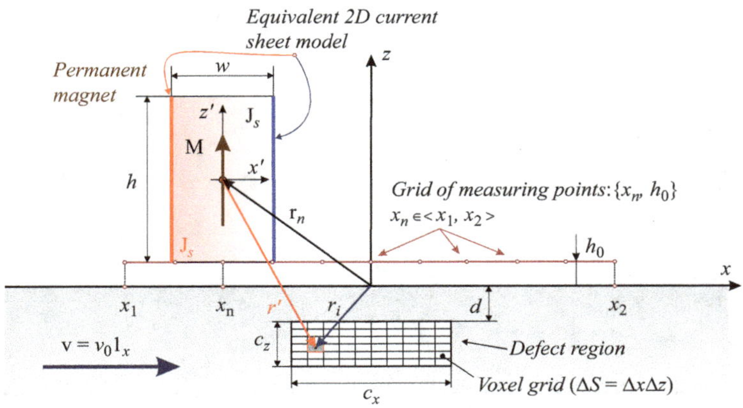

Figure 2.12 Setup for DRS calculations using 2D current sheet model of a long rectangular permanent magnet together with a grid of voxels for a defect modeling

and

$$\Delta \widetilde{\mathbf{F}}^{(n)} = l \Delta S \sigma_0 v_0 \sum_{i=1}^{N_V^D} \left[-\left(B_{z,i}^{(n)} \right)^2 \mathbf{1}_x + B_{x,i}^{(n)} B_{z,i}^{(n)} \mathbf{1}_z \right], \tag{2.39}$$

where N_V^D is the number of voxels in S_d region, $B_{x,i}^{(n)} = B_x(x_i - x_n, z_i - z_n)$, $B_{z,i}^{(n)} = B_z(x_i - x_n, z_i - z_n)$, $z_n = h_0 + h/2$, and B_x, B_z are defined by (2.36) and (2.37), respectively.

The total Lorentz force $\widetilde{\mathbf{F}}^{(n)}$ exerted on a rectangular permanent magnet located above the moving conducting plate $(L \times D)$ with K ideal defects can be calculated as

$$\begin{aligned} \widetilde{\mathbf{F}}^{(n)} &= l \Delta S_0 \sigma_0 v_0 \sum_{i=1}^{N_0} \left[\left(B_{z,i}^{(n)} \right)^2 \mathbf{1}_x - B_{x,i}^{(n)} B_{z,i}^{(n)} \mathbf{1}_z \right] \\ &+ l \sigma_0 v_0 \sum_{k=1}^{K} \Delta S_k \sum_{j=1}^{N_k^D} \left[-\left(B_{z,i}^{(n)} \right)^2 \mathbf{1}_x + B_{x,i}^{(n)} B_{z,i}^{(n)} \mathbf{1}_z \right], \end{aligned} \tag{2.40}$$

where N_0 and ΔS_0 are the number of voxels and the element size in the conductor without defects, and N_k^D, ΔS_k are the number of voxels and the element size used for the k-th defect.

Figure 2.13 shows results of simulations using the same 2D LET configuration as in the previous example. In the calculation of force profiles, a voxel grid with elements of size $\Delta S_0 = \Delta x_0 \times \Delta z_0 = 1$ mm $\times$ 1 mm for a conductor without a defect was used. For calculating the DRS, the voxel grid size in the area of the defect was $\Delta S_1 = \Delta x_1 \times \Delta z_1 = 1$ mm $\times$ 1 mm.

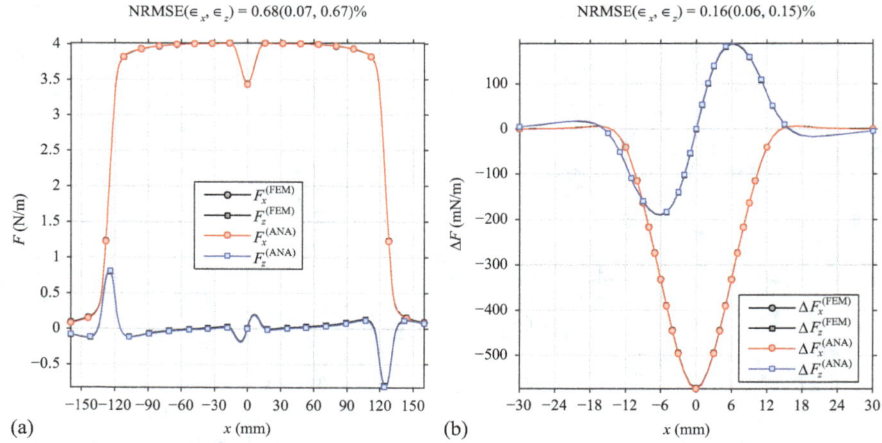

Figure 2.13 Long rectangular permanent magnet—Lorentz force and DRS profiles calculated with exact magnetic field formulas (2.30)–(2.30) *and the voxel grid density* $\Delta x \times \Delta z = 1\,mm \times 1\,mm$*. (a) Lorentz force profiles. (b) DRS profiles*

The results of semianalytic simulations have negligible errors compared to FEM, i.e., 0.68% and 0.16% for Lorentz force and DRS profiles, respectively.

In this section, two analytical approaches of Lorentz force calculations in 2D LET systems were described. In both approaches, the weak reaction formulation of the LET problem is applied, i.e., the corresponding magnetic Reynolds number is much less than 1. Both approaches are based on the principle of superposition to calculate the DRSs. In the first approach, the permanent magnet is replaced by a single equivalent magnetic l-dipole located inside the permanent magnet. This enables to find analytical formulas of DRS for one or more rectangular defects as well as the global Lorentz force profile exerted on the permanent magnet above the moving plate. The optimal position of the l_α-dipole can be found using procedure which minimizes the NRMSE between the Lorentz force profile calculated based on the l_α-model and the reference solution. In the second approach, the exact analytical formulas describing the magnetic field generated by a rectangular permanent magnet were used. For this case, it was not possible to find analytical expressions describing DRS. Therefore, a semianalytical approach based on regular grids of voxels replacing the conductor as well as defects was introduced. Sample profiles of DRS and Lorentz forces calculated by both methods together with reference solutions obtained by FEM were also shown.

2.2.2 *Lorentz forces acting on 3D permanent magnets above moving conducting plate without defects*

Analytical calculation of Lorentz force exerted on a permanent magnet located above a conducting plate without any defects moving at a constant velocity **v** is only possible

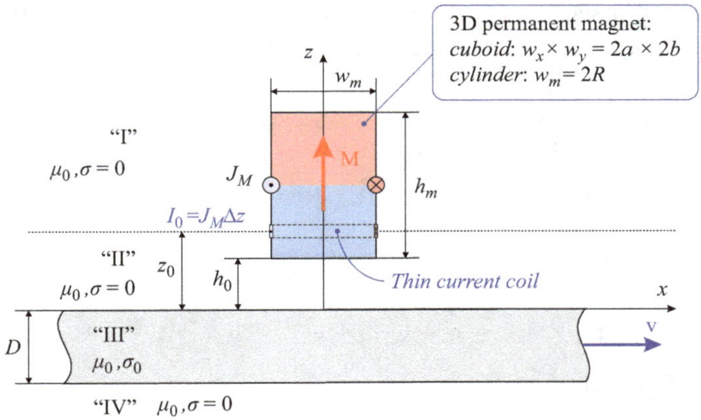

Figure 2.14 3D permanent magnet (cuboid/cylinder) above a moving nonmagnetic, conducting wide plate

when it is assumed that the plate is infinitely wide. Additionally, it is assumed that the plate is nonmagnetic ($\mu = \mu_0$) and its electrical conductivity is constant and equals σ_0. In the following, only cuboidal/cylindrical permanent magnets with constant magnetization vector **M** will be considered. In this case, the permanent magnet used in the analysis can be replaced by an infinitely thin solenoid of height h_m with a surface current density $\mathbf{J}_M$, where $J_M = M$ (see Figure 2.14).

First, the Lorentz force exerted on the infinitely thin coil with current $I_0 = J_M \Delta z$ located at $z = z_0$ above the moving conducting plate of thickness D is calculated. Using approach presented in [61], the problem in the coordinate system fixed to the coil can be described by the following set of equations:

$$\nabla^2 \mathbf{H} = 0, \qquad \text{Region: I, II, IV,} \tag{2.41}$$

$$\nabla^2 \mathbf{H} = \mu_0 \sigma_0 v_0 \frac{\partial \mathbf{H}}{\partial x}, \qquad \text{Region: III,} \tag{2.42}$$

$$\nabla \cdot \mathbf{H} = 0, \qquad \text{Region: all,} \tag{2.43}$$

where regions I–IV are defined in Figure 2.14.

For solving (2.41)–(2.43), 2D spatial Fourier transform is applied

$$\widehat{\mathbf{F}} = \widehat{\mathbf{F}}(k_x, k_y) = \int_{-\infty}^{\infty} \int_{-\infty}^{\infty} \mathbf{F}(\mathbf{r}) e^{-j(k_x x + k_y y)} \mathrm{d}x \ \mathrm{d}y, \tag{2.44}$$

where $\mathbf{r} = x\mathbf{1}_x + y\mathbf{1}_y + z\mathbf{1}_z$, and k_x, k_y are spatial frequencies. The inverse 2D spatial Fourier transform is defined as

$$\mathbf{F}(\mathbf{r}) = \left(\frac{1}{2\pi}\right)^2 \int_{-\infty}^{\infty} \int_{-\infty}^{\infty} \widehat{\mathbf{F}}(k_x, k_y) e^{j(k_x x + k_y y)} \mathrm{d}k_x \ \mathrm{d}k_y. \tag{2.45}$$

For simplicity of notation, $\widehat{\mathbf{F}}(k_x, k_y)$ is replaced with the symbol $\widehat{\mathbf{F}}$. After applying (2.44) to (2.41)–(2.43), the following equations are obtained

$$-(k_x^2 + k_y^2)\widehat{\mathbf{H}} + \frac{d^2\widehat{\mathbf{H}}}{dz^2} = 0, \qquad \text{Region: I, II, IV,} \tag{2.46}$$

$$-(k_x^2 + k_y^2)\widehat{\mathbf{H}} + \frac{d^2\widehat{\mathbf{H}}}{dz^2} = jk_x\mu_0\sigma_0 v\widehat{\mathbf{H}}, \quad \text{Region: III,} \tag{2.47}$$

$$jk_x\widehat{H}_x + jk_y\widehat{H}_y + \frac{d\widehat{H}_z}{dz} = 0, \qquad \text{Region: all.} \tag{2.48}$$

The solution of (2.46)–(2.48) can be written as:

$$\widehat{\mathbf{H}}_1 = \widehat{\mathbf{H}}^{(i)} + \widehat{\mathbf{H}}^{(e)} = \widehat{\mathbf{H}}^{(i)} + \widehat{a}e^{-kz}, \quad \text{Region: I, II,} \tag{2.49}$$

$$\widehat{\mathbf{H}}_2 = \widehat{b}e^{-\alpha z} + \widehat{c}e^{\alpha z}, \qquad \text{Region: III,} \tag{2.50}$$

$$\widehat{\mathbf{H}}_3 = \widehat{d}e^{kz}, \qquad \text{Region: IV,} \tag{2.51}$$

where $\alpha^2 = j\mu_0\sigma_0 v_0 k_x + k^2$, $k^2 = k_x^2 + k_y^2$, and $\widehat{\boldsymbol{a}}, \widehat{\boldsymbol{b}}, \widehat{\boldsymbol{c}}, \widehat{\boldsymbol{d}}$ are unknown complex constants defined as $\widehat{\boldsymbol{a}} = [\widehat{a}_x, \widehat{a}_y, \widehat{a}_z]^{\mathrm{T}}, \widehat{\boldsymbol{b}} = [\widehat{b}_x, \widehat{b}_y, \widehat{b}_z]^{\mathrm{T}}, \widehat{\boldsymbol{c}} = [\widehat{c}_x, \widehat{c}_y, \widehat{c}_z]^{\mathrm{T}}, \widehat{\boldsymbol{d}} = [\widehat{d}_x, \widehat{d}_y, \widehat{d}_z]^{\mathrm{T}}$, respectively. $\widehat{\mathbf{H}}_i = [\widehat{H}_{x,i}, \widehat{H}_{y,i}, \widehat{H}_{z,i}]^{\mathrm{T}}$ denotes the resultant magnetic field in the ith region ($i = 1, 2, 3$) while $\widehat{\mathbf{H}}^{(i)}$ is the primary (incident) magnetic field produced by the current coil in absence of the plate and $\widehat{\mathbf{H}}^{(e)}$ is the magnetic field produced by eddy currents induced in the moving conducting plate.

The spatial Fourier transform of the primary magnetic field $\widehat{\mathbf{H}}^{(i)}$ produced by a thin coil located at $z = z_0$ and carrying the current I_0 can be derived in regions I and II as

$$\text{Region I}: \begin{cases} \widehat{\mathbf{H}}^{(i)}_{||,1} = -\frac{1}{2}\left(\mathbf{1}_z \times \widehat{\mathbf{J}}_s\right) e^{-k(z-z_0)} \\ \widehat{H}^{(i)}_{z,1} = j\frac{k_x}{k}\widehat{H}^{(i)}_{x,1} + j\frac{k_y}{k}\widehat{H}^{(i)}_{y,1}, \end{cases} \tag{2.52}$$

$$\text{Region II}: \begin{cases} \widehat{\mathbf{H}}^{(i)}_{||,2} = \frac{1}{2}\left(\mathbf{1}_z \times \widehat{\mathbf{J}}_s\right) e^{k(z-z_0)} \\ \widehat{H}^{(i)}_{z,2} = -j\frac{k_x}{k}\widehat{H}^{(i)}_{x,2} - j\frac{k_y}{k}\widehat{H}^{(i)}_{y,2}, \end{cases} \tag{2.53}$$

where $\widehat{\mathbf{H}}^{(i)}_{||,n} = \widehat{H}^{(i)}_{x,n}\mathbf{1}_x + \widehat{H}^{(i)}_{y,n}\mathbf{1}_y$, $n = 1, 2$ and $\widehat{\mathbf{J}}_s$ is 2D Fourier transform of the coil with current. To complete formulas (2.52) and (2.53), it is necessary to find the 2D Fourier transform $\widehat{\mathbf{J}}_s$.

At the beginning, it is analyzed an infinitely thin rectangular coil $2a \times 2b$ with a current given as

$$\mathbf{J}_s(x, y) = I_0 \begin{bmatrix} \Pi\left(\frac{x}{2a}\right)\delta(y+b) - \Pi\left(\frac{x}{2a}\right)\delta(y-b) \\ \Pi\left(\frac{y}{2b}\right)\delta(x-a) - \Pi\left(\frac{y}{2b}\right)\delta(x+a) \end{bmatrix}, \tag{2.54}$$

where $I_0 = M\Delta z$, $\delta(\,\cdot\,)$ is the Dirac function and $\Pi(\,\cdot\,)$ is the rectangle function [113] defined as

$$\Pi(x) = \begin{cases} 1, & |x| < \frac{1}{2} \\ \frac{1}{2}, & |x| = \frac{1}{2} \\ 0, & |x| > \frac{1}{2} \end{cases} \tag{2.55}$$

The 2D spatial Fourier transform $\widehat{\mathbf{J}}_s$ of (2.54) can be expressed as

$$\widehat{\mathbf{J}}_s = \begin{bmatrix} \widehat{J}_{sx} \\ \widehat{J}_{sy} \end{bmatrix} = j\,4I_0 \sin(k_x a)\sin(k_y b) \begin{bmatrix} \dfrac{1}{k_x} \\ -\dfrac{1}{k_y} \end{bmatrix}. \tag{2.56}$$

Similarly, a current flowing in an infinitely thin loop of radius R can be defined as

$$\mathbf{J}_s(x,y) = I_0 \begin{bmatrix} -\delta(r-R)\sin\varphi \\ \delta(r-R)\cos\varphi \end{bmatrix}. \tag{2.57}$$

The corresponding 2D spatial Fourier transform can be found as

$$\widehat{\mathbf{J}}_s = \begin{bmatrix} \widehat{J}_{sx} \\ \widehat{J}_{sy} \end{bmatrix} = j\,2\pi R I_0 \mathrm{J}_1(kR) \begin{bmatrix} \dfrac{k_y}{k} \\ -\dfrac{k_x}{k} \end{bmatrix}, \tag{2.58}$$

where $r^2 = x^2 + y^2$, $\sin\varphi = y/r$, $\cos\varphi = x/r$, and $\mathrm{J}_1(\,\cdot\,)$ is the first order Bessel function of the first kind [114]

$$\mathrm{J}_1(z) = \frac{j}{\pi}\int_0^{\pi} e^{-jz\cos\theta}\cos\theta\,\mathrm{d}\theta. \tag{2.59}$$

Unknown constants $\widehat{\boldsymbol{a}}, \widehat{\boldsymbol{b}}, \widehat{\boldsymbol{c}}, \widehat{\boldsymbol{d}}$ in (2.49)–(2.51) can be determined from the continuity conditions at $z = 0$ $(i = 2)$ and $z = -D$ $(i = 3)$

$$\widehat{\mathbf{H}}_{t,i} = \widehat{\mathbf{H}}_{t,i+1}, \tag{2.60}$$

$$\widehat{\mathbf{B}}_{n,i} = \widehat{\mathbf{B}}_{n,i+1}. \tag{2.61}$$

The indexes t and n denote tangential and normal components of vectors $\mathbf{H}$ and $\mathbf{B}$, respectively. Additionally to (2.60)–(2.61), the condition $\nabla \times \mathbf{B}_{|z} = 0$ in the region III has to be taken into account because the induced eddy currents in the conducting plate flow only in x-y–planes ($J_z = 0$). After a few elementary transformations, the constant $\widehat{\boldsymbol{a}}$ can be found as

$$\widehat{a} = \begin{bmatrix} \widehat{a}_x = T(k,\beta)\,\widehat{H}^{(i)}_{x,2}\Big|_{z=0} = \frac{1}{2}T(k,\beta)\left(\mathbf{1}_z \times \widehat{\mathbf{J}}_s\right)e^{-kz_0} \\ \widehat{a}_y = \dfrac{k_y}{k_x}\widehat{a}_x \\ \widehat{a}_z = j\dfrac{k}{k_x}\widehat{a}_x \end{bmatrix}, \tag{2.62}$$

where $T(k, \beta)$ is given by the expression

$$T(k,\beta) = \frac{(\beta^2 - 1)\tanh \beta kD}{2\beta + (1+\beta^2)\tanh \beta kD}, \tag{2.63}$$

where $\beta = \alpha/k$, $k^2 = k_x^2 + k_y^2$, and $\alpha^2 = j\mu_0\sigma_0 v_0 k_x + k^2$. The coefficient $T(k, \beta)$ can be treated as a *reflection coefficient* of the incident magnetic field *reflected* from the moving conducting plate [61,66,115]. Finally, using (2.53) and (2.62), the magnetic field $\widehat{\mathbf{H}}^{(e)}$ in regions I and II produced by eddy currents in the moving plate can be obtained as

$$\widehat{\mathbf{H}}_{||}^{(e)} = T(k,\beta)\,\widehat{\mathbf{H}}_{||,2}^{(i)}\Big|_{z=0} e^{-kz} = \tfrac{1}{2}T(k,\beta)\left(\mathbf{1}_z \times \widehat{\mathbf{J}}_s\right)e^{-k(z+z_0)}, \tag{2.64}$$

$$\widehat{H}_z^{(e)} = -j\tfrac{1}{2}T(k,\beta)\left(\frac{k_x}{k}\widehat{J}_{sy}e^{-kz_0} - \frac{k_y}{k}\widehat{J}_{sx}e^{-kz_0}\right)e^{-kz}. \tag{2.65}$$

The Lorentz force exerted on a thin coil above a moving conducting plate [66] can be determined using the following formula

$$\mathbf{F} = \mu_0 \int_{-\infty}^{\infty}\int_{-\infty}^{\infty} \mathbf{J}_s(\mathbf{r}) \times \mathbf{H}^{(e)}(\mathbf{r})\Big|_{z=z_0} \mathrm{d}x\, \mathrm{d}y. \tag{2.66}$$

According to Parseval's theorem [113], equation (2.66) can be written as

$$\mathbf{F} = \frac{\mu_0}{4\pi^2}\int_{-\infty}^{\infty}\int_{-\infty}^{\infty} \widehat{\mathbf{J}}_s^* \times \widehat{\mathbf{H}}^{(e)}\Big|_{z=z_0} \mathrm{d}k_x\, \mathrm{d}k_y. \tag{2.67}$$

After substituting (2.64)–(2.65) into (2.67), the following expressions are given to the Lorentz force components

$$F_x = -j\frac{\mu_0}{8\pi^2}\int_{-\infty}^{\infty}\int_{-\infty}^{\infty} T(k,\beta)\left(\frac{k_x}{k}\left|\widehat{J}_{sy}\right|^2 - \frac{k_y}{k}\widehat{J}_{sy}^*\widehat{J}_{sx}\right)e^{-2kz_0}\mathrm{d}k_x\, \mathrm{d}k_y, \tag{2.68}$$

$$F_y = 0, \tag{2.69}$$

$$F_z = \frac{\mu_0}{8\pi^2}\int_{-\infty}^{\infty}\int_{-\infty}^{\infty} T(k,\beta)\left(\left|\widehat{J}_{sx}\right|^2 + \left|\widehat{J}_{sy}\right|^2\right)e^{-2kz_0}\mathrm{d}k_x\, \mathrm{d}k_y. \tag{2.70}$$

The side force F_y equals 0 due to the x-symmetry of the coil. Using the symmetry of $\widehat{\mathbf{J}}_s$ in the spatial Fourier domain [see (2.56) and (2.58)] defined as

$$\widehat{J}_{sx}(k_x) = \widehat{J}_{sx}(-k_x), \qquad \widehat{J}_{sx}(k_y) = -\widehat{J}_{sx}(-k_y),$$
$$\widehat{J}_{sy}(k_x) = -\widehat{J}_{sy}(-k_x), \quad \widehat{J}_{sy}(k_y) = \widehat{J}_{sy}(-k_y),$$

equations (2.68)–(2.70) can be simplified to the following form

$$F_x = \frac{\mu_0}{2\pi^2}\int_0^{\infty}\int_0^{\infty} \Im[T(k,\beta)]\left(\frac{k_x}{k}|\widehat{J}_{sy}|^2 - \frac{k_y}{k}\widehat{J}_{sy}^*\widehat{J}_{sx}\right)e^{-2kz_0}\,\mathrm{d}k_x\, \mathrm{d}k_y, \tag{2.71}$$

$$F_z = \frac{\mu_0}{2\pi^2}\int_0^{\infty}\int_0^{\infty} \Re[T(k,\beta)]\left(|\widehat{J}_{sx}|^2 + |\widehat{J}_{sy}|^2\right)e^{-2kz_0}\,\mathrm{d}k_x\, \mathrm{d}k_y. \tag{2.72}$$

In order to calculate the Lorentz force $\mathbf{F}^{PM}$ acting on the infinitely thin solenoid of the height h_m shown in Figure 2.14, and thus the permanent magnet, it is necessary first to integrate the incident magnetic field contributions at $z = 0$ from infinitely thin coils evenly distributed along the solenoid. The resultant field $\widehat{\mathbf{H}}_2$ after integration takes the following form

$$\widehat{\mathbf{H}}^{(i)}_{||,2}\Big|_{z=0} = \tfrac{1}{2}\left(\mathbf{1}_z \times \widehat{\mathbf{J}}_s\right)\frac{e^{-kh_0}}{k}\left(1 - e^{-kH_m}\right), \tag{2.73}$$

$$\widehat{H}^{(i)}_{z,2}\Big|_{z=0} = -j\frac{k_x}{k}\,\widehat{H}^{(i)}_{x,2}\Big|_{z=0} - j\frac{k_y}{k}\,\widehat{H}^{(i)}_{y,2}\Big|_{z=0}. \tag{2.74}$$

Next, the magnetic field $\widehat{\mathbf{H}}^{(e)}$ produced by induced eddy currents in the plate (2.64)–(2.65) has to be integrated along the solenoid height [115]. Finally, the following expressions for the Lorentz force exerted on a permanent magnet above moving conducting plate can be obtained as

$$\begin{aligned} F_x^{PM} &= \frac{\mu_0}{2\pi^2}\int_0^\infty\int_0^\infty \Im[T(k,\beta)]\left(k_x|\widehat{J}_{sy})|^2 - k_y\widehat{J}^*_{sy}\widehat{J}_{sx}\right) \\ &\qquad\qquad \times\frac{\left(1 - e^{-kH_m}\right)^2}{k^3}e^{-2kh_0}\,\mathrm{d}k_x\mathrm{d}k_y, \end{aligned} \tag{2.75}$$

$$F_y^{PM} = 0 \tag{2.76}$$

$$\begin{aligned} F_z^{PM} &= \frac{\mu_0}{2\pi^2}\int_0^\infty\int_0^\infty \Re[T(k,\beta)]\left(|\widehat{J}_{sx}|^2 + |\widehat{J}_{sy}|^2\right) \\ &\qquad\qquad \times\frac{\left(1 - e^{-kH_m}\right)^2}{k^2}e^{-2kh_0}\,\mathrm{d}k_x\mathrm{d}k_y. \end{aligned} \tag{2.77}$$

Formulas (2.75) and (2.77) are verified using the FEM applied to similar configuration setup as for the analytical method. Figure 2.15 shows normalized Lorentz forces calculated using (2.75)–(2.77) and the FEM. The reference velocity v_0 is the velocity at which the lift force is equal to the drag force ($F_x = F_z$). The reference force F_0 is an asymptotic limit of (2.77) for $\sigma_0 v_0 \to \infty$ equals

$$F_0 = \frac{\mu_0}{2\pi^2}\int_0^\infty\int_0^\infty\left(|\widehat{J}_{sx}|^2 + |\widehat{J}_{sy}|^2\right)\frac{\left(1 - e^{-kH_m}\right)^2}{k^2}e^{-2kh_0}\,\mathrm{d}k_x\,\mathrm{d}k_y. \tag{2.78}$$

It can be observed a very good agreement between forces calculated analytically and obtained from FEM simulations. The normalized root mean deviation between ANA and FEM is equal to 0.42% for both simulated cases. The presented approach enables also to calculate induced eddy currents in the moving plate in easy way. Using $\widehat{\mathbf{J}} = \nabla \times \widehat{\mathbf{H}}_3$, the 2D spatial Fourier transform of the eddy current density $\widehat{\mathbf{J}}$ takes the form

$$\widehat{\mathbf{J}} = \begin{bmatrix}\widehat{J}_x\\ \widehat{J}_y\end{bmatrix} = (1-\beta)\frac{1 - \frac{1-\beta}{1+\beta}e^{-2\beta k(D+z)}}{1 - \left(\frac{1-\beta}{1+\beta}\right)^2 e^{-2\beta kD}}e^{\beta kz}\left(1 - e^{-kH_m}\right)e^{-kh_0}\begin{bmatrix}\widehat{J}_{sx}\\ \widehat{J}_{sy}\end{bmatrix}. \tag{2.79}$$

Applying symmetry properties of $\widehat{\mathbf{J}}_s$ in the spatial Fourier domain, the eddy current density $\mathbf{J}$ at any point $\mathbf{r}$ located in the plate can be determined from the inverse Fourier

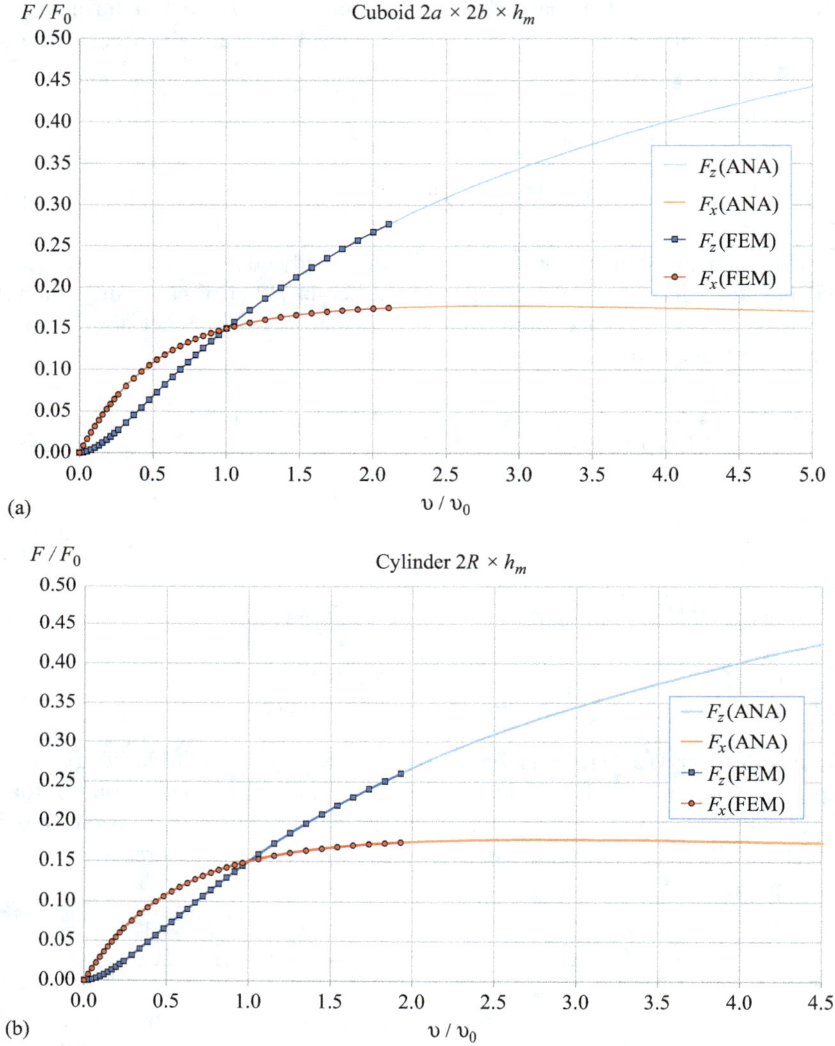

Figure 2.15 Normalized Lorentz force acting on a permanent magnet placed $h_0 = 1$ mm above a moving conductive plate with a thickness of $D = 100$ mm ($\sigma_0 = 30.61$ MS/m). (a) Cuboid: 15 mm × 15 mm × 25 mm, $F_0 = 65.6$ N, $v_0 = 9.48$ m/s. (b) Cylinder: 15 mm × 25 mm, $F_0 = 51.1$ N, $v_0 = 10.37$ m/s. (ANA) analytical solutions, (FEM) finite element method

transform as

$$J_x(\mathbf{r}) = \frac{1}{\pi^2}\int_0^\infty\int_0^\infty \sin(k_y y)\left\{\sin(k_x x)\Re[\widehat{J_x}] + \cos(k_x x)\Im[\widehat{J_x}]\right\} \mathrm{d}k_x\ \mathrm{d}k_y, \tag{2.80}$$

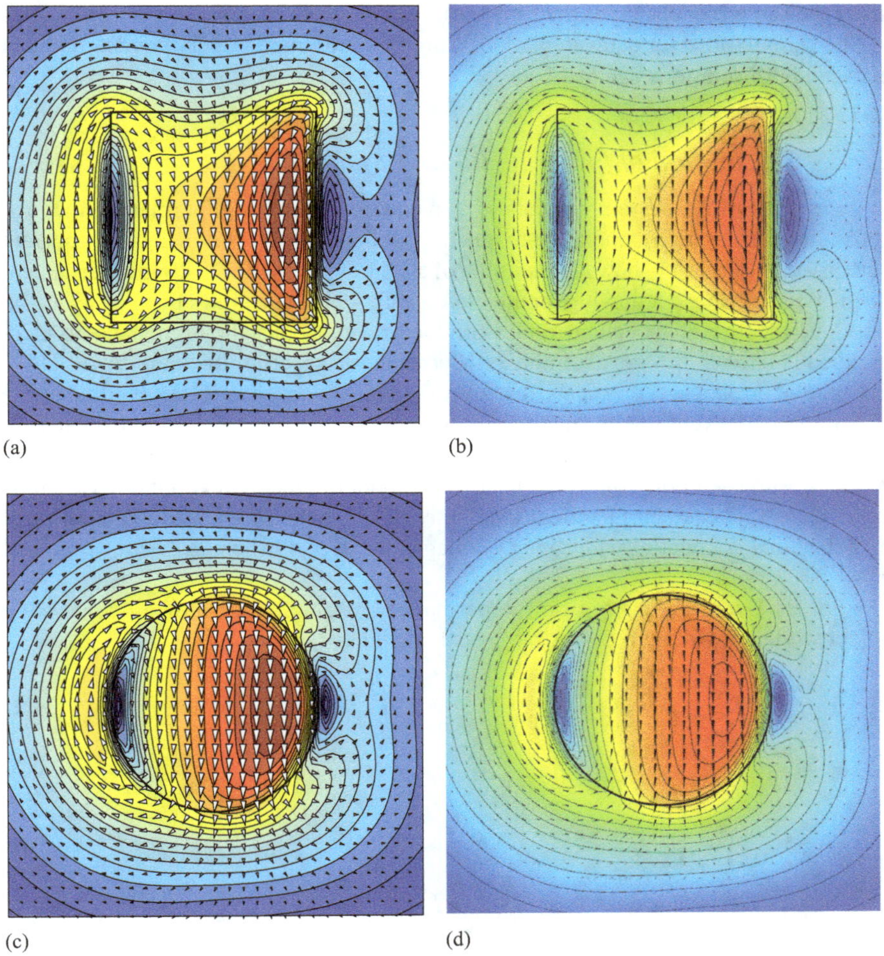

Figure 2.16 Eddy currents on the upper surface of a conductive plate moving at a velocity $v_0 = 10$ m/s under a rectangular (a–b) or cylindrical (c–d) permanent magnet. Analytical solutions: (a), (c). FEM: (b), (d)

$$J_y(\mathbf{r}) = \frac{1}{\pi^2} \int_0^\infty \int_0^\infty \cos(k_y y) \left\{ \sin(k_x x) \Im[\widehat{J_y}] - \cos(k_x x) \Re[\widehat{J_y}] \right\} \mathrm{d}k_x \, \mathrm{d}k_y, \tag{2.81}$$

where $\mathbf{r} = x\mathbf{1}_x + y\mathbf{1}_y + z\mathbf{1}_z$.

Figure 2.16 shows sample stationary eddy current distributions on the surface $z = 0$ of the plate moving at a velocity $v_0 = 10$ m/s. The distributions are calculated using the same models as for the Lorentz force calculations. For comparison, results of FEM simulations for equivalent models are also presented. A very good agreement between both solutions can be observed.

2.2.3 Calculation of forces in 3D LET systems

In general, 3D LET system consists of a magnetic system (e.g. permanent magnet, Halbach configuration), and a nonmagnetic, conducting object moving at a constant velocity **v**. Figure 2.17 shows an example of a LET system with a cuboidal permanent magnet placed above a moving, conductive block with a surface defect.

A conductive block can be a solid conductor with a constant, homogeneous electrical conductivity σ_0 or an assembly of N conducting sheets of thickness $\Delta h = D/N$. The use of a package of sheets instead of a solid block decisively facilitates the preparation of experiments involving studies on the same defects but located at different depths [91,102]. If thickness Δh of sheets is small enough and sheets are isolated from each other then an anisotropic model of electrical conductivity can be applied. In this case an electrical conductivity of stacked sheets can be described by a diagonal conductivity tensor $[\boldsymbol{\sigma}] = \mathrm{diag}(\sigma_{xx}, \sigma_{yy}, \sigma_{zz})$, where $\sigma_{xx} = \sigma_{yy} = \sigma_0$ and $\sigma_{zz} = 0$.

Forces **F** and $\mathbf{F}_0$ exerted on the permanent magnet above moving conducting object with and without a defect can be calculated as

$$\mathbf{F} = -\int_{V-V_D} \mathbf{J} \times \mathbf{B}\, \mathrm{d}V \quad \text{and} \quad \mathbf{F}_0 = -\int_V \mathbf{J}_0 \times \mathbf{B}\, \mathrm{d}V, \tag{2.82}$$

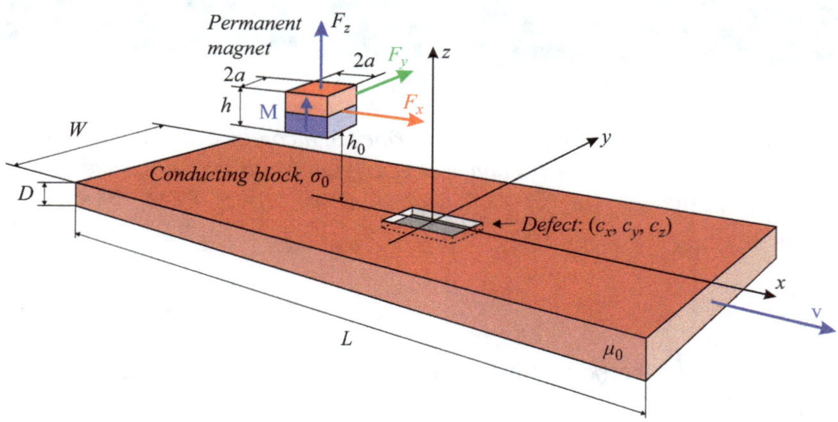

Figure 2.17 An exemplary 3D LET configuration consisting of a cuboidal permanent magnet placed above a moving nonmagnetic, conductive block with a cuboidal surface defect ($c_x \times c_y \times c_z, d = 0$)

where V and V_D denote the volumes of the object and the defect, respectively. $\mathbf{J}$ describes eddy currents induced in the defective object while $\mathbf{J}_0$ in the object without the defect. DRSs are defined analogically to the 2D case as

$$\Delta\mathbf{F} = \mathbf{F} - \mathbf{F}_0, \quad \Delta\mathbf{F} = [\Delta F_x, \Delta F_y, \Delta F_z]^{\mathrm{T}}. \tag{2.83}$$

After substituting formulas (2.82) into equation (2.83) one obtains

$$\begin{aligned}
\Delta\mathbf{F} &= \int_V \mathbf{J}_0 \times \mathbf{B}\,\mathrm{d}V - \int_{V-V_D} \mathbf{J} \times \mathbf{B}\,\mathrm{d}V \\
&= \int_{V-V_D} (\mathbf{J}_0 - \mathbf{J}) \times \mathbf{B}\,\mathrm{d}V + \int_{V_D} \mathbf{J}_0 \times \mathbf{B}\,\mathrm{d}V \\
&= \int_{V-V_D} \mathbf{j}_E \times \mathbf{B}\,\mathrm{d}V + \int_{V_D} \mathbf{j}_D \times \mathbf{B}\,\mathrm{d}V = \int_V (\mathbf{j}_E + \mathbf{j}_D) \times \mathbf{B}\,\mathrm{d}V \\
&= \int_V \mathbf{j} \times \mathbf{B}\,\mathrm{d}V\,,
\end{aligned} \tag{2.84}$$

where $\mathbf{j}_E = \mathbf{J}_0 - \mathbf{J}$ is the density of distortion currents caused by the defect flowing outside the defect and $\mathbf{j}_D = \mathbf{J}_0$ is the density of eddy currents without defect truncated to the defect (so-called defect eddy currents). The sum $\mathbf{j} = \mathbf{j}_E + \mathbf{j}_D$ is called the total defect distortion current. The analytical calculation of DRS (2.84) in the general case is not possible. However, the problem can be significantly simplified if the magnetic Reynolds number R_m is much smaller than 1 and the approach of the weak reaction of eddy currents can be applied. In this case, the DRSs from the defect located in the conductive object can be determined as

$$\Delta\mathbf{F} = \int_V \mathbf{j} \times \mathbf{B}_0\,\mathrm{d}V = \int_V (\mathbf{j}_E + \mathbf{j}_D) \times \mathbf{B}_0\,\mathrm{d}V\,, \tag{2.85}$$

where $\mathbf{B}_0$ is the magnetic flux density produced by a permanent magnet.

Most of the magnetic configurations used in LET systems can be modeled using the appropriate sets of magnetic dipoles [100]. If a permanent magnet is modeled using N_D magnetic dipoles with magnetic moments $\mathbf{m}_i = [m_{x,i}, m_{y,i}, m_{z,i}]^{\mathrm{T}}$ located in points $\mathbf{r}_i = [x_i, y_i, z_i]^{\mathrm{T}}$, its magnetic field $\mathbf{B}_{0,k}$ at any point $\mathbf{r}_k = [x_k, y_k, z_k]^{\mathrm{T}}$ can be determined as

$$\mathbf{B}_{0,k} = \sum_{i=1}^{N_D} \mathbf{b}_{ki}, \tag{2.86}$$

and

$$\mathbf{b}_{ki} = \frac{\mu_0}{4\pi}\begin{bmatrix} 3\dfrac{x_{ki}^2}{R_{ki}^5} - \dfrac{1}{R_{ki}^3} & 3\dfrac{x_{ki}y_{ki}}{R_{ki}^5} & 3\dfrac{x_{ki}z_{ki}}{R_{ki}^5} \\ 3\dfrac{x_{ki}y_{ki}}{R_{ki}^5} & 3\dfrac{y_{ki}^2}{R_{ki}^5} - \dfrac{1}{R_{ki}^3} & 3\dfrac{y_{ki}z_{ki}}{R_{ki}^5} \\ 3\dfrac{x_{ki}z_{ki}}{R_{ki}^5} & 3\dfrac{y_{ki}z_{ki}}{R_{ki}^5} & 3\dfrac{z_{ki}^2}{R_{ki}^5} - \dfrac{1}{R_{ki}^3} \end{bmatrix} \cdot \mathbf{m}_i^{\mathrm{T}}, \tag{2.87}$$

where

$$x_{ki} = x_k - x_i, \quad y_{ki} = y_k - y_i, \quad z_{ki} = z_k - z_i,$$
$$R_{ki}^2 = r_{ki}^2 + z_{ki}^2, \quad r_{ki}^2 = x_{ki}^2 + y_{ki}^2.$$

Induced eddy currents $\mathbf{J}_0$ in a moving conducting object ($\mathbf{v} = v_0\mathbf{1}_x$) can be directly calculated using Ohm's law (2.19)

$$\mathbf{J}_0 = \sigma_0(\mathbf{E} + \mathbf{v} \times \mathbf{B}_0) = \sigma_0(-\nabla\varphi + \mathbf{v} \times \mathbf{B}_0) = -\sigma_0\begin{bmatrix} \dfrac{\partial\varphi}{\partial x} \\ \dfrac{\partial\varphi}{\partial y} + v_0B_{z,0} \\ \dfrac{\partial\varphi}{\partial z} - v_0B_{y,0} \end{bmatrix}, \tag{2.88}$$

where φ is the electric scalar potential fulfilling Laplace equation $\nabla^2\varphi = 0$ with appropriate boundary conditions regarding the behavior of eddy currents at the boundaries of the object and defects.

In the case of a conductive block $L \times W \times D$ whose L and W dimensions are much larger than the magnets' dimensions (infinitely wide plate model), the following expressions describing induced eddy currents evoked by the magnetic dipole $\mathbf{m}_i$ above the moving plate [116] can be used for calculation of $\mathbf{J}_0$

$$\mathbf{J}_{0,ki} = \frac{\mu_0\sigma_0v_0}{4\pi}\begin{bmatrix} C_1g_1 + C_3g_3 & C_1f_1 - C_3f_3 & C_2g_2 \\ -3C_1f_1 - C_3f_3 & -C_1g_1 - C_3g_3 & C_0 - C_2f_2 \\ 0 & 0 & 0 \end{bmatrix} \cdot \mathbf{m}_i^{\mathrm{T}}, \tag{2.89}$$

where

$$\mathbf{J}_{0,ki} = \mathbf{J}_{0,i}(x_k, y_k, z_k) = [J_{x0,i}, J_{y0,i}, J_{z0,i}]^{\mathrm{T}},$$

$$f_1 = \frac{x_{ki}}{r_{ki}}, \quad f_2 = \frac{x_{ki}^2 - y_{ki}^2}{r_{ki}^2}, \quad f_3 = \frac{x_{ki}(x_{ki}^2 - 3y_{ki}^2)}{r_{ki}^3},$$

$$g_1 = \frac{y_{ki}}{r_{ki}}, \quad g_2 = 2\frac{x_{ki}y_{ki}}{r_{ki}^2}, \quad g_3 = \frac{y_{ki}(3x_{ki}^2 - y_{ki}^2)}{r_{ki}^3},$$

and

$$C_0 = \frac{3r_{ki}^2}{2R_{ki}^5} - \frac{1}{R_{ki}^3}, \quad C_1 = \mathrm{sign}(z_{ki})\frac{3r_{ki}|z_{ki}|}{4R_{ki}^5}, \quad C_2 = -\frac{3r_{ki}^2}{2R_{ki}^5},$$

$$C_3 = \mathrm{sign}(z_{ki})\left[\frac{|z_{ki}|}{R_{ki}^3}\left(\frac{3r_{ki}}{4R_{ki}^2} + \frac{1}{r_{ki}}\right) + \frac{2}{r_{ki}^3}\left(\frac{|z_{ki}|}{R_{ki}} - 1\right)\right].$$

It should be noted that the component J_{z0} disappears which means that the induced eddy currents flow only in planes parallel to the surface of the plate.

For cuboidal permanent magnets, it is possible to find analytical expressions which enable to determine the magnetic field at any point around the magnet. According to [110], the formulas take the following form

$$\mathbf{B}_0(x,y,z) = \begin{bmatrix} B_{x,0} \\ B_{y,0} \\ B_{z,0} \end{bmatrix} = \frac{\mu_0 M}{4\pi} \begin{bmatrix} \sum_{k=1}^{2}\sum_{n=1}^{2}(-1)^{k+n}\ln\left[F_{nk}(x,y,z)\right] \\ \sum_{k=1}^{2}\sum_{m=1}^{2}(-1)^{k+m}\ln\left[H_{mk}(x,y,z)\right] \\ \sum_{k=1}^{2}\sum_{m=1}^{2}\sum_{n=1}^{2}(-1)^{k+m+n}\arctan\left[G_{nmk}(x,y,z)\right] \end{bmatrix}, \quad (2.90)$$

where

$$F_{nk}(x,y,z) = \frac{R_{n1k}(x,y,z) + \Delta y_1}{R_{n2k}(x,y,z) + \Delta y_2}, \quad (2.91)$$

$$G_{nmk}(x,y,z) = \frac{\Delta x_n \Delta y_m}{\Delta z_k R_{nmk}(x,y,z)}, \quad (2.92)$$

$$H_{mk}(x,y,z) = \frac{R_{1mk}(x,y,z) + \Delta x_1}{R_{2mk}(x,y,z) + \Delta x_2} \quad (2.93)$$

and

$$\begin{aligned} R_{nmk}(x,y,z) &= \sqrt{\Delta x_n^2 + \Delta y_m^2 + \Delta z_k^2}, \\ \Delta x_n &= x - x_n, \\ \Delta y_m &= y - y_m, \\ \Delta z_k &= z - z_k. \end{aligned}$$

The parameters x_1, x_2, y_1, y_2, and z_1, z_2 are defined in Figure 2.18.

If a cuboidal permanent magnet is located above a large conducting plate moving at a speed small enough to allow the use of WRA, it is possible to find the flow of induced eddy currents analytically.

The electric scalar potential φ at any point $\mathbf{r} = [x,y,z]^{\mathrm{T}}$ in the moving plate can be found as

$$\begin{aligned} \varphi(x,y,z) = \frac{\mu_0 M v_0}{4\pi} \sum_{k=1}^{2}\sum_{m=1}^{2}\sum_{n=1}^{2}(-1)^{k+m+n}\big\{\Delta x_n \ln\left[R_{nmk}(x,y,z) - \Delta z_k\right] \\ + \Delta y_m \arctan\frac{\Delta x_n \Delta z_k}{\Delta y_m R_{nmk}(x,y,z)} - \Delta z_k \ln\left[R_{nmk}(x,y,z) + \Delta x_n\right]\big\}. \quad (2.94) \end{aligned}$$

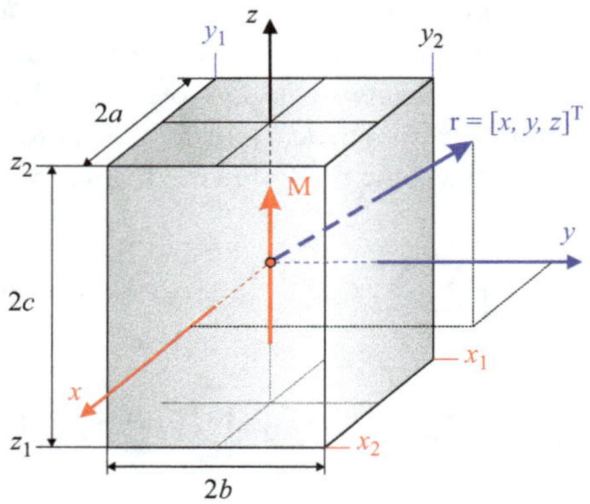

Figure 2.18 Cuboidal permanent magnet—parameter definitions according to [110]

The eddy current density $\mathbf{J}_0$ in the plate is calculated from:

$$\mathbf{J}_0(x, y, z) = -\sigma_0 \begin{bmatrix} \dfrac{\partial \varphi}{\partial x} \\ \dfrac{\partial \varphi}{\partial y} + vB_{z,0} \\ 0 \end{bmatrix}. \tag{2.95}$$

where $B_{z,0}$ is given by (2.90) and gradient components are as follows:

$$\frac{\partial \varphi}{\partial x} = \frac{\mu_0 M v_0}{4\pi} \sum_{k=1}^{2} \sum_{m=1}^{2} \left\{ (-1)^{k+m} \ln \frac{R_{2mk}(x, y, z) - \Delta z_k}{R_{1mk}(x, y, z) - \Delta z_k} \right.$$
$$\left. + \sum_{n=1}^{2} \frac{(-1)^{k+m+n}}{R_{nmk}(x, y, z)} \left[\frac{\Delta y_m^2 \Delta z_k}{\Delta x_n^2 + \Delta y_m^2} + \frac{\Delta x_n^2}{R_{nmk}(x, y, z) - \Delta z_k} - \Delta z_k \right] \right\}, \tag{2.96}$$

$$\frac{\partial \varphi}{\partial y} = \frac{\mu_0 M v_0}{4\pi} \sum_{k=1}^{2} \sum_{m=1}^{2} \sum_{n=1}^{2} (-1)^{k+m+n} \left\{ \arctan \frac{\Delta x_n \Delta z_k}{\Delta y_m R_{nmk}(x, y, z)} \right.$$
$$+ \frac{\Delta y_m}{R_{nmk}(x, y, z)} \left[\frac{\Delta x_n}{R_{nmk}(x, y, z) - \Delta z_k} - \frac{\Delta z_k}{R_{nmk}(x, y, z) + \Delta x_n} \right.$$
$$\left. \left. - \frac{\Delta x_n \Delta z_k \left(\Delta x_n^2 + 2\Delta y_m^2 + \Delta z_k^2 \right)}{\left(\Delta x_n^2 + \Delta y_m^2 \right) \left(\Delta y_m^2 + \Delta z_k^2 \right)} \right] \right\}. \tag{2.97}$$

It is also possible to determine analytical formulas for the calculation of the magnetic field produced by a cylindrical permanent magnet [117]. Unfortunately, in this case, it is not possible to provide analytical dependencies allowing the calculation of induced currents in the moving conductive plate.

2.2.4 Oscillatory motion of permanent magnets above a conducting plate

2.2.4.1 Introduction and motivation

This chapter addresses the analysis of motion-induced eddy currents in the case of harmonic motion of current carrying coils or permanent magnets in the vicinity of electrical conductors. The survey in the framework of MIECT is motivated by the observation of velocity oscillations of the specimen, resulting for example from the control circuit of the linear drive. Three normalized velocity profiles, which are obtained during operation, are shown in Figure 2.19. The velocities are determined using the incremental position encoder TONiC T1000 (Renishaw plc, www.renishaw.com). It has been observed that the relative oscillation amplitudes reach up to 3% in the present case. The time-dependent velocity influences the induced eddy current density together with the total magnetic field. In consequence, the Lorentz force is influenced as well. The goal is to determine this electromagnetic force $\mathbf{F}^{(\mathrm{EM})}$ resulting from the observed oscillations. This quantity serves as an input of the force sensing unit. An ideal sensor would convert this force into another physical quantity, which is proportional to the measured force without any alteration such that $\mathbf{F}^{(\mathrm{EM})} = \mathbf{F}^{(\mathrm{MEAS})}$. However, in reality, every sensor obeys a distinct characteristic, which can be described by its transfer function. As a result, the output of the sensor differs with respect to its input. Despite of the academic nature of this problem and its particular observation pursuant to MIECT, analogies can be found in many modern engineering problems where oscillations occur and motion-induced eddy currents, together with the associated forces, are utilized. A typical example are eddy current brakes [118,119], magnetic levitation [59,120], electromagnetic damping [121–124], electromagnetic coupling [125], electromagnetic vibration isolation systems or suspension systems [126,127], or even very recently, energy harvesting [128]. Several analytical approaches considering constant motion, especially in case of high speeds,

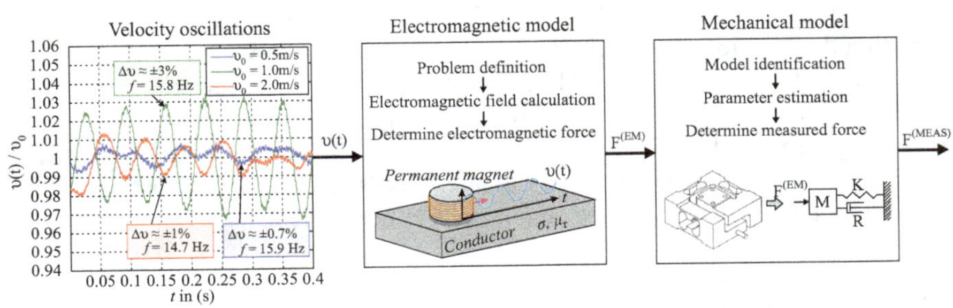

Figure 2.19 Problem under investigation. A permanent magnet moves with a sinusoidal velocity profile relative to a conductive slab. (left) Observed velocity oscillations in the experimental LET setup; (middle) Electromagnetic model including time-dependent velocities; (right) Mechanical model of the force sensor unit altering the electromagnetic input force

can be found in the literature [58,60,61,66,68,69,129–131]. Linear electromagnetic transducers are either analyzed with simplified analytical approximations neglecting the reactance of the conductor [132] or by means of time-consuming numerical simulations [133]. Together with MIECT, all these applications have in common that they involve mechanical systems which obey eigenfrequencies and the tendency to oscillate, either intended or parasitic, in form of undesired vibrations of industrial or laboratory setups. A typical example of the latter was also observed by Ramos *et al.* [77] in the framework of NDT.

The time-dependent velocity accounts for a more intricate eddy current profile inside the conductor. As a consequence, the problem cannot be treated as stationary or quasi-static anymore. The resistive and inductive nature of the conductor implies a complex interaction between the primary magnetic field generated by the magnetic field source, which oscillates at a given frequency, and the temporally as well as spatially pulsating secondary magnetic field generated by the induced eddy currents. During analytical analysis of oscillating systems, the back reaction of the conductor was merely taken into account [134]. Admittedly, Amati *et al.* [135] addressed the question of sinusoidal speed variations in torsional eddy current dampers but neglected the inductive character of the conductor supported by the assumption of small amplitude oscillations. Considerable studies from Ooi *et al.*[136–140] tackled the analysis of several transient problems using the concept of dynamic circuit theory. The method is based on the evaluation of lumped parameter matrices which are determined by the stored magnetic energy. Besides the good agreement to experimental results, the modeling of source- and induced-eddy currents as pure surface currents can be disadvantageous. Recently, Weidermann *et al.* [141] addressed the problem of time-dependent velocities and the associated Lorentz forces in the framework of LFV in 2014. However, they assumed a homogeneous external magnetic field and simplified the problem to the 1D case. Besides the intelligible and descriptive nature of this study, it is far from a realistic scenario. Its actuality indicates a lack of knowledge in this field and confirms the need for more advanced solutions. In the following, an analytical approach is presented to model this kind of electromagnetic field problems bypassing the mentioned simplifications.

At first, the problem is formulated and the solution of the governing equations is presented in Section 2.2.4.2. The obtained analytical expressions are verified by comparing them to FEM simulations in Section 2.2.4.3. Subsequently, the influence of oscillatory motion on the resulting Lorentz force is investigated in Section 2.2.4.4. The chapter closes by drawing the conclusions in Section 2.2.4.4.

2.2.4.2 Mathematical formulation of the problem

2.2.4.2.1 The governing equations and its solutions

The investigated problem is sketched in Figure 2.20. The computational domain is divided into four sub domains. Domains I, II, and IV are air domains and domain III represents the conductor. The origin of the Cartesian coordinate system is defined at the surface of the conductor. The conducting slab is infinitely extended in the xy-plane but has a finite thickness d. The magnetic field source is modeled by an arrangement

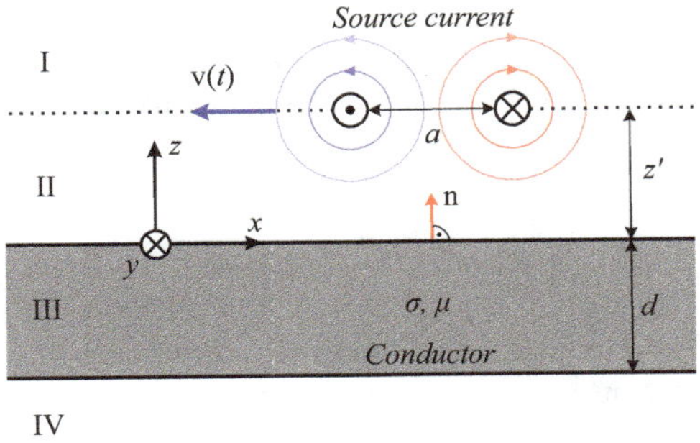

Figure 2.20 Cross-section of the problem under investigation. A configuration of current carrying coils is moving over a conducting slab, infinitely extended in the xy-plane but of finite thickness d [143]

of current carrying wires located at a height z'. It moves along the x-axis with a time-dependent velocity $v(t) = v_0 + v_1 \cos \omega_v t$. The problem is addressed in rest frame of reference K', considering that the magnetic field source moves and the conductor is at rest. Hence, motion-induced eddy currents are induced in consequence of a time-dependent magnetic field $\mathbf{B}(t)$. Displacement currents are neglected under the assumption that the velocity is much smaller than the speed of light and that the oscillation frequency is moderate such that $\omega\varepsilon/\sigma \ll 1$. If bodies are accelerated, stresses can arise which could alter the material properties. However, the present analysis is based on the hypothesis that the electrical properties in the instantaneous rest frame K' are unaffected by the acceleration [142]. The negligence of acceleration effects can be motivated by an electron-theoretical viewpoint considering an example of rotary motion. This particular effect was studied by Shiozawa [144]. It is stated that acceleration effects can be neglected if the angular frequency of the solid body ω_v is much smaller than the angular frequency of the mass m_e and the electron-nucleus spring ω_0, which lies in the range of infrared and ultraviolet light. Because moderate velocities are assumed in the present analysis, it is possible to continue with the phenomenological viewpoint, where electromagnetic effects of matter are solely described by the material properties σ, μ, and ε, without taking into account its internal structure. In this sense, the constitutive equations derived in Section 2.1 are still valid and hold [145,146].

In short, the governing equations are derived from Maxwell's equations in the rest frame K', omitting primed quantities to simplify matters:

$$\nabla \times \mathbf{H} = \mathbf{J}, \tag{2.98a}$$

$$\nabla \times \mathbf{E} = -\frac{\partial \mathbf{B}}{\partial t}. \tag{2.98b}$$

The magnetic flux density $\mathbf{B} = \mu\mathbf{H}$ is expressed by the magnetic vector potential $\mathbf{A}$ ($\mathbf{B} = \nabla \times \mathbf{A}$, $\nabla \cdot \mathbf{A} = 0$) under the assumption that the magnetic material is linear and obeys a constant relative permeability μ_r. In the present case, the induced eddy currents are given by $\mathbf{J} = -\sigma(\nabla\phi + \partial\mathbf{A}/\partial t)$. Because the source currents are assumed to flow solely in the xy-plane, the scalar electric potential ϕ is constant and can be omitted in further calculations. Additionally, $A_z = 0$ and the calculations are simplified to the components A_x and A_y. As a result, (2.98a) and (2.98b) can be written as:

$$\frac{\partial^2 A_{x|y}}{\partial x^2} + \frac{\partial^2 A_{x|y}}{\partial y^2} + \frac{\partial^2 A_{x|y}}{\partial z^2} = \mu\sigma \frac{\partial A_{x|y}}{\partial t}. \tag{2.99}$$

The variable separation method is applied by taking the Fourier transform of $\mathbf{A}$ with respect to the spatial coordinates x and y as well as with respect to the time t such that $\tilde{\mathbf{A}} = \mathscr{F}_x \mathscr{F}_y \mathscr{F}_t\{\mathbf{A}\}$. In consequence, the problem is described in the aforementioned dimensions by the spatial angular frequencies k_x and k_y and the temporal angular frequency ω. In the following, the x- and y-component of the magnetic vector potential are termed as $\tilde{A}_{x|y}$.

The governing equations in the respective domains are then given by:

$$\frac{\partial^2 \tilde{A}_{x|y}^{\mathrm{I,II,IV}}}{\partial z^2} = k^2 \tilde{A}_{x|y}^{\mathrm{I,II,IV}}, \tag{2.100a}$$

$$\frac{\partial^2 \tilde{A}_{x|y}^{\mathrm{III}}}{\partial z^2} = \gamma^2 \tilde{A}_{x|y}^{\mathrm{III}}, \tag{2.100b}$$

with

$$k = \sqrt{k_x^2 + k_y^2}, \tag{2.101a}$$

$$\gamma = \sqrt{k_x^2 + k_y^2 + j\omega\mu\sigma}. \tag{2.101b}$$

The corresponding solutions of (2.100a) and (2.100b) are:

$$\tilde{A}_{x|y}^{\mathrm{I,II,IV}} = \tilde{\mathrm{C}}_{x|y}^{\mathrm{I,II,IV}} \mathrm{e}^{kz} + \tilde{\mathrm{D}}_{x|y}^{\mathrm{I,II,IV}} \mathrm{e}^{-kz}, \tag{2.102a}$$

$$\tilde{A}_{x|y}^{\mathrm{III}} = \tilde{\mathrm{C}}_{x|y}^{\mathrm{III}} \mathrm{e}^{\gamma z} + \tilde{\mathrm{D}}_{x|y}^{\mathrm{III}} \mathrm{e}^{-\gamma z}. \tag{2.102b}$$

This results in a total number of 16 constants $\tilde{\mathrm{C}}_{x|y}^{\text{I-IV}}$ and $\tilde{\mathrm{D}}_{x|y}^{\text{I-IV}}$, which are determined by means of the underlying boundary conditions.

The magnetic vector potential vanishes at infinity, such that:

$$\tilde{A}_{x|y}^{\mathrm{I}} = 0|_{z\to\infty}, \tag{2.103a}$$

$$\tilde{A}_{x|y}^{\mathrm{IV}} = 0|_{z\to-\infty}. \tag{2.103b}$$

Hence, $\tilde{\mathrm{C}}_{x|y}^{\mathrm{I}} = 0$ and $\tilde{\mathrm{D}}_{x|y}^{\mathrm{IV}} = 0$.

The normal component of the magnetic flux density $\tilde{\mathbf{B}}$ is continuous across the interfaces such that $\tilde{B}_z^i = \tilde{B}_z^{i+1}$, where $i = \{\mathrm{I, II, III}\}$. This condition is fulfilled by ensuring the continuity of $\tilde{A}_{x|y}$ across the interfaces:

$$\tilde{A}_{x|y}^{\mathrm{I}} = \tilde{A}_{x|y}^{\mathrm{II}}|_{z=z'}, \tag{2.104a}$$

$$\tilde{A}_{x|y}^{\mathrm{II}} = \tilde{A}_{x|y}^{\mathrm{III}}|_{z=0}, \tag{2.104b}$$

$$\tilde{A}_{x|y}^{\mathrm{III}} = \tilde{A}_{x|y}^{\mathrm{IV}}|_{z=-d}. \tag{2.104c}$$

The remaining constants are determined by ensuring the continuity of the tangential component of the magnetic field $\tilde{\mathbf{H}}$ across the interfaces, such that $\tilde{H}^{i}_{x|y} = \tilde{H}^{i+1}_{x|y}$:

$$\frac{\partial \tilde{A}^{\mathrm{II}}_{x|y}}{\partial z} - \frac{\partial \tilde{A}^{\mathrm{I}}_{x|y}}{\partial z} = \mu_0 \tilde{I}_{x|y}|_{z=z'}, \tag{2.105a}$$

$$\frac{\partial \tilde{A}^{\mathrm{II}}_{x|y}}{\partial z} - \frac{1}{\mu_r}\frac{\partial \tilde{A}^{\mathrm{III}}_{x|y}}{\partial z} = 0|_{z=0}, \tag{2.105b}$$

$$\frac{\partial \tilde{A}^{\mathrm{IV}}_{x|y}}{\partial z} - \frac{1}{\mu_r}\frac{\partial \tilde{A}^{\mathrm{III}}_{x|y}}{\partial z} = 0|_{z=-d}. \tag{2.105c}$$

The function $\tilde{I}_{x|y} = \mathscr{F}_x\mathscr{F}_y\mathscr{F}_t\{I_{x|y}\}$ in (2.105a) is the Fourier transform of the x- and y-component of the oscillating source current above the conductor.

Using the boundary conditions together with the ansatz from (2.102a) and (2.102b), a system of equations (2.106) can be derived to determine the unknown constants. The solutions are given in (2.107a)–(2.107f). The expressions for $\tilde{A}_x$ and $\tilde{A}_y$ are similar, with the only difference regarding the source current $\tilde{I}_x$ and $\tilde{I}_y$.

$$\begin{bmatrix} \mathrm{e}^{-kz'} & -\mathrm{e}^{kz'} & -\mathrm{e}^{-kz'} & 0 & 0 & 0 \\ \mathrm{e}^{-kz'} & \mathrm{e}^{kz'} & -\mathrm{e}^{-kz'} & 0 & 0 & 0 \\ 0 & 1 & 1 & -1 & -1 & 0 \\ 0 & k & -k & -\frac{\gamma}{\mu_r} & \frac{\gamma}{\mu_r} & 0 \\ 0 & 0 & 0 & \mathrm{e}^{-\gamma d} & \mathrm{e}^{\gamma d} & -\mathrm{e}^{-\gamma d} \\ 0 & 0 & 0 & \frac{\gamma}{\mu_r}\mathrm{e}^{-\gamma d} & -\frac{\gamma}{\mu_r}\mathrm{e}^{\gamma d} & -k\mathrm{e}^{-kd} \end{bmatrix} \begin{bmatrix} \tilde{\mathrm{D}}^{\mathrm{I}}_{x|y} \\ \tilde{\mathrm{C}}^{\mathrm{II}}_{x|y} \\ \tilde{\mathrm{D}}^{\mathrm{II}}_{x|y} \\ \tilde{\mathrm{C}}^{\mathrm{III}}_{x|y} \\ \tilde{\mathrm{D}}^{\mathrm{III}}_{x|y} \\ \tilde{\mathrm{C}}^{\mathrm{IV}}_{x|y} \end{bmatrix} = \begin{bmatrix} 0 \\ \frac{\mu_0}{k}\tilde{I}_{x|y} \\ 0 \\ 0 \\ 0 \\ 0 \end{bmatrix}. \tag{2.106}$$

$$\tilde{\mathrm{D}}^{\mathrm{I}}_{x|y} = \mu_0 \tilde{\mathrm{I}}_{x|y} \frac{\frac{\gamma}{k}\sinh(kz')[k\mu_r + \gamma\tanh(\gamma d)] + \mu_r\cosh(kz')[\gamma + k\mu_r\tanh(\gamma d)]}{(\gamma^2 + k^2\mu_r^2)\tanh(\gamma d) + 2k\gamma\mu_r}, \tag{2.107a}$$

$$\tilde{\mathrm{C}}^{\mathrm{II}}_{x|y} = \mu_0 \tilde{\mathrm{I}}_{x|y} \frac{\mathrm{e}^{-kz'}}{2k}, \tag{2.107b}$$

$$\tilde{\mathrm{D}}^{\mathrm{II}}_{x|y} = \mu_0 \tilde{\mathrm{I}}_{x|y} \frac{k^2\mu_r^2 - \gamma^2}{(\gamma^2 + k^2\mu_r^2) + 2k\gamma\mu_r\tanh^{-1}(\gamma d)} \frac{\mathrm{e}^{-kz'}}{2k}, \tag{2.107c}$$

$$\tilde{\mathrm{C}}^{\mathrm{III}}_{x|y} = \mu_0 \tilde{\mathrm{I}}_{x|y} \frac{\mu_r(\gamma + k\mu_r)}{(\gamma + k\mu_r)^2 - (\gamma - k\mu_r)^2\mathrm{e}^{-2d\gamma}}\mathrm{e}^{-kz'}, \tag{2.107d}$$

$$\tilde{\mathrm{D}}^{\mathrm{III}}_{x|y} = \mu_0 \tilde{\mathrm{I}}_{x|y} \frac{\mu_r(\gamma - k\mu_r)}{(\gamma + k\mu_r)^2\mathrm{e}^{2d\gamma} - (\gamma - k\mu_r)^2}\mathrm{e}^{-kz'}, \tag{2.107e}$$

$$\tilde{\mathrm{C}}^{\mathrm{IV}}_{x|y} = \mu_0 \tilde{\mathrm{I}}_{x|y} \frac{\gamma\mu_r\mathrm{e}^{(d-z')k}}{2k\gamma\mu_r\cosh(\gamma d) + (\gamma^2 + k^2\mu_r^2)\sinh(\gamma d)}. \tag{2.107f}$$

2.2.4.2.2 Fourier transform of the source current

The Fourier transformation with respect to the spatial coordinates x and y as well as the time t of the x- and y-component of the source current is required to evaluate the magnetic vector potential. It is assumed that the magnetic field source moves along the negative x-direction with a periodically changing velocity $\mathbf{v} = -v(t)\mathbf{e}_x$. The velocity oscillates with a frequency of $\omega_v = 2\pi f_v$ around a nominal value of v_0 with an amplitude of v_1:

$$v(t) = v_0 + v_1 \cos \omega_v t. \tag{2.108}$$

Hence, the time-dependent position of the magnetic field source is given by:

$$\xi(t) = v_0 t + \frac{v_1}{\omega_v} \sin \omega_v t. \tag{2.109}$$

The ratio v_1/ω_v can be interpreted as a displacement amplitude of the magnetic field source.

In the present context, rectangular and circular coil shapes are considered. In the further course of this work, the planar cases are extended to stacked configurations as they are shown in Figure 2.21. This is equivalent to the current model of permanent magnets [110]. Hence, cuboidal and cylindrical permanent magnets are included in the present analysis.

Rectangular current sources: The x- and y-component of the source current of a rectangular coil in the spatial domain are given by:

$$I_x^{(\mathrm{rect})} = I_s \left[\mathrm{rect}\left(\frac{x-\xi(t)}{a}\right) \delta\left(y + \frac{b}{2}\right) - \mathrm{rect}\left(\frac{x-\xi(t)}{a}\right) \delta\left(y - \frac{b}{2}\right) \right], \tag{2.110a}$$

$$I_y^{(\mathrm{rect})} = I_s \left[\mathrm{rect}\left(\frac{y}{b}\right) \delta\left(x - \frac{a}{2} - \xi(t)\right) - \mathrm{rect}\left(\frac{y}{b}\right) \delta\left(x + \frac{a}{2} - \xi(t)\right) \right], \tag{2.110b}$$

where rect(·) is the rectangular function, $\delta(\,\cdot\,)$ is the Dirac delta distribution, and I_s is the current strength. In the case of permanent magnets, I_s is the outer surface current density in A/m determined by $\mathbf{J}_s = I_s \mathbf{t} = \mathbf{M} \times \mathbf{n}$, where $\mathbf{M}$ is the magnetization, and

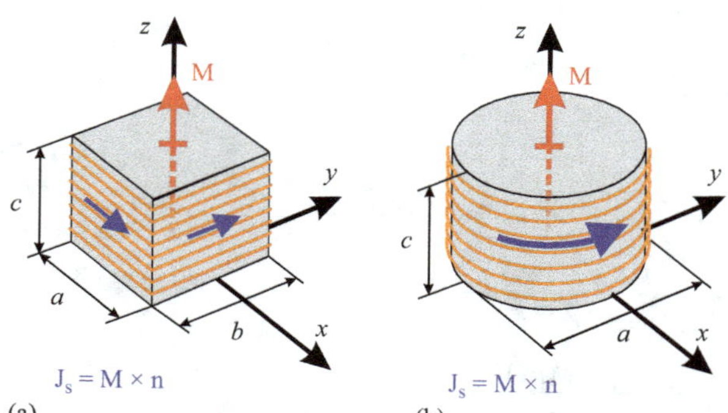

Figure 2.21 Current models used to model permanent magnets [143]. (a) Cuboidal magnet. (b) Cylindrical magnet

n and **t** are the unit normal and tangential vectors of the outer surface of the magnet, respectively. Their spatial Fourier transforms with respect to the coordinates x and y are:

$$\mathscr{F}_x\mathscr{F}_y\{I_x^{(\mathrm{rect})}\} = \underbrace{I_s\frac{4j}{k_x}\sin\left(k_x\frac{a}{2}\right)\sin\left(k_y\frac{b}{2}\right)}_{\hat{I}_x^{(\mathrm{rect})}}\bar{f}(t), \tag{2.111a}$$

$$\mathscr{F}_x\mathscr{F}_y\{I_y^{(\mathrm{rect})}\} = \underbrace{-I_s\frac{4j}{k_y}\sin\left(k_x\frac{a}{2}\right)\sin\left(k_y\frac{b}{2}\right)}_{\hat{I}_y^{(\mathrm{rect})}}\bar{f}(t). \tag{2.111b}$$

The terms which solely depend on the spatial frequencies k_x and k_y are indicated by $\hat{I}_{x|y}$ in further calculations. The complex function $\bar{f}(t)$ includes the time-dependency of the vibrating source current. Since the magnetic field source moves as a whole, it does not depend on the explicit geometry of the current carrying wire and can be treated independently. The calculation of the Fourier transform will be treated separately in the further course.

Circular current sources: In case of circular coils, the x- and y-component of the source current can be described by:

$$I_x^{(\mathrm{circ})} = -2I_s\frac{y}{a}\delta\left(\sqrt{(x-\xi(t))^2+y^2}-\frac{a}{2}\right), \tag{2.112a}$$

$$I_y^{(\mathrm{circ})} = 2I_s\frac{x}{a}\delta\left(\sqrt{(x-\xi(t))^2+y^2}-\frac{a}{2}\right). \tag{2.112b}$$

Their Fourier transforms with respect to x and y are:

$$\mathscr{F}_x\mathscr{F}_y\{I_x^{(\mathrm{circ})}\} = \underbrace{ja\pi I_s\frac{k_y}{k}\mathrm{J}_1\left(k\frac{a}{2}\right)}_{\hat{I}_x^{(\mathrm{circ})}}\bar{f}(t), \tag{2.113a}$$

$$\mathscr{F}_x\mathscr{F}_y\{I_y^{(\mathrm{circ})}\} = \underbrace{-ja\pi I_s\frac{k_x}{k}\mathrm{J}_1\left(k\frac{a}{2}\right)}_{\hat{I}_y^{(\mathrm{circ})}}\bar{f}(t), \tag{2.113b}$$

where $\mathrm{J}_1(\cdot)$ is the first order Bessel function of the first kind. The calculation of the Fourier transform is explained in more detail in the following.

2D Fourier transform circulating currents: The time dependency of its position is omitted in the first place, without loss of generality. The current loop in the spatial domain, considering a Cartesian coordinate system, is described by:

$$I_x^{(\mathrm{circ})} = -2I_s\frac{y}{a}\delta\left(\sqrt{x^2+y^2}-\frac{a}{2}\right), \tag{2.114a}$$

$$I_y^{(\mathrm{circ})} = 2I_s\frac{x}{a}\delta\left(\sqrt{x^2+y^2}-\frac{a}{2}\right). \tag{2.114b}$$

The parameter a denotes the diameter of the current loop. In analogy, the expressions can be described in the cylindrical coordinate system:

$$I_x^{(\text{circ})} = -I_s \sin\varphi\delta\left(r - \frac{a}{2}\right), \tag{2.115a}$$

$$I_y^{(\text{circ})} = I_s \cos\varphi\delta\left(r - \frac{a}{2}\right). \tag{2.115b}$$

The Fourier transform of a 2D function $f(x, y)$ is given by:

$$\mathscr{F}_x\mathscr{F}_y\{f(x,y)\} = \int_{-\infty}^{\infty}\int_{-\infty}^{\infty} f(x,y)e^{-j(xk_x+yk_y)}\,\mathrm{d}x\mathrm{d}y, \tag{2.116}$$

with the spatial frequencies k_x and k_y. This expression can be transformed into the cylindrical coordinate system by considering the following identities:

$$ke^{j\beta} = k_x + jk_y \quad\rightarrow\quad \sin\beta = \frac{k_y}{k},\quad \cos\beta = \frac{k_x}{k},\quad k = \sqrt{k_x^2 + k_y^2}, \tag{2.117}$$

$$re^{j\varphi} = x + jy \quad\rightarrow\quad \sin\varphi = \frac{y}{r},\quad \cos\varphi = \frac{x}{r},\quad r = \sqrt{x^2 + y^2}, \tag{2.118}$$

$$xk_x = kr\cos\beta\cos\varphi, \tag{2.119}$$

$$yk_y = kr\sin\beta\sin\varphi, \tag{2.120}$$

$$xk_x + yk_y = kr(\cos\beta\cos\varphi + \sin\beta\sin\varphi), \tag{2.121}$$

$$= kr\cos(\varphi - \beta). \tag{2.122}$$

Substituting (2.122) in (2.116) and performing the integration with respect to the radial coordinate, r, and the azimuthal angle, φ, yields:

$$\int_{-\infty}^{\infty}\int_{-\infty}^{\infty} f(x,y)e^{-j(xk_x+yk_y)}\,\mathrm{d}x\mathrm{d}y = \int_0^{\infty}\int_0^{2\pi} f(r,\varphi)e^{-jkr\cos(\varphi-\beta)}r\,\mathrm{d}r\mathrm{d}\varphi. \tag{2.123}$$

By substituting $\theta = \varphi - \beta$ one gets:

$$\int_0^{\infty}\int_{-\beta}^{2\pi-\beta} f(r,\theta+\beta)e^{-jkr\cos\theta}r\,\mathrm{d}r\mathrm{d}\theta. \tag{2.124}$$

Substituting the expressions of the current loop (2.115a) and (2.115b) in (2.124) for the function $f(r, \theta + \beta)$ leads to:

$$I_s\int_0^{\infty}\int_{-\beta}^{2\pi-\beta}\begin{bmatrix}-\sin(\theta+\beta)\\ \cos(\theta+\beta)\end{bmatrix}\delta\left(r - \frac{a}{2}\right)e^{-jkr\cos\theta}r\,\mathrm{d}r\mathrm{d}\theta. \tag{2.125}$$

The integral over the radial coordinate can be solved by using the sifting property of the Dirac delta function at $r = \frac{a}{2}$.

$$I_s\frac{a}{2}\int_{-\beta}^{2\pi-\beta}\begin{bmatrix}-\sin(\theta+\beta)\\ \cos(\theta+\beta)\end{bmatrix}e^{-jk\frac{a}{2}\cos\theta}\,\mathrm{d}\theta. \tag{2.126}$$

By reformulating the expressions in brackets by means of the following trigonometric identities:

$$\sin(\theta + \beta) = \sin\theta\cos\beta + \cos\theta\sin\beta = \frac{1}{k}\left(k_x \sin\theta + k_y \cos\theta\right), \tag{2.127}$$

$$\cos(\theta + \beta) = \cos\theta\cos\beta - \sin\theta\sin\beta = \frac{1}{k}\left(k_x \cos\theta - k_y \sin\theta\right), \tag{2.128}$$

Eq. (2.126) can be written as:

$$I_s \frac{a}{2k} \int_{-\beta}^{2\pi-\beta} \begin{bmatrix} -k_x \sin\theta - k_y \cos\theta \\ -k_y \sin\theta + k_x \cos\theta \end{bmatrix} e^{-jk\frac{a}{2}\cos\theta}\, \mathrm{d}\theta. \tag{2.129}$$

In the following, only the first component is considered since the considerations do also apply for the second one.

Splitting up the integral involving the sine and the cosine term, one can identify the following relationship using Eulers identity:

$$\begin{aligned} &\int_{-\beta}^{2\pi-\beta} \sin\theta e^{-jk\frac{a}{2}\cos\theta}\, \mathrm{d}\theta \\ &= \int_{-\beta}^{2\pi-\beta} \sin\theta \cos(k\frac{a}{2}\cos\theta)\, \mathrm{d}\theta - j\int_{-\beta}^{2\pi-\beta} \sin\theta \sin(k\frac{a}{2}\cos\theta)\, \mathrm{d}\theta \\ &= -\underbrace{\left[\frac{\sin(k\frac{a}{2}\cos\theta)}{k\frac{a}{2}}\right]_{-\beta}^{2\pi-\beta}}_{=0} - j\underbrace{\left[\frac{\cos(k\frac{a}{2}\cos\theta)}{k\frac{a}{2}}\right]_{-\beta}^{2\pi-\beta}}_{=0} = 0. \end{aligned} \tag{2.130}$$

As a consequence, the integrals involving the sine terms vanish. Next, the cosine terms are considered, taking into account the following identities of the Bessel functions of the first kind [114]:

$$\mathrm{J}_n(z) = \frac{i^{-n}}{\pi} \int_0^{\pi} \cos(n\theta) e^{jz\cos\theta}, \tag{2.131}$$

$$\mathrm{J}_n(ze^{jm\pi}) = e^{jm\pi n}\mathrm{J}_n(z) \xrightarrow{m=n=1} \mathrm{J}_1(-z) = -\mathrm{J}_1(z). \tag{2.132}$$

Substituting $z = -k\frac{a}{2}$ and considering the symmetry of the function, one gets:

$$-2\pi j \mathrm{J}_1\left(k\frac{a}{2}\right) = \int_0^{2\pi} \cos\theta e^{-jk\frac{a}{2}\cos\theta}. \tag{2.133}$$

Finally, by using this analogy in (2.129), the 2D Fourier transforms of both current components are given by:

$$\mathscr{F}_x\mathscr{F}_y\{I_x^{(\mathrm{circ})}\} = ja\pi I_s \frac{k_y}{k}\mathrm{J}_1\left(k\frac{a}{2}\right), \tag{2.134a}$$

$$\mathscr{F}_x\mathscr{F}_y\{I_y^{(\mathrm{circ})}\} = -ja\pi I_s \frac{k_x}{k}\mathrm{J}_1\left(k\frac{a}{2}\right). \tag{2.134b}$$

Mathematical modeling of oscillating current sources: The expressions in (2.111a) and (2.111b) as well as in (2.113a) and (2.113b) have to be additionally transformed with respect to the time t. The complex function $\bar{f}(t)$ results from applying the law of displacement to the $x - \xi(t)$ terms in (2.110a), (2.110b) and (2.112a), (2.112b), and is given by:

$$\bar{f}(t) = \exp\left[-jk_x\left(v_0 t + \frac{v_1}{\omega_v}\sin\omega_v t\right)\right]. \tag{2.135}$$

It is noted, that it contains a harmonic function in the exponent due to the periodically changing velocity. This function bears analogies to phase modulated signals, which are well known in communication engineering [147]. Its Fourier transform is obtained by applying the Jacobi-Anger expansion [114] of the nested harmonics using the following identities:

$$\cos(\alpha\sin\beta) = \mathrm{J}_0(\alpha) + 2\sum_{n=1}^{\infty}\mathrm{J}_{2n}(\alpha)\cos(2n\beta), \tag{2.136a}$$

$$\sin(\alpha\sin\beta) = 2\sum_{n=1}^{\infty}\mathrm{J}_{2n-1}(\alpha)\sin[(2n-1)\beta], \tag{2.136b}$$

where $\mathrm{J}_n(\cdot)$ are the nth order Bessel functions. In this sense, $\bar{f}(t)$ is decomposed to an infinite sum of harmonics:

$$\begin{aligned}\bar{f}(t) = {} & [\cos(k_x v_0 t) - j\sin(k_x v_0 t)] \times \left\{\mathrm{J}_0\left(k_x\frac{v_1}{\omega_v}\right) + 2\sum_{n=1}^{\infty}\mathrm{J}_{2n}\left(k_x\frac{v_1}{\omega_v}\right)\right. \\ & \left.\times\cos(2n\omega_v t) - j\mathrm{J}_{2n-1}\left(k_x\frac{v_1}{\omega_v}\right)\sin[(2n-1)\omega_v t]\right\}.\end{aligned} \tag{2.137}$$

In consequence, its Fourier transform can be represented by a sum of Dirac delta distributions:

$$\begin{aligned}\mathscr{F}_t\{\bar{f}(t)\} = {} & \mathrm{J}_0\left(k_x\frac{v_1}{\omega_v}\right)\delta(\omega + k_x v_0) + \sum_{n=1}^{\infty}(-1)^n\mathrm{J}_n\left(k_x\frac{v_1}{\omega_v}\right) \\ & \times[\delta(\omega - n\omega_v + k_x v_0) + (-1)^n\delta(\omega + n\omega_v + k_x v_0)].\end{aligned} \tag{2.138}$$

A graphical representation of the signal in the frequency domain is shown in Figure 2.22. The main component of the spectrum is determined by the nominal velocity v_0. However, the oscillatory motion introduces additional side components, displaced by multiples of the oscillation frequency ω_v. Their magnitudes are determined by the corresponding n-th order Bessel functions of the first kind $\mathrm{J}_n(\cdot)$. The higher order harmonics vanish if the magnetic field source moves at a constant velocity v_0 with $v_1 = 0$ since $\mathrm{J}_n(0) = 0$ for $n > 0$ and $\mathrm{J}_0(0) = 1$.

2.2.4.2.3 Force calculation

As a result of the harmonic oscillation, the forces which act on the current carrying wires are time-dependent. They can be determined directly in the Fourier domain by applying the theorem of Parseval [148]:

$$\mathbf{F}(t) = \frac{1}{8\pi^3}\int_{-\infty}^{\infty}\int_{-\infty}^{\infty}\int_{-\infty}^{\infty}\tilde{\mathbf{I}}^* \times \tilde{\mathbf{B}}^{\mathrm{I}(d)}e^{j\omega t}\,\mathrm{d}\omega\mathrm{d}k_x\mathrm{d}k_y, \tag{2.139}$$

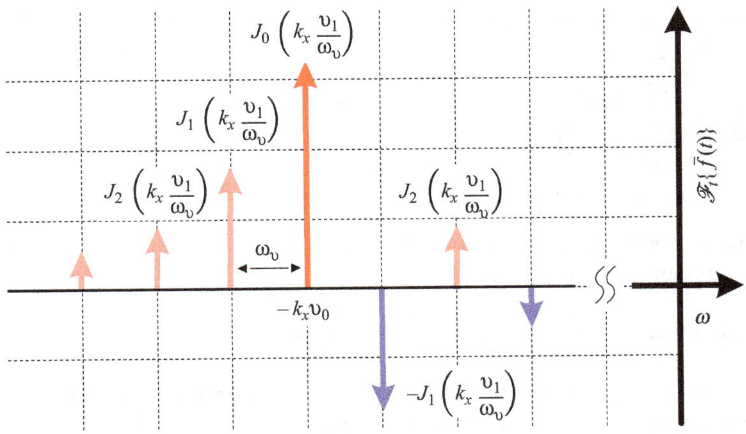

Figure 2.22 Harmonics of Fourier transformed oscillating magnetic field sources [143]

where $\tilde{\mathbf{I}}$ is the Fourier transformed source current vector and the superscript * indicates complex conjugation. The distorted magnetic flux $\tilde{\mathbf{B}}^{\mathrm{I}(d)}$ can be interpreted as the part of the total magnetic field, $\tilde{\mathbf{B}}^{\mathrm{I}}$, in the air domain I, which is affected by the magnetic flux of the induced eddy currents and the magnetic properties of the conductor. The drag force F_x and the lift force F_z are then:

$$F_x(t) = \frac{1}{8\pi^3}\int_{-\infty}^{\infty}\int_{-\infty}^{\infty}\int_{-\infty}^{\infty} \tilde{I}_y^{*}\tilde{B}_z^{\mathrm{I}(d)}\mathrm{e}^{j\omega t}\,\mathrm{d}\omega\mathrm{d}k_x\mathrm{d}k_y, \tag{2.140a}$$

$$F_z(t) = \frac{1}{8\pi^3}\int_{-\infty}^{\infty}\int_{-\infty}^{\infty}\int_{-\infty}^{\infty} \left(\tilde{I}_x^{*}\tilde{B}_y^{\mathrm{I}(d)} - \tilde{I}_y^{*}\tilde{B}_x^{\mathrm{I}(d)}\right)\mathrm{e}^{j\omega t}\mathrm{d}\omega\mathrm{d}k_x\mathrm{d}k_y. \tag{2.140b}$$

The components of $\tilde{\mathbf{B}}^{\mathrm{I}(d)}$ can be determined by subtracting the primary field $\tilde{\mathbf{B}}^{\mathrm{I}(p)}$, generated by the magnetic field source, from the total field:

$$\tilde{\mathbf{B}}^{\mathrm{I}(d)} = \tilde{\mathbf{B}}^{\mathrm{I}} - \tilde{\mathbf{B}}^{\mathrm{I}(p)} = \nabla \times \left(\tilde{\mathbf{A}}^{\mathrm{I}}\Big|_{\sigma}^{\mu_r} - \tilde{\mathbf{A}}^{\mathrm{I}}\Big|_{\sigma=0}^{\mu_r=1}\right). \tag{2.141}$$

For $\tilde{B}_x^{\mathrm{I}(d)}$, one gets:

$$\tilde{B}_x^{\mathrm{I}(d)} = \left(\tilde{\mathrm{D}}_y^{\mathrm{I}} - \mu_0\tilde{I}_y\frac{\mathrm{e}^{kz'}}{2k}\right)k\mathrm{e}^{-kz}. \tag{2.142}$$

Since the constant $\tilde{\mathrm{D}}_y^{\mathrm{I}}$ contains $\tilde{I}_y$, the following expressions can be simplified by using the identity $\tilde{\mathrm{D}}_y^{\mathrm{I}} = \tilde{I}_y\hat{\mathrm{D}}^{\mathrm{I}}$:

$$\tilde{B}_x^{\mathrm{I}(d)} = k\tilde{I}_y\underbrace{\left(\hat{\mathrm{D}}^{\mathrm{I}} - \mu_0\frac{\mathrm{e}^{kz'}}{2k}\right)}_{\mathrm{G}(\omega,\mathbf{k})\mathrm{e}^{-kz'}}\mathrm{e}^{-kz}, \tag{2.143}$$

with $\mathbf{k} = [k_x, k_y]$. The remaining components of the distorted magnetic field can be determined in a similar way:

$$\tilde{B}_x^{\mathrm{I}(d)} = k\tilde{I}_y \mathrm{G}(\omega, \mathbf{k})\mathrm{e}^{-k(z+z')}, \tag{2.144a}$$

$$\tilde{B}_y^{\mathrm{I}(d)} = -k\tilde{I}_x \mathrm{G}(\omega, \mathbf{k})\mathrm{e}^{-k(z+z')}, \tag{2.144b}$$

$$\tilde{B}_z^{\mathrm{I}(d)} = j(k_y\tilde{I}_x - k_x\tilde{I}_y)\mathrm{G}(\omega, \mathbf{k})\mathrm{e}^{-k(z+z')}. \tag{2.144c}$$

The function $\mathrm{G}(\omega, \mathbf{k})$ contains properties of the conductor such as its thickness, conductivity and relative permeability:

$$\mathrm{G}(\omega, \mathbf{k}) = \frac{\mu_0(\mathrm{e}^{2d\gamma} - 1)(k^2\mu_r^2 - \gamma^2)}{2k[(\gamma + k\mu_r)^2\mathrm{e}^{2d\gamma} - (\gamma - k\mu_r)^2]}. \tag{2.145}$$

The final expression for the drag- and lift force in case of a single winding can be determined by combining the following expressions in (2.140a) and (2.140b):

- The Fourier transformed source currents, i.e., the functions describing the spatial transformation from (2.111a) and (2.111b) or (2.113a) and (2.113b).
- The transformed time-dependent part of the source currents from (2.138).
- The distorted magnetic field from (2.144a)–(2.144c) at the location $z = z'$ of the current.

Additionally, an inverse Fourier transform from frequency to time domain is applied in order to evaluate the waveforms of both force components. As a consequence of the symmetry in the spatial Fourier domain, the integration limits can be changed to 0 and ∞. Finally, $F_{x|z}(t)$ is then given by:

$$F_{x|z}(t) = \frac{1}{\pi^2}\int_0^\infty \int_0^\infty \mathscr{I}_{x|z}(\mathbf{k})\mathscr{G}(t, \mathbf{k})\mathscr{S}_{w|m}(\mathbf{k})\bar{f}(t)^{-1}\, \mathrm{d}k_x \mathrm{d}k_y, \tag{2.146}$$

with the auxiliary functions:

$$\mathscr{I}_x(\mathbf{k}) = -j\hat{I}_y^* \left(k_y\hat{I}_x - k_x\hat{I}_y\right), \tag{2.147}$$

$$\mathscr{I}_z(\mathbf{k}) = -k\left(\hat{I}_x^*\hat{I}_x + \hat{I}_y^*\hat{I}_y\right), \tag{2.148}$$

$$\begin{aligned}\mathscr{G}(t, \mathbf{k}) = {} & \mathrm{J}_0\left(k_x\frac{v_1}{\omega_v}\right)\mathrm{G}(-k_x v_0, \mathbf{k})\mathrm{e}^{-jk_x v_0 t} + \sum_{n=1}^{N}(-1)^n \mathrm{J}_n\left(k_x\frac{v_1}{\omega_v}\right) \\ & \times [\mathrm{G}(-k_x v_0 + n\omega_v, \mathbf{k})\mathrm{e}^{-j(v_0 k_x - n\omega_v)t} \\ & + (-1)^n \mathrm{G}(-k_x v_0 - n\omega_v, \mathbf{k})\mathrm{e}^{-j(v_0 k_x + n\omega_v)t}],\end{aligned} \tag{2.149}$$

$$\mathscr{S}_w(\mathbf{k}) = \mathrm{e}^{-2kz'}. \tag{2.150}$$

The functions $\mathscr{I}_x(\mathbf{k})$ and $\mathscr{I}_z(\mathbf{k})$ include information about the shape of the source current in the *xy*-plane and have to be used depending on the force component to be calculated. They can be determined using (2.111a) and (2.111b) for rectangular current sources or (2.113a) and (2.113b) for circular current sources, respectively. The function $\mathscr{G}(t, \mathbf{k})$ includes the time-dependence of the magnetic field, originating from the oscillating current source. The infinite bandwidth of the function leads to a sum of harmonics with their amplitudes determined by the velocity oscillation amplitude v_1 and the frequencies ω_v and k_x. In numerical calculations, the sum in (2.150) is truncated after N terms. Finally, the function $\mathscr{S}_{w,m}(\mathbf{k})$ includes the height information of the magnetic field source. The index w and m distinguishes between the single wire with $c = 0$ or a magnet of height $c > 0$, respectively.

The solution of a single wire can be extended to permanent magnets by means of the surface current model shown in Figure 2.21. It is assumed that the permanent magnets have a height of c and their lower surface is located at a lift-off distance h above the plate. The total force is determined by integrating the force on each wire over the z-coordinate, resulting from the superposition of the total distorted magnetic flux density generated from all wires. Hence, (2.146) becomes:

$$F_{x|z}(t) = \frac{1}{\pi^2}\int_{-\infty}^{\infty}\int_{-\infty}^{\infty}\mathscr{I}_{x|z}\mathscr{G}(t,\mathbf{k})f(t)^{-1}\underbrace{\int_{h}^{h+c}\int_{h}^{h+c}\mathrm{e}^{-k(z+z')}\,\mathrm{d}z\mathrm{d}z'}_{\mathscr{S}_m(\mathbf{k})}\,\mathrm{d}k_x\mathrm{d}k_y. \tag{2.151}$$

Carrying out the integration over z and z' yields the expression of the function $\mathscr{S}_m(\mathbf{k})$ valid as long as $c > 0$:

$$\mathscr{S}_m(\mathbf{k}) = \frac{1}{k^2}\left(\mathrm{e}^{-kh} - \mathrm{e}^{-k(h+c)}\right)^2. \tag{2.152}$$

By inserting (2.152) in (2.146), one gets the final expression for the drag- and lift force for permanent magnets. In order to evaluate (2.146) at a certain point in time, an adaptive 2D numerical integration technique based on a weighted quadrature approach is applied [149,150].

2.2.4.3 Comparison to numerical simulations

The analytical solution is compared to numerical simulations obtained using the FEM. The numerical simulation of the complete time-dependent 3D problem is not trivial. Difficulties are arising in the case of high velocities, when the skin-effect becomes dominant. Hence, the element size has to be reduced, which results in large system matrices. Moreover, the time-step has to be chosen appropriately, depending on the oscillation frequency ω_v, and the conducting slab has to be truncated at a sufficient distance to avoid edge effects. In view of those circumstances, the decision was taken to perform the verification in a simplified 2D model, which can be solved numerically with a high accuracy and within reasonable time. At first, the numerical model is described, which is followed by the derivation of the analytical solution in 2D. It is noted that the basic concept of the analytical approach is unaltered by the dimensional reduction and that the solutions of the 2D and 3D case are similar.

2.2.4.3.1 The 2D numerical model

The FEM model is set up using the software environment Comsol Multiphysics [151]. The problem is defined in the same way as in the 3D case from Figure 2.20, on the assumption that the setup has an infinite extend along the y-axis. The geometrical parameters of the model are given in Table 2.1. The problem can be completely described by the y-component of the magnetic vector potential A_y. Efficient numerical modeling is realized by changing the frame of reference from the conductor to

Table 2.1 Parameters of the exemplary problem used in the analysis of harmonic motion [143]

Parameter	Value	Description
a	15 mm	Length of the magnet
b	15 mm	Width of the magnet
c	25 mm	Height of the magnet
d	50 mm	Thickness of the plate
h	1 mm	Lift-off distance
B_r	1.17 T	Remanence of the magnet (NdFeB-N35)
σ_{Al}	30.66 MSm^{-1}	Electrical conductivity of aluminum
μ_r	1	Relative permeability

the magnet. In this way, the time-dependent velocity is modeled by the $\mathbf{v} \times \mathbf{B}$ term rather than the $\partial \mathbf{B}/\partial t$ term used in the analytical approach. In this way, it is possible to use a stationary mesh, avoiding the time consuming re-meshing procedure after every time-step. The governing equation for A_y in the whole computational domain, including the conductor, the magnet, and the surrounding air region, is given by:

$$\sigma\left(\frac{\partial A_y}{\partial t} + v_x\frac{\partial A_y}{\partial x}\right) + \frac{\partial}{\partial x}\left(\frac{1}{\mu}\frac{\partial A_y}{\partial x}\right) + \frac{\partial}{\partial z}\left(\frac{1}{\mu}\frac{\partial A_y}{\partial z}\right) = 0. \tag{2.153}$$

Equation (2.153) is solved by the FEM in the time-domain using a fifth order backward differentiation formula. The time-step is chosen such that every oscillation period contains 100 steps $\Delta t = 1/(100 f_v)$.

2.2.4.3.2 The 2D analytical model

The evaluation of the 2D analytical solution follows the same procedure as in the former case. The geometry of the problem is unaltered and already given in Figure 2.20. The governing equation (2.99) is simplified to determine A_y since $A_x = 0$. As a consequence of the infinite extend in the y-direction, the spatial frequency k_y vanishes, which simplifies the expressions for k and γ from (2.101a) and (2.101b), respectively. Applying the boundary conditions (2.103a)–(2.105c) results in the same coefficients $\tilde{\mathrm{C}}^{\text{I-IV}}$ and $\tilde{\mathrm{D}}^{\text{I-IV}}$ from (2.107a)–(2.107f) as in the 3-D case.

The major difference between the 2D and 3D solution arises considering the source current above the conducting slab, which is now described by:

$$I_y^{(2D)} = I_s\left[\delta\left(x + \frac{a}{2} - \xi(t)\right) - \delta\left(x - \frac{a}{2} - \xi(t)\right)\right]. \tag{2.154}$$

Its Fourier transform with respect to x is:

$$\mathscr{F}_x\{I_y^{(2D)}\} = \underbrace{2jI_s \sin\left(k_x\frac{a}{2}\right)\bar{f}(t)}_{\hat{I}_y^{(2D)}}. \tag{2.155}$$

The expressions for the magnetic field can be readily derived from the 3D case considering $\tilde{I}_x = 0$ and $k_y = 0$ in (2.144a)–(2.144c). The forces are calculated by integrating (2.146), but now only with respect to the spatial frequency k_x. It is noted that the pre-factor of $1/\pi^2$ becomes $1/\pi$ by omitting the integration with respect to k_y. This confirms the analogy between the 2D and 3D solution necessary to perform a meaningful verification.

2.2.4.3.3 Comparison of analytical and numerical results

The verification problem is defined using the parameters from Table 2.1. The rectilinear part of the velocity is $v_0^{(\max)} = 14.5$ m/s. The velocity is chosen such that the system operates at the point of maximal drag force, i.e. in the nonlinear regime. This characteristic velocity will be addressed and explained in more detail in Section 2.2.4.4 for the 3D case. The velocity oscillates at a frequency of $f_v = 100$ Hz considering a rather high oscillation amplitude of $v_1 = 0.5v_0$. The direction of motion is chosen such that the drag force is positive (see Figure 2.20). The parameters of the verification are defined to correspond to a numerically challenging benchmark problem. The relative difference between analytical (ANA) and numerical (FEM) results are quantified using the NRMSD over one oscillation period T:

$$\mathrm{NRMSD}_{x|z} = \frac{100\%}{\max\left(F_{x|z}^{\mathrm{FEM}}(t)\right) - \min\left(F_{x|z}^{\mathrm{FEM}}(t)\right)} \times \sqrt{\frac{1}{T}\int_0^T \left(F_{x|z}^{\mathrm{FEM}}(t) - F_{x|z}^{\mathrm{ANA}}(t)\right)^2 \mathrm{d}t. \tag{2.156}$$

The results obtained by both approaches are shown in Figure 2.23. They show an almost perfect agreement with an NRMSD of only 0.051% and 0.049% for the drag force F_x and lift force F_z, respectively. It can be clearly seen that both force

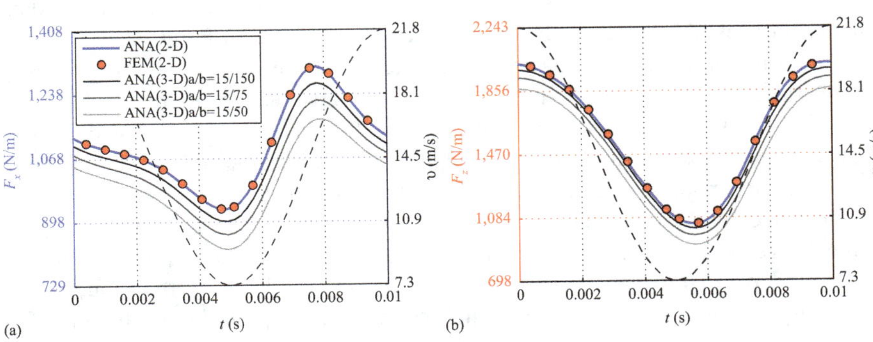

Figure 2.23 Comparison of the time-dependent drag- and lift-force density (N/m) acting on a moving rectangular magnet evaluated by the analytical and numerical approaches. The dashed line indicates the underlying velocity profile considering a rectilinear velocity, superimposed by an oscillation with a frequency of $f_v = 100$ Hz, and an amplitude of $v_1 = 0.5v_0$. The different 3D analytical solutions are normalized with respect to the individual y-extensions of the magnet b [143]. (a) Drag force F_x. (b) Lift force F_z

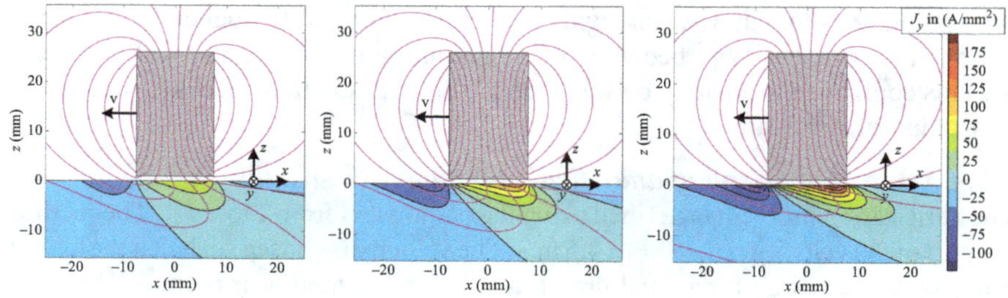

(a) $v(t) = 7.25$m/s, $t = 5$ms (b) $v(t) = 14.5$m/s, $t = 7.5$ms (c) $v(t) = 21.75$m/s, $t = 10$ms

Figure 2.24 Induced-eddy current density J_y and streamlines of the total magnetic flux density **B** *for three different instances, namely at (a) the minimal velocity, (b) the mean velocity, and (c) the maximal velocity [143]*

components are delayed and do not follow the underlying velocity profile. Especially the drag force is heavily distorted, illustrating the presence of higher order harmonics. A more comprehensive analysis on the amplitude and phase-shift of the Lorentz force will be given in Section 2.2.4.4 for the 3D case.

If the y-extension of the magnet is sufficiently large compared to the other dimensions, the 3D problem can be approximated by the 2D model. Hence, in order to compare the results of the 3D analytical solution, the force profiles are evaluated for different y-extensions b of the magnet. By normalizing the forces with respect to the individual b values, it is possible to determine the equivalent force densities. The expected convergence towards the 2D case can be clearly seen in Figure 2.23, which proves the validity of the 3D solution.

In order to illustrate the impact of skin-effect and the influence of time-dependent velocity profiles, the induced eddy current density is shown in Figure 2.24 together with the streamlines of the magnetic flux density. The first illustration in Figure 2.24(a) exemplifies the situation when the velocity reaches its minimum value of $v(t) = 7.5$ m/s at $t = 5$ ms (see Figure 2.23). At this stage, field suppression can be already observed. With progression in time, the velocity is rising and the induced eddy current density increases. As a consequence, the secondary magnetic field increases and further expels the primary field out of the conductor. This procedure is harmonically repeating. The time-dependent current density fluctuations generate an also time-dependent magnetic flux density. This in turn leads to the induction of (secondary) induced eddy currents, which counteract the periodic oscillation. This particular effect explains the inductive character of the conductor and the observed phase-shift between the force and the velocity oscillation.

2.2.4.4 Results and discussion

As mentioned previously, the parameters influencing the time-dependent force profiles are v_0, v_1, and f_v. In this section, the underlying effects are described, differentiating between three different kinds of motion: (i) Constant rectilinear motion without oscillation ($v_0 \neq 0$, $v_1 = 0$) similar to those already reported in the literature [61,66], (ii) pure harmonic motion ($v_0 = 0$, $v_1 \neq 0$) such that the magnet is vibrating over the conductive slab and (iii), the most complicated case, a mixture of both ($v_0 \neq 0$, $v_1 \neq 0$).

2.2.4.4.1 Constant rectilinear motion

At first, the case of constant rectilinear motion is considered, such that the oscillation amplitude of the velocity is zero ($v_1 = 0$). This case is in analogy to electromagnetic damping or magnetic levitation where the speed is assumed to be constant or slowly varying. The results are obtained considering an axially magnetized cylindrical permanent magnet with a diameter of 15 mm and a height of 25 mm and a cuboidal permanent magnet of size $[a, b, c] = [15\ \text{mm}, 15\ \text{mm}, 25\ \text{mm}]$. The remaining parameters of the magnet and the conducting slab are the same used during the verification and are given in Table 2.1.

The drag- and lift-force as a function of the nominal velocity v_0 is shown in Figure 2.25. As expected from Lenz's law, the drag-component acts against the direction of motion and the lift-component intends to push the magnet away from the conductor. It can be observed that the cuboidal magnet generates higher Lorentz forces than the cylindrical one. On the one hand, this originates from the difference in magnet volume. On the other hand, it results from the difference in area of the magnet surface close to the specimen. Both are higher in case of the cuboidal magnet. In general, three characteristic velocity regions indicated by ①–③ can be identified.

Region ① corresponds to the low velocity regime which extends in this particular example up to 5 m/s. In this regime, diffusion effects dominate over advection phenomena and the resistive nature of the conductor is prevalent. The secondary magnetic field from the induced eddy currents is much weaker than the primary magnetic field from the source ($B^{(s)} \ll B^{(p)}$). Numerical methods dealing with computational expensive problems benefit from this circumstance by simplifying the numerical

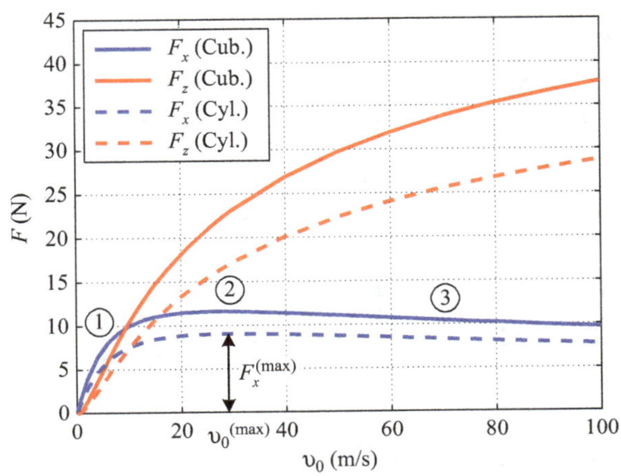

Figure 2.25 Stationary drag- (F_x) and lift-force (F_z) acting on a moving cuboidal or cylindrical permanent magnet as a function of constant velocity v_0 for three characteristic regions without oscillations ($v_1 = 0$) [143]

model (see the e/dWRA in Section 2.4). Typical examples can be found in magneto-hydro dynamics [90]. As already observed by Reitz [60] and others, the drag- and lift force show a linear and quadratic proportionality with respect to the σv product. It can be stated that the reaction from the conductor is low compared to the primary field of the magnetic field source.

In region ②, advection phenomena become more important and inductive effects can be observed. For example, the drag force generated by the cylindrical permanent magnet reaches its maximum at a velocity of $v_0^{(\max)} = 29.8$ m/s, which is higher compared to the 2D case, where the maximum was already reached at 14.5 m/s (see Section 2.2.4.3). This can be explained by emerging edge effects originating from the finite y-extension of the magnet. In contrast to the drag force, the lift force behaves differently and starts to saturate.

Finally, regime ③ is characterized by the precedence of advection phenomena. The secondary magnetic field, generated by the induced eddy currents, is as strong as the primary field inside the conductor. Hence, the effect of field suppression is distinctive and cannot be neglected. As a consequence, Joule losses, which are directly proportional to the drag force, decrease while the lift force continues to saturate.

2.2.4.4.2 Harmonic motion

In the following, the force on a periodically oscillating cylindrical permanent magnet is investigated. This case corresponds to the damping mode in electromagnetic damping. The same geometrical parameters as in the previous section are used for the calculations. It is assumed that the magnet moves with a nominal velocity of $v_0 = 0$ but oscillates with a frequency of $f_v = 100$ Hz. The oscillation amplitude is varied between $v_1 = 0.1$–0.5 m/s, which is typical for a shock absorber [152].

The drag- and lift-force over one period are shown in Figure 2.26(a) and (b), respectively. The dashed line indicates the normalized velocity profile for the purpose

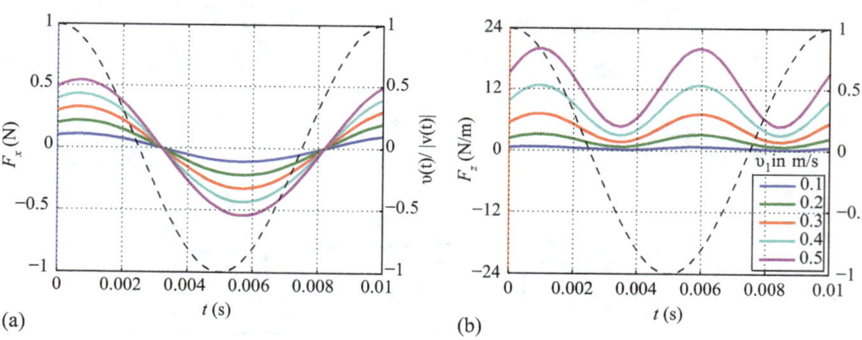

Figure 2.26 Time-dependent drag- and lift-force acting on a cylindrical permanent magnet purely harmonically oscillating ($v_0 = 0$) with a frequency of $f_v = 100$ Hz for different oscillation amplitudes v_1. The dashed line indicates the normalized velocity profile [143]. (a) Drag force F_x. (b) Lift force F_z

of illustration. As expected, it can be observed that the drag force changes the direction with respect to the underlying velocity. In contrast, the lift force stays positive over the whole period, resulting in a doubled frequency. When the oscillation amplitude v_1 increases, the lift force does not become zero at the reversal point. However, if only the maxima are considered, the linear and quadratic behavior of both components can still be observed in this velocity regime. A clear phase shift between the velocity profile and the forces is visible, resulting from the inductive nature of the conductor, which nicely illustrates Lenz's law of induction. It is observed that the lift force is influenced slightly more by this effect than the drag force.

Representing the force profiles as a function of velocity nicely illustrates the fundamental differences compared to constant rectilinear motion. The results for both the drag- and the lift-force are shown in Figure 2.27 considering three different oscillation frequencies. The dashed line shows the force profile in case of constant rectilinear motion to pin down the differences more easily. In case of harmonic motion, the drag force in Figure 2.27(a) shows a hysteretic behavior. With increasing frequency, the hysteresis is getting wider and the force reaches its zero-point when the velocities is close to its peak. Interestingly, in the transition between $f_v = 1$ Hz and $f_v = 100$ Hz (here $f_v = 10$ Hz, cf. red curve), the drag force may exceed the maximum observed in the rectilinear case ($\pm$ 38 mN) and reaches values of up to $\pm$47 mN. The corresponding lift force profile is shown in Figure 2.27(b). Since the lift force is always positive, its profile has a similar shape to a butterfly. With increasing frequency, to e.g. $f_v = 10$ Hz, the lift force is not reaching the zero-point anymore, i.e. its mean increases. However, increasing the frequency further, to e.g. $f_v = 100$ Hz, considerably changes the whole

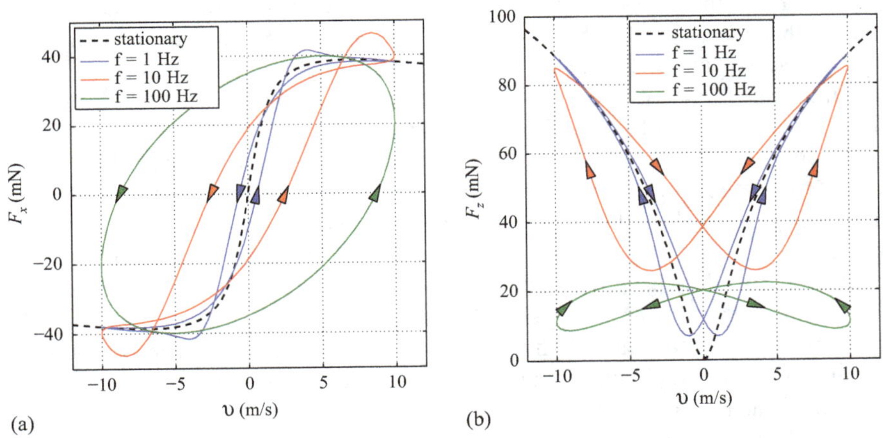

Figure 2.27 Dependency between drag force, lift force and velocity acting on a cylindrical permanent magnet purely harmonically oscillating ($v_0 = 0$) for three different frequencies considering an oscillation amplitude of $v_1 = 10$ m/s. The dashed lines indicate the force profile for constant rectilinear motion. (a) Drag force F_x. (b) Lift force F_z

profile and even flips the *wings* downward. Induction effects due to the harmonic part of the motion are thus dominating.

2.2.4.4.3 Constant rectilinear motion superimposed by harmonic oscillations

The present study is extended to the case of mixed motion. In analogy to electromagnetic damping, this corresponds to the coupler operation. In this sense, the cylindrical permanent magnet moves rectilinear with a constant velocity v_0 superimposed by an oscillation of $v_1 \cos \omega_v t$. The forces are calculated as a function of the oscillation frequency f_v and the nominal velocity v_0. The oscillation amplitude v_1 is chosen to be 3% of v_0 such that $v_1/v_0 = 0.03$. In this regard, it corresponds to the observations from the laboratory LET setup shown in Figure 2.19.

At very low frequencies, i.e. $f_v < 10$ Hz, it is observed that a high approximation order of around $N = 200$ in (2.150) is needed in order to evaluate a converged solution of (2.146). Lower approximation orders lead to noisy signals in the frequency domain which complicated the numerical integration. In order to provide compact information about the magnitude of the time-dependent forces, the relative oscillation amplitude $\Delta F_{x|z}^{(\mathrm{rel})}$ with respect to the stationary force $F_{x|z}^{(0)}$, without any oscillation ($v_1 = 0$), is calculated as:

$$\Delta F_{x|z}^{(\mathrm{rel})} = \frac{\max\left(F_{x|z}(t)\right) - \min\left(F_{x|z}(t)\right)}{F_{x|z}^{(0)}} 100\%. \tag{2.157}$$

It provides a relative measure of the force perturbation at a given working point. The corresponding results for the cylindrical permanent magnet, analyzed in the previous section, are shown in Figure 2.28(a) and (b) for the drag- and lift-component of the Lorentz force, respectively. It can be observed that both components show diverse characteristics. The highest relative oscillation amplitude of the drag force can be observed at low frequencies and when the system operates at low velocities, i.e., in the linear regime (see ① in Figure 2.28(a)). As expected from the linear relationship between the velocity and the drag force, slow velocity oscillations are directly projected onto the drag force, i.e., oscillations of about $v_1/v_0 = 0.03$ result in relative force perturbations of $\Delta F_x^{(\mathrm{rel})} \approx 6\%$. The force oscillations are damped by the secondary magnetic field of the induced eddy currents when the frequency of the velocity oscillation increases. In this sense, the conductor acts like a nonlinear inductance and similarities to a low pass filter can be observed. In consequence, high-frequency oscillations are weakly projected onto the Lorentz force. On the other hand, the relative oscillation amplitudes of both forces also decrease, when increasing the nominal velocity v_0. A local minimum can be observed in region ⓐ in Figure 2.28(a), where the drag force reaches its maximum at $v_0^{(\max)} = 29.8$ m/s [see ② in Figure 2.28(a)]. Reason for this effect is the small gradient of the drag force with respect to the velocity in this region. However, with increasing frequency, the oscillation amplitude increases before high-frequency damping effects become inherent. This can be explained by the time-dependent secondary induced eddy currents, emerging from the oscillating velocity.

The relative oscillation amplitude of the lift force is shown in Figure 2.28(b). Compared to the drag force, similar characteristics regarding the oscillation frequency

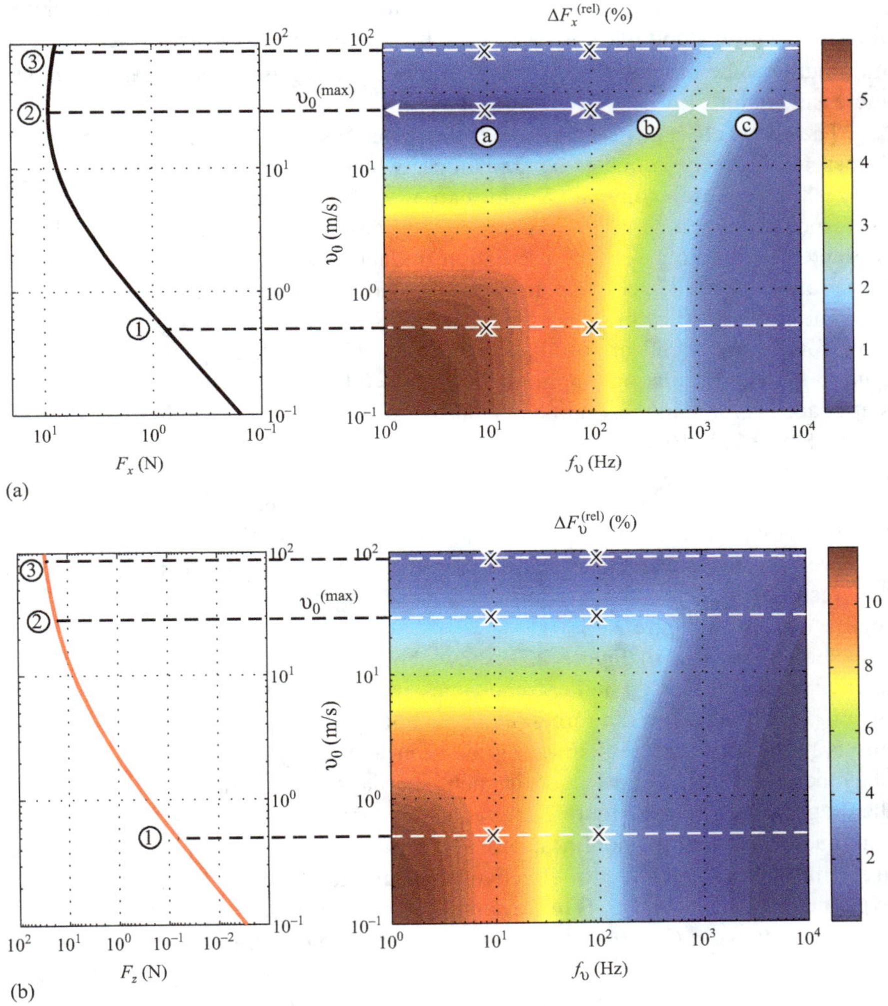

Figure 2.28 *Relative force perturbations defined by* (2.157) *as a function of the oscillation frequency* f_v *and the nominal velocity* v_0. *The plots on the left correspond to the forces for constant rectilinear motion in logarithmic scale (see Figure 2.25). All calculations are performed with a cylindrical magnet considering a velocity oscillation amplitude of* $v_1/v_0 = 0.03$. *The markers indicate the parameters used in the calculations shown in Figure 2.29 [143]. (a) Relative force perturbation of the drag force* $\Delta F_x^{(rel)}$. *(b) Relative force perturbation of the lift force* $\Delta F_z^{(rel)}$

f_v can be observed. However, its quadratic behavior at low velocities leads to relative force oscillation amplitudes of $\Delta F_z^{(\mathrm{rel})} > 11\%$. In contrast to the drag force, no local minimum can be observed at $v_0^{(\max)} = 29.8\,\mathrm{m/s}$, resulting from its overall monotonic increasing nature.

The above observations are further illustrated by showing the waveforms of the forces for distinct parameter combinations, which are marked by crosses in Figure 2.28. The chosen nominal velocities are $v_0 = \{0.5\,\mathrm{m/s},\ 29.8\,\mathrm{m/s},\ 100\,\mathrm{m/s}\}$, which correspond to the three velocity regions ①–③ described previously. The results are presented considering velocity oscillation frequencies of $f_v = \{10\,\mathrm{Hz},\ 100\,\mathrm{Hz}\}$. The forces are calculated over one period, and are shown in Figure 2.29.

The force profiles of $v_0 = 0.5\,\mathrm{m/s}$ and $f_v = 10\,\mathrm{Hz}$ are shown in Figure 2.29(a). Both force components are oscillating almost in phase with respect to the velocity. Due to the low oscillation amplitude of $v_1/v_0 = 0.03$, the quadratic behavior of the stationary lift force in this regime is weakly projected to the oscillating case.

In Figure 2.29(b), the velocity is increased up to the turning point of the drag force ($v_0^{(\max)} = 29.8\,\mathrm{m/s}$). The perturbed waveform indicates the superposition of multiple harmonics. As shown before, the oscillation amplitude of the drag force is very small in this region. Finally, the case of $v_0 = 100\,\mathrm{m/s}$ is shown in Figure 2.29(c). It can be observed that the waveform of the drag force is inverted with respect to the velocity as a result of its degressive proportionality and the regressive characteristic in this regime (see ③ in Figure 2.28(a)).

The bottom row of Figure 2.29 shows both forces considering a velocity frequency of $f_v = 100\,\mathrm{Hz}$. At higher frequencies, inductive effects become prevalent and the phase shift between the forces and the velocity increases. Concomitant with the observations from Figure 2.26, it can be seen that the drag- and lift-forces do not obey the same delay. The effect of magnetic inertia is well exemplified by comparing the drag forces between Figure 2.29(b) and (e). In a figurative sense, the drag force is not able to follow the nonlinear profile anymore. Hence, higher order harmonics are damped and the number of effective harmonics is reduced. Formally, this can be explained by the fact that ω_v is in the denominator of the argument of the higher order Bessel functions in (2.150). As a consequence, the nonlinearities are linearized and (2.150) can be truncated after a few terms, e.g. $N = 10$. Considering the lift force at $f_v = 100\,\mathrm{Hz}$ for different values of v_0, it can be seen that the phase delay decreases when the nominal velocity v_0 increases.

The behavior of the time-dependent drag force in the nonlinear region ② around the point of maximal drag force is further illustrated in Figure 2.30. The results correspond to the highlighted intervals in Figure 2.28(a). The abscissae in Figure 2.30 are normalized with respect to f_v to compare the different waveforms to each other. It can be seen that the explicit waveform strongly depends on the velocity oscillation frequency f_v. Figure 2.30(a) illustrates the non-harmonic character of the drag force in the low frequency regime. The time-dependent results are compared to the stationary case obtained by sampling the force-velocity-curve from Figure 2.25 according to the harmonic oscillation. The stationary maximum $F_x^{(0)}$ is indicated by the horizontal solid line. It can be observed that the amplitude of the drag force increases with increasing frequency and even exceeds the maximum from the stationary study in the second

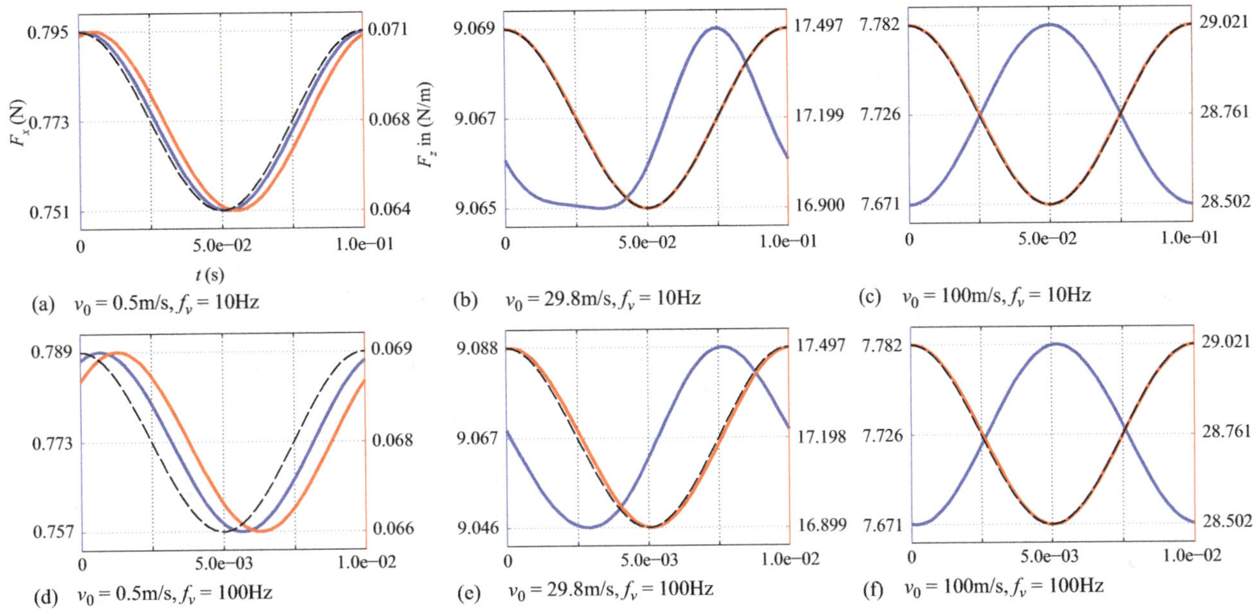

Figure 2.29 *Time-dependent drag- and lift-force waveforms in case of (a) and (d) low velocities ($v_0 = 0.5$ m/s), (b) and (e) moderate velocities ($v_0 = 29.8$ m/s) and (c) and (f) high velocities ($v_0 = 100$ m/s), considering a relative velocity oscillation amplitude of $v_1/v_0 = 0.03$. The top row corresponds to an oscillation frequency of $f_v = 10$ Hz and the bottom row to $f_v = 100$ Hz. The results correspond to the working points marked with crosses in Figure 2.28. The blue and red lines correspond to the drag-force and lift-force, respectively. The dotted line indicates the principal velocity profile of the permanent magnet for visual orientation [143]*

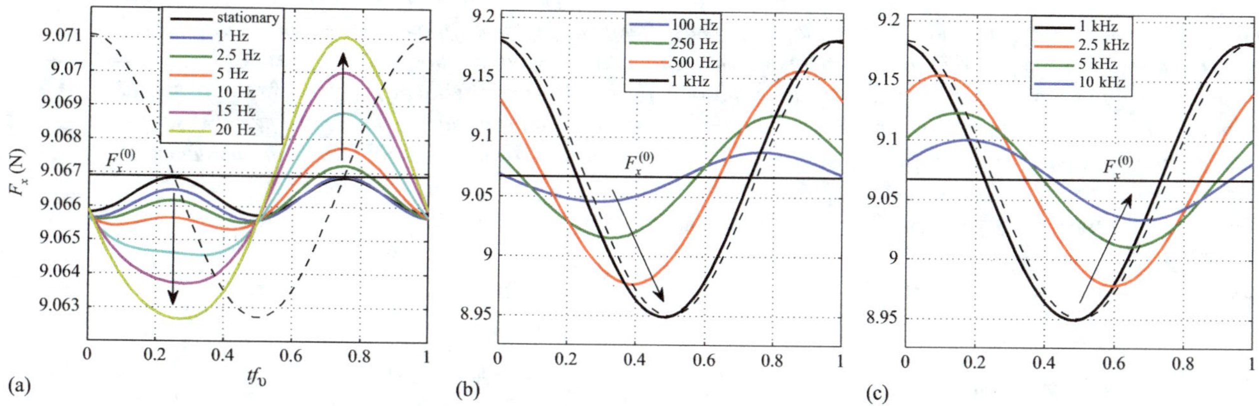

Figure 2.30 *Time-dependent drag force at the nominal velocity of* $v_0^{(\max)} = 29.8$ *m/s for different velocity oscillation frequencies* f_v*, considering an oscillation amplitude of* $v_1/v_0 = 0.03$*. The dashed line indicates the underlying velocity profile of the cylindrical permanent magnet; (a)–(c) correspond to the highlighted regions in Figure 2.28 [143]. (a) Low* f_v*. (b) Medium* f_v*. (c) High* f_v

half of the period. Figure 2.30(b) shows the drag force for frequencies up to 1 kHz. It can be seen, that the force perturbation amplitude continuously increases together with the phase-shift. At $f_v = 1$ kHz, the time-dependent drag force has its maximum and exceeds $F_x^{(0)}$ by up to 1.3%. The waveforms for frequencies up to 10 kHz are shown in Figure 2.30(c). It can be observed that the low-pass character of the conductor becomes predominant and the amplitudes decrease, while the phase-shift still increases.

The former observations can be further illustrated by the corresponding force–velocity curves shown exemplary in Figure 2.31 for the drag and lift forces. The permanent magnet was oscillating with a frequency of $f_v = 10$ Hz at an amplitude of $v_1 = 2$ m/s while he was moving with different velocities v_0. The drag force in Figure 2.31(a) shows characteristic sub-hystereses depending on the rectilinear part of the velocity v_0. The case of $v_0 = 2$ m/s (c.f. red curve) is a special case, because the magnet passes with its velocity different regimes of the magnetic Reynolds number R_m. In the transition between low and medium R_m, the drag force overshoots the maximum force, which can be reached in case of purely rectilinear motion (dotted curve). Similar observations were also made for higher velocities, i.e. at $v_0 = 6$ m/s (c.f. blue curve). Similar observations can be also made for the lift force in Figure 2.31(b).

In summary, the present study demonstrates the complexity of the problem and underlines the necessity to consider the reactance of the conductor in the case of time-dependent velocity profiles during the development of new measurement systems or devices making use of oscillating magnetic field sources.

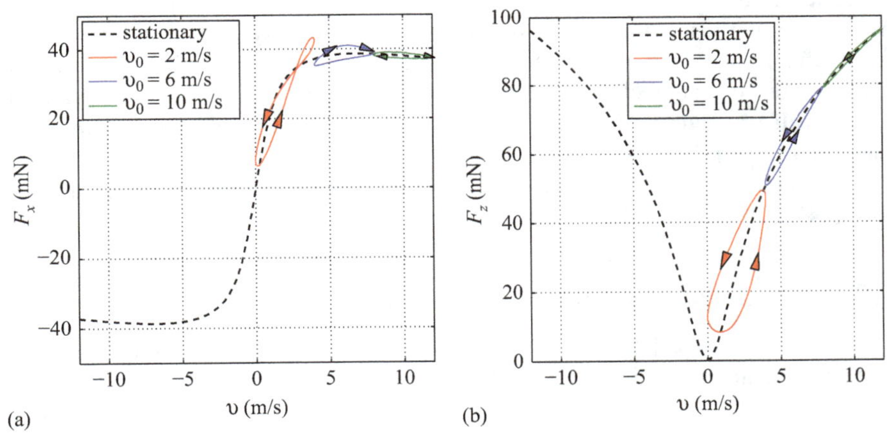

Figure 2.31 *Dependency between drag force, lift force and velocity acting on a cylindrical permanent magnet harmonically oscillating at three different offset velocities v_0 at a frequency of $f_v = 10$ Hz considering an oscillation amplitude of $v_1 = 2$ m/s. The dashed lines indicate the force profile for constant rectilinear motion. (a) Drag force F_x. (b) Lift force F_z*

2.2.4.4.4 Conclusions

Vibrations play a major role in industrial and laboratory MIECT setups. The present study shows that the force profiles strongly depend on the operating point of the system under investigation, which is determined by the level of constant motion together with the oscillation amplitude and frequency. The complex interaction between induced eddy currents, resulting from the constant part and the time-dependent part of the velocity is exemplified. The back reaction of the conductor and its reactance cause phase shifts and lower damping forces. This should be taken into account when evaluating the dynamic characteristics of oscillating systems in the future. Due to the nonlinear characteristic of the drag force as a function of velocity v_0, higher order harmonics are emerging in the oscillatory case.

In contrast, the gradual nature of the lift force mitigates this effect. It seems inherent that these effects should be considered in future developments of new systems involving time-dependent motion in order to provide more accurate predictions. Current research for example is devoted to apply ball screw mechanisms to energy harvesters which can significantly magnify the vibrational motion [126] and in turn increase the performance of such devices.

The analytical integral expressions can be easily modified according to different coil geometries by replacing the corresponding Fourier transforms of the source current in (2.147) and (2.148), respectively. Note that the presented approach could also be adopted to other application scenarios as for example to analyze the dynamics of linear or rotating eddy current couplers [153] or MAGLEV systems. Moreover, the presented approach can be used as a reference during the development of more advanced numerical models.

The phase and amplitude of the drag- and lift force provide information about the material properties of the conductor. Oscillating magnet systems could be applied in the future within the context of MIECT as an alternative to systems which are based on constant rectilinear motion. By this analysis, it is possible to provide the required electromagnetic force $\mathbf{F}^{(EM)}$, acting on a mechanical system to determine the measured force $\mathbf{F}^{(MEAS)}$ (see Figure 2.19).

Subsequently, several semianalytical methods for calculating DRSs for 3D LET systems will be discussed.

2.2.5 The simplest approach to calculate DRS

The simplest way to calculate DRSs is to omit distortion current $\mathbf{j}_E$ in equation (2.85). In this case, the determination of DRS reduces to the calculation of the following integral

$$\Delta\mathbf{F} \simeq \int_{V_D} \mathbf{j}_D \times \mathbf{B}_0 \, \mathrm{d}V = \int_{V_D} \mathbf{J}_0 \times \mathbf{B}_0 \, \mathrm{d}V, \tag{2.158}$$

where $\mathbf{B}_0$ is the primary field produced by a magnet and $\mathbf{J}_0$ are induced eddy currents in the conductor without any defect. This approach was used for the first time in [98]. Assuming a weak reaction of eddy currents, it uses a superposition of eddy currents induced in a block without a defect with eddy currents occurring in the virtual area covering the defect. The permanent magnet is replaced by a single magnetic dipole. In addition, it is assumed that the dimensions of the block are much larger than the permanent magnet and the block itself is represented by a package of thin conducting sheets isolated from each other. The paper [98] describes a defect reconstruction procedure using normalized DRS. The main advantage of this method is the speed of calculation. However, a direct comparison of the non-normalized DRS with the signals calculated by FEM shows that the approximate signals differ significantly from the reference signals. Figure 2.32 shows an example of DRS profiles near the defect calculated according to the procedure described in [98] for $y = 0$. The following system LET configuration was used in the calculations. The cuboidal permanent magnet with dimensions $w^2 \times h = 15\,\text{mm} \times 15\,\text{mm} \times 25\,\text{mm}$, magnetization $\mathbf{M} = M\mathbf{1}_z$, $M = 931\,\text{A/mm}$ is located at a lift-off $h_0 = 1\,\text{mm}$ above the conductive block with a thickness of $D = 100\,\text{mm}$. The block moves with the velocity $v = 1\,\text{cm/s}$ along the x-axis. Since in experiments a set of insulated thin plates instead of solid material was used, the electrical conductivity of the block is described by a diagonal tensor with parameters $\sigma_{xx} = \sigma_{yy} = \sigma_0 = 30.61\,\text{MS/m}$, $\sigma_{zz} = 0$. The rectangular defect $c_x \times c_y \times c_z = 12\,\text{mm} \times 2\,\text{mm} \times 2\,\text{mm}$ is at depth $d = 4\,\text{mm}$. The permanent magnet is modeled by a single equivalent magnetic m_α-dipole with a magnetic moment $\mathbf{m}_\alpha = m_\alpha \mathbf{1}_z = Mw^2h\mathbf{1}_z$ placed inside the permanent magnet at $\mathbf{r}_\alpha = [0.5w, 0.5w, \alpha h]^{\mathrm{T}}$, where $\alpha = 0.418$. The parameter α was determined in a similar way to the procedure from Section 2.2.1.

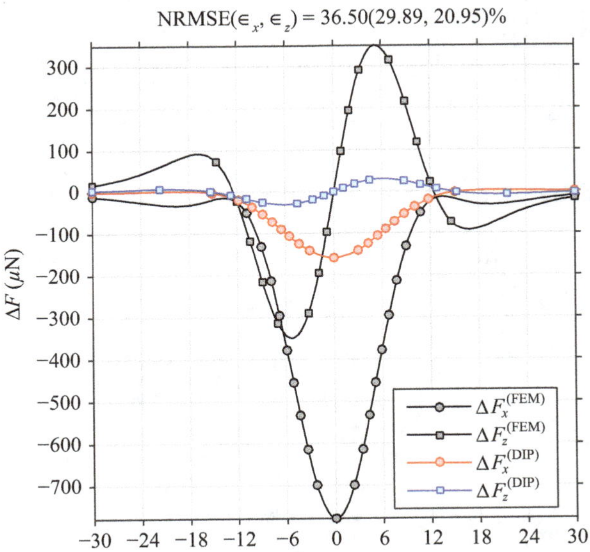

Figure 2.32 DRS profiles calculated using the method presented in [98] (DIP) together with the reference signals (FEM)

The calculated NRMSE (see (2.34)) for the above setup is equal to 36.5%. Although the use of more accurate permanent magnet modeling improves the approximated DRS, the improvement is relatively small and NRMSE remains at around 30% [112].

To explain the reason of large errors in the calculation of DRS, eddy currents near the defect are determined. Figure 2.33 shows the distribution of eddy currents in the *x*-*y*–plane crossing the defect. The distributions are plotted for the moment when the permanent magnet is just above the center of the defect. $\mathbf{J}_0$ denotes the density of eddy currents induced in a moving block without a defect. $\mathbf{j}_D$ is the current density vector describing induced eddy currents in the defect filled with a material with electrical conductivity σ_0. According to [98], eddy currents flowing around the defect are described by the superposition of the $\mathbf{J}_0$ and $\mathbf{j}_D$ vectors as $\mathbf{J} = \mathbf{J}_0 - \mathbf{j}_D$. As can be seen from Figure 2.33c, the obtained distribution of eddy currents $\mathbf{J}$ is not correct because the continuity of currents on the defect boundary is not satisfied. The correct eddy current distribution can be calculated by FEM and it is shown in Figure 2.34.

In conclusion, the correct description of eddy currents in 3D LET systems requires in the superposition process to take into account not only the area of the

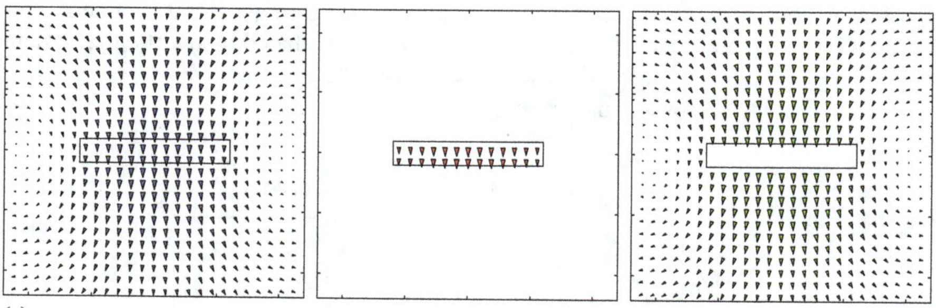

(a) (b) (c)

Figure 2.33 Eddy currents in the x-y–plane crossing the defect for the moment when the permanent magnet is above the defect center (according to [98]). (a) $\mathbf{J}_0$*—without defect. (b)* $\mathbf{j}_D$*—in defect region. (c)* $\mathbf{J}_0$*—*$\mathbf{j}_D$

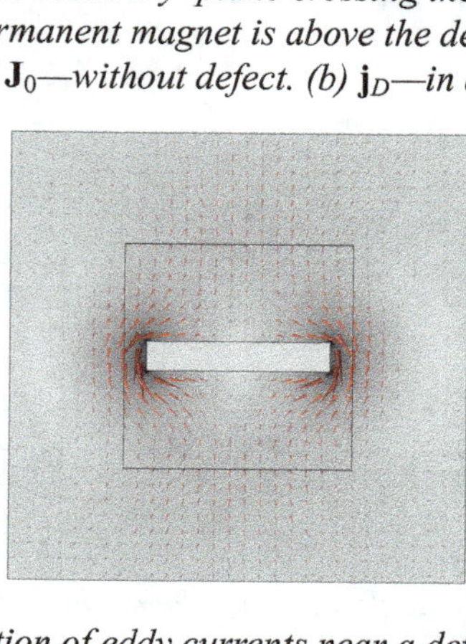

Figure 2.34 Distribution of eddy currents near a defect calculated by FEM

defect itself, as was the case in 2D systems, but also the area outside the defect. This also explains the large errors in the calculation of DRS using the approach given in [98].

In the following sections, several methods will be presented that will enable more accurate calculation of DRSs.

2.2.6 A hole in a thin, large, conductive sheet

Before presenting the next method of calculating the DRS, the distribution of currents around the circular hole in a thin conductive plate will be analyzed. The uniform electric current with density $\mathbf{J}_0$ flows in a thin conductive plate with an homogeneous electrical conductivity σ_0. A hole with a radius R is drilled in the plate (Figure 2.35). The purpose of the analysis is to find a distribution of currents around the hollow hole.

Far away from the origin, the current density is uniform with $\mathbf{J} = \mathbf{J_0} = -J_0\mathbf{1}_y$. Due to the presence of the hole, the current distribution is disturbed. Introducing the electric scalar potential φ defined as $\mathbf{E} = -\nabla\varphi$, the problem can be described using Laplace equation in a cylindrical coordinate system (r, ψ, z)

$$\nabla^2\varphi = \frac{\partial^2\varphi}{\partial r^2} + \frac{1}{r}\frac{\partial\varphi}{\partial r} + \frac{1}{r^2}\frac{\partial^2\varphi}{\partial\psi^2} = 0, \tag{2.159}$$

together with the boundary conditions

$$r = R: \quad J_n = 0, \quad \frac{\partial\varphi}{\partial r} = 0, \tag{2.160}$$

$$r \to \infty: \quad \mathbf{J} = -J_0\mathbf{1}_y, \quad V = \frac{1}{\sigma_0}J_0 r \sin\psi. \tag{2.161}$$

The solution of equation (2.159) can be obtained as

$$\varphi = \varphi_0 + \frac{J_0}{\sigma_0}r\left[1 + \left(\frac{R}{r}\right)^2\right]\sin\psi, \tag{2.162}$$

where $\varphi_0 = \text{const}$ is a potential with any fixed value.

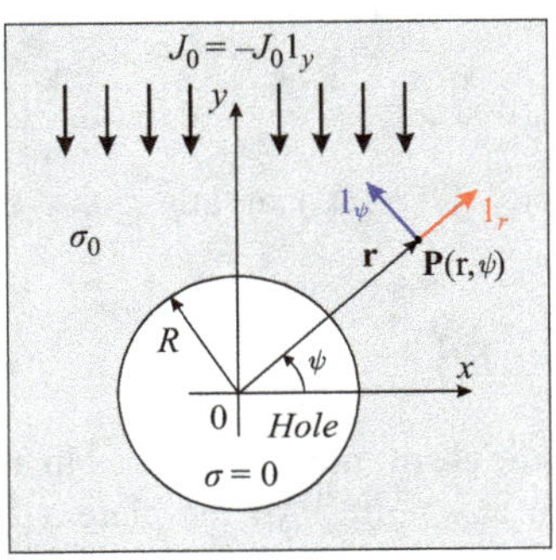

Figure 2.35 Setup to solve the problem of electric current flowing in a thin conducting sheet with a circular hollow hole

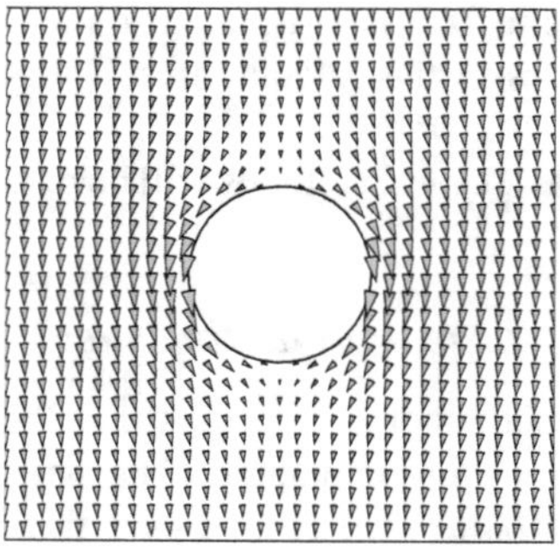

Figure 2.36 Current density distribution in the vicinity of a circular hole

The current density **J** can be calculated following from

$$\mathbf{J} = \sigma_0 \mathbf{E} = -\sigma_0 \nabla \varphi = -\sigma_0 \left(\frac{\partial \varphi}{\partial r} \mathbf{1}_r + \frac{1}{r} \frac{\partial \varphi}{\partial \psi} \mathbf{1}_\psi + \frac{\partial \varphi}{\partial z} \mathbf{1}_z \right) \tag{2.163}$$

and finally takes the form

$$\mathbf{J} = -J_0 \left[1 - \left(\frac{R}{r} \right)^2 \right] \sin \psi \mathbf{1}_r - J_0 \left[1 + \left(\frac{R}{r} \right)^2 \right] \cos \psi \mathbf{1}_\psi . \tag{2.164}$$

An exemplary current distribution around a circular hole is shown in Figure 2.36. The current density **J** can be replaced by the superposition of two terms, namely the primary input current density $\mathbf{J}_0$ and the distortion current density **j** evoked by the hole

$$\mathbf{J} = \mathbf{J}_0 - \mathbf{j}, \tag{2.165}$$

with **j** given by

$$\mathbf{j} = \left(\frac{R}{r} \right)^2 \left(-J_0 \sin \psi \mathbf{1}_r + J_0 \cos \psi \mathbf{1}_\psi \right) . \tag{2.166}$$

Using the vector identity

$$\mathbf{J}_0 = -J_0 \sin \psi \mathbf{1}_r - J_0 \cos \psi \mathbf{1}_\psi = \frac{\mathbf{J}_0 \cdot \mathbf{r}}{r^2} \mathbf{r} + (\mathbf{J}_0 \cdot \mathbf{1}_\psi) \mathbf{1}_\psi$$

the distortion current density **j** (2.166) at any point **r** outside the hole can be expressed as

$$\mathbf{j} = \left(\frac{R}{r} \right)^2 \left(2 \frac{\mathbf{J}_0 \cdot \mathbf{r}}{r^2} \mathbf{r} - \mathbf{J}_0 \right) . \tag{2.167}$$

A current dipole with a dipole moment $\mathbf{p}_0$, placed in a large thin conducting plate of thickness δz, produces at any point **r** of the plate a 2D flow of electric current described by

$$\mathbf{j}_p = \frac{1}{2\pi r^2 \delta z} \left(2 \frac{\mathbf{p}_0 \cdot \mathbf{r}}{r^2} \mathbf{r} - \mathbf{p}_0 \right) . \tag{2.168}$$

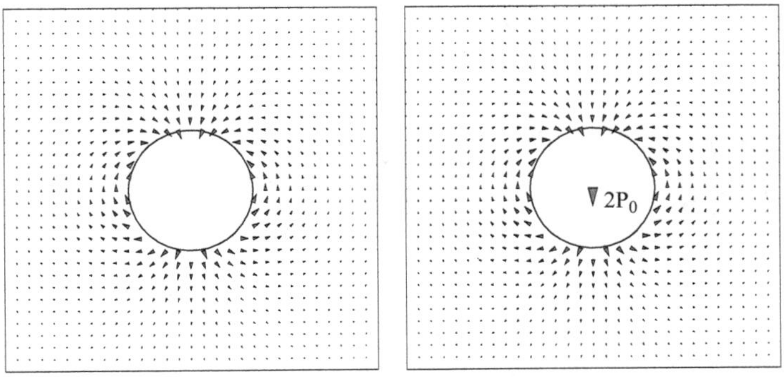

(a) (b)

Figure 2.37 Distortion current near a hollow hole in a large thin conducting plate. (a) Distortion current $\mathbf{j}$*. (b) Current generated by an equivalent current dipole* $2\mathbf{p}_0$

By introducing an equivalent current dipole with a dipole moment $\mathbf{p}_0 = V_0\,\mathbf{J}_0 = \pi R^2\delta z\,\mathbf{J}_0$, equation (2.167) can be replaced by

$$\mathbf{j} = 2\frac{1}{2\pi r^2\delta z}\left(2\frac{\mathbf{p}_0\cdot\mathbf{r}}{r^2}\mathbf{r} - \mathbf{p}_0\right) = 2\mathbf{j}_p, \tag{2.169}$$

where V_0 is the volume of the hole. Thus, it can be seen that a distortion current caused by a circular hole in a large thin conducting plate in which a uniform flow of electric current was forced, can be replaced by the current produced by a single current dipole located at the center of the hole. Figure 2.37 shows the distortion current around a circular hollow hole and the current produced by the dipole $2\mathbf{p}_0$ clipped through the hole.

2.2.7 An extended area approach in the calculation of DRS

Inspired by the results presented in the previous section, a semianalytical approach, called an extended area approach (EAA) [112], for calculating the flow of distortion currents around a hollow hole in a thin conductive plate has been developed.

It is considered a plate moving with a velocity $\mathbf{v} = v_0\mathbf{1}_x$. The plate with the thickness Δz is parallel to the XOY plane. The hole region and the region around is replaced by a regular mesh of volumetric elements (voxels) with a volume $V_0 = \Delta x\Delta y\Delta z$, where Δx and Δy are grid densities in x- and y-directions, respectively. In the center $\mathbf{r}_i$ of each voxel located inside the hole a source current dipole with a dipole moment $\mathbf{p}_i = V_0\,\mathbf{j}_{D,i}$ is placed, where $\mathbf{j}_{D,i} = \mathbf{J}_{0,i}$ is the density of induced eddy currents in the plate without any defect truncated to the defect region.

The distortion current density $\mathbf{j}_E$ at any point $\mathbf{r}$ outside the hole can be approximated by

$$\mathbf{j}_E(\mathbf{r}) = \chi_{DCF}\frac{V_0}{2\pi\delta z}\sum_{i=1}^{N_V^D}\left[2\frac{\mathbf{j}_{D,i}\cdot(\mathbf{r}-\mathbf{r}_{D,i})}{|\mathbf{r}-\mathbf{r}_{D,i}|^4}(\mathbf{r}-\mathbf{r}_{D,i}) - \frac{\mathbf{j}_{D,i}}{|\mathbf{r}-\mathbf{r}_{D,i}|^2}\right], \tag{2.170}$$

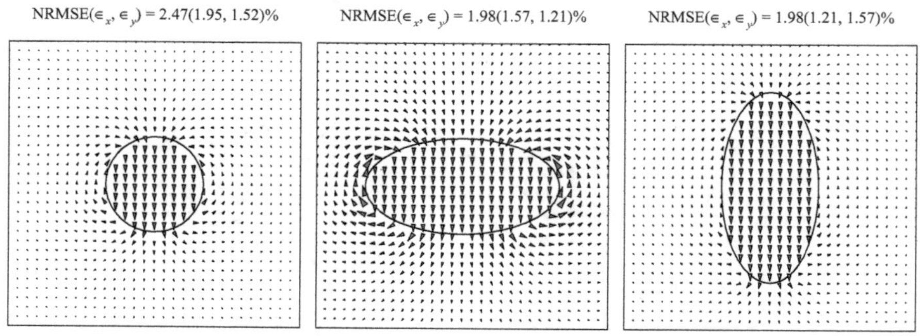

Figure 2.38 Distortion currents around hollow holes in a thin conducting plate together with $\mathbf{J}_0$ *currents in the hole area*

where N_V^D is the number of voxels inside the hole region, $\mathbf{r}_{D,i}$ is the center of the i-th voxel, and χ_{DCF} denotes a dipolar correction factor (DCF).

The dipolar correction factor χ_{DCF} for a hole defined by parameters c_x, c_y, e.g. a circular, elliptical or rectangular defect, is generally defined as

$$\chi_{DCF} = 1 + \frac{c_x}{c_y}. \tag{2.171}$$

Although the dipolar correction factor is only an indicative factor it enables to model the distortion field around holes of various shapes with a sufficient accuracy.

In order to calculate DRS profiles, a rectangular extended region defined as $e_{\text{ext}} \times e_{\text{ext}} \times \Delta z$ around the hollow hole $c_x \times c_y \times \Delta z$ is introduced. The size e_{ext} of the extended region is defined as

$$e_{\text{ext}} = (s_{\text{ext}} + 1)\max(c_x, c_y), \tag{2.172}$$

where s_{ext} is an arbitrary chosen scaling factor greater or equal 0. If $s_{\text{ext}} = 0$ then the extended region does not exist and the model is reduced to the simplest approach presented in Section 2.2.5.

Figure 2.38 shows sample distributions of distortion currents evoked by holes of various shapes located in a thin conductive plate in which flows homogeneous current $\mathbf{J}_0 = -J_0\mathbf{1}_y$. The NRMSEs of the EAA in Figure 2.38 were calculated in the reference to analytical solutions.

The method EAA can be directly used to calculate DRS from a defect (c_x, c_y, c_z) located in a conductive block of $L \times W \times D$ dimensions much larger than the dimensions of the magnet and anisotropic conductivity of $[\boldsymbol{\sigma}] = \text{diag}(\sigma_0, \sigma_0, 0)$ $(J_z = 0)$. In this case, the extended area consists of N_L x-y–layers Δz thick ($\Delta z = c_z/N_L$) in which regular voxel grids have been defined, each with a volume of $V_E = \Delta x \times \Delta y \times \Delta z$ (Figure 2.39).

Finally, the DRS formula (2.85) takes the following discrete form

$$\Delta\mathbf{F}^{(n)} \cong V_E \sum_{l=1}^{N_L} \left[\sum_{j=1}^{N_V^D} \mathbf{j}_{D,j}^{l,(n)} \times \mathbf{B}_{0,j}^{l,(n)} + \sum_{i=1}^{N_V^E} \mathbf{j}_{E,i}^{l,(n)} \times \mathbf{B}_{0,i}^{l,(n)} \right], \tag{2.173}$$

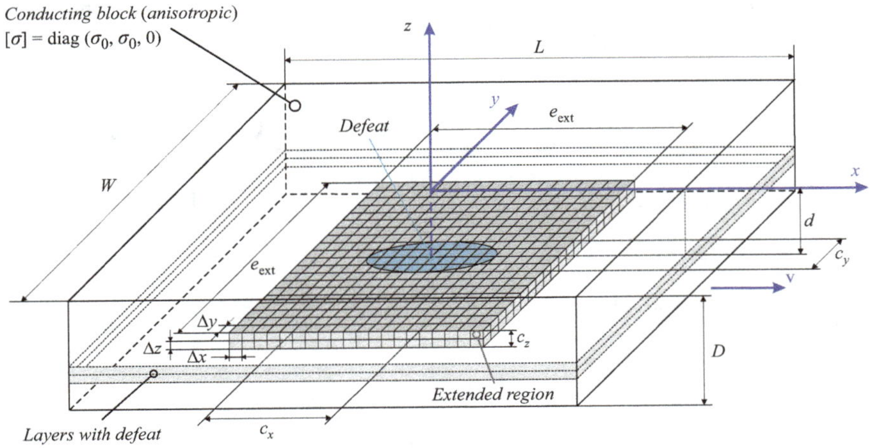

Figure 2.39 Extended region for $s_{ext} = 1$ around a sample cylindrical defect located in an anisotropic conducting block

where the distortion current $\mathbf{j}_{E,i}^{l,(n)}$ is expressed by (2.170). The defect current $\mathbf{j}_{D,j}^{l,(n)} = \mathbf{J}_{0,j}^{l,(n)}$ can be calculated according to (2.89) when a multi-dipole permanent magnet model is used or (2.95) for rectangular permanent magnets. Respectively, the primary magnetic flux density $\mathbf{B}_{0,k}^{l,(n)}$ can be calculated using (2.87) or (2.90).

Figure 2.40 shows DRS profiles for $y = 0$ determined according to (2.173) for an artificial cuboidal defect $c_x \times c_y \times c_z = 12\,\text{mm} \times 2\,\text{mm} \times 2\,\text{mm}$ placed at a depth $d = 4\,\text{mm}$ in a conductive plate with a thickness $D = 50\,\text{mm}$ moving at a velocity $v_0 = 1\,\text{cm/s}$ under a rectangular permanent magnet with dimensions of $w \times w \times h = 15\,\text{mm} \times 15\,\text{mm} \times 25\,\text{mm}$ and magnetization $M = 931\,\text{A/mm}$ suspended at a lift-off $h_0 = 1\,\text{mm}$ above the plate. The anisotropic conductivity $[\boldsymbol{\sigma}] = \text{diag}(\sigma_0, \sigma_0, 0)$ is equal to $[\boldsymbol{\sigma}] = \text{diag}\ (30.61\,\text{MS/m}, 30.61\,\text{MS/m}, 0)$. The voxel size used for construction of various extended areas is defined as follows: $\Delta x = 1\,\text{mm}$, $\Delta y = 1\,\text{mm}$, and $\Delta z = 2\,\text{mm}$. The profiles were calculated using in (2.173) the analytical formulas (2.95) and (2.90) for the description of $\mathbf{J}_0$ and $\mathbf{B}_0$, respectively. For comparison, the profiles calculated with the simplest approach from Section 2.2.5 ($s_{ext} = 0$) and using the extended region with $s_{ext} = 4$ are presented.

Assuming a weak reaction of eddy currents, the EAA method can also be used to calculate DRS from defects located near the surface of the thick solid plate with a uniform electrical conductivity σ_0. In this case, the extended region around the defect (c_x, c_y, c_z) is defined as a cuboid with dimensions of $e_{ext} \times e_{ext} \times (e_{ext}/2 + c_z + d)$, where e_{ext} is given by (2.172) and d is the defect depth. For the defined extended region, formula (2.173) can be used to calculate the DRS.

As in the case of an anisotropic plate, eddy currents $\mathbf{J}_0$ induced in an isotropic plate can be determined using equation (2.89) when the magnet is modeled using magnetic dipoles or analytical formulas (2.95) in the case of a rectangular magnet.

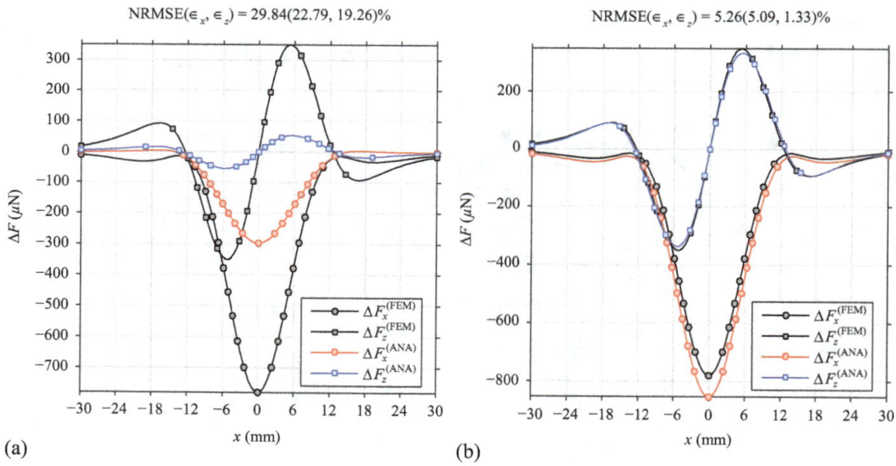

Figure 2.40 Anisotropic conductor—DRS profiles obtained for a defect $c_x \times c_y \times c_z = 12\,\text{mm} \times 2\,\text{mm} \times 2\,\text{mm}$ *located at a depth* $d = 4\,\text{mm}$ *in a conductive plate with a thickness* $D = 50\,\text{mm}$ *moving at a velocity* $v_0 = 1\,\text{cm/s}$. *(ANA)* (2.173) *with analytical formulas* (2.95), (2.170) *and* (2.90), *(FEM) finite element method. (a)* $s_{ext} = 0$. *(b)* $s_{ext} = 4$

As well, the primary field $\mathbf{B}_0$ produced by a permanent magnet is calculated according to (2.87) or (2.90).

Calculation of the distortion current $\mathbf{j}_{E,k}^{(n)}$ at point $\mathbf{r}_k$ in an isotropic environment requires a different formula than (2.170), namely

$$\mathbf{j}_{E,k}^{(n)} = \chi_{DCF} \frac{V_E}{4\pi} \sum_{i=1}^{N_V^D} \left[3 \frac{\mathbf{j}_{D,i}^{(n)} \cdot (\mathbf{r}_k - \mathbf{r}_{D,i})}{|\mathbf{r}_k - \mathbf{r}_{D,i}|^5} (\mathbf{r}_k - \mathbf{r}_{D,i}) - \frac{\mathbf{j}_{D,i}^{(n)}}{|\mathbf{r}_k - \mathbf{r}_{D,i}|^3} \right], \tag{2.174}$$

where N_V^D is the number of voxels inside the defect region, $\mathbf{r}_{D,i}$ is the center of the i-th voxel, and χ_{DCF} is a dipolar correction factor. It should be noted that the estimation (2.171) cannot be further used for defects in isotropic conductors. For regular defects, such as a sphere or a cube, the χ_{DCF} can be determined analytically as $\chi_{DCF} = 3/2$ [154]. To find the DCF for an defect defined by the outline $c_x \times c_y \times c_z$ distortion currents around the idealized cavity in a form of oblate/prolate spheroid have been analyzed [154]. The cavity is located in a conductor in which homogeneous flow of electric current is forced. The distribution of distortion currents can be found analytically using a separation of variables method for solving Laplace equation $\nabla^2\varphi = 0$ in the appropriate coordinate system (oblate/prolate spheroidal coordinate system) [155]. The performed analysis allows to formulate the following approximate formula for the dipolar correction factor

$$\chi_{DCF} = 1 + \frac{1}{4}\left(\frac{c_x}{c_y} + \frac{c_z}{c_y}\right). \tag{2.175}$$

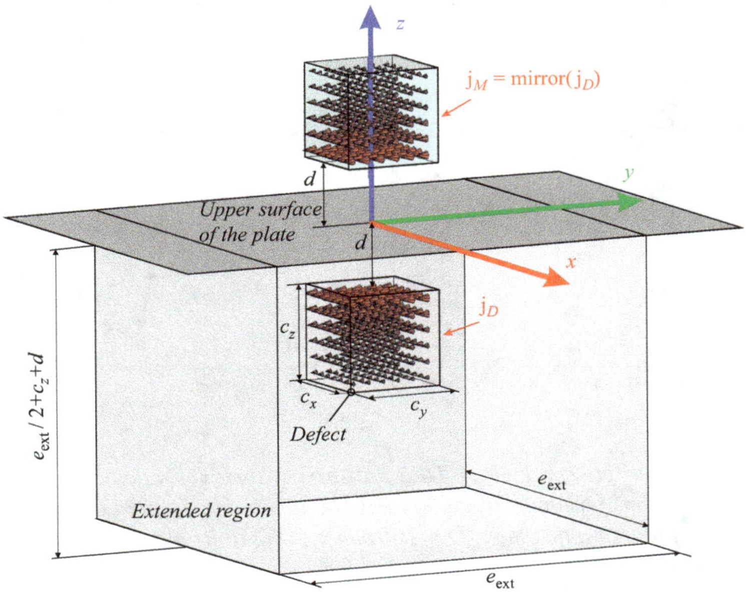

Figure 2.41 Extended region around a cuboidal defect in a solid isotropic conductive plate together with mirrored defect currents

However, a simple application of (2.174) in the calculation of DRSs is not correct because the conditions $j_z = 0|_{z=0}$ and $j_z = 0|_{z=-D}$ at the upper and bottom surfaces of the plate will be not satisfied. If the thickness of the plate is much greater than the depth and size of the defect, then the condition $j_z = 0$ can be ensured by introducing additional source currents $\mathbf{j}_M = \text{mirror}(\mathbf{j}_D)$ obtained as a result of the mirror reflection of $\mathbf{j}_D$ currents relative to the upper surface of the plate (see Figure 2.41).

Finally, the corrected distortion currents $\mathbf{j}_{E,k}^{(n)}$ can be obtained as

$$\mathbf{j}_{E,k}^{(n)} = \chi_{DCF}\frac{V_E}{4\pi}\left\{\sum_{i=1}^{N_V^D}\left[3\frac{\mathbf{j}_{D,i}^{(n)}\cdot(\mathbf{r}_k-\mathbf{r}_{D,i})}{|\mathbf{r}_k-\mathbf{r}_{D,i}|^5}(\mathbf{r}_k-\mathbf{r}_{D,i})-\frac{\mathbf{j}_{D,i}^{(n)}}{|\mathbf{r}_k-\mathbf{r}_{D,i}|^3}\right]\right.$$
$$\left.+\sum_{i=1}^{N_V^D}\left[3\frac{\mathbf{j}_{D,i}^{(n)}\cdot(\mathbf{r}_k-\mathbf{r}_{M,i})}{|\mathbf{r}_k-\mathbf{r}_{M,i}|^5}(\mathbf{r}_k-\mathbf{r}_{M,i})-\frac{\mathbf{j}_{D,i}^{(n)}}{|\mathbf{r}_k-\mathbf{r}_{M,i}|^3}\right]\right\}, \quad (2.176)$$

where $\mathbf{r}_{D,i} = [x_{D,i}, y_{D,i}, z_{D,i}]^{\mathrm{T}}$ is the center of the i-th voxel and $\mathbf{r}_{M,i} = [x_{D,i}, y_{D,i}, -z_{D,i}]^{\mathrm{T}}$ and $\mathbf{j}_{D,i}^{(n)} = \mathbf{J}_{0,i}^{(n)}$ are eddy currents induced in the plate without defects truncated to the defect region.

Figure 2.42 shows DRS profiles for $y = 0$ determined according to (2.173) for an artificial cuboidal defect $c_x \times c_y \times c_z = 6\,\text{mm} \times 6\,\text{mm} \times 6\,\text{mm}$ placed at a depth $d = 4\,\text{mm}$ in an isotropic conductive plate with a thickness $D = 100\,\text{mm}$ moving

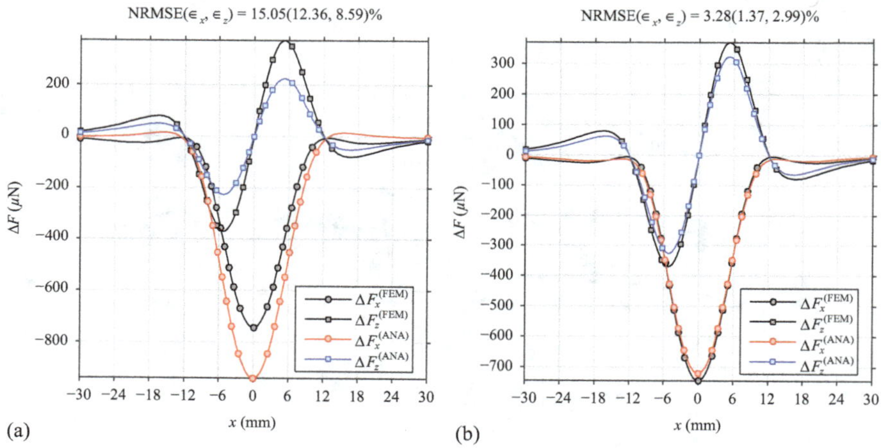

Figure 2.42 Isotropic conductor—DRS profiles obtained for a defect $c_x \times c_y \times c_z = 6\,\text{mm} \times 6\,\text{mm} \times 6\,\text{mm}$ *located at a depth* $d = 4\,\text{mm}$ *in a conductive plate of a thickness* $D = 100\,\text{mm}$ *moving at a velocity* $v_0 = 1\,\text{cm/s}$. *(ANA)* (2.173) *with analytical formulas* (2.95), (2.176) *and* (2.90), *(FEM) finite element method. (a)* $s_{ext} = 0$. *(b)* $s_{ext} = 5$

at a velocity $v_0 = 1$ cm/s under a rectangular permanent magnet with dimensions of $w \times w \times h = 15\,\text{mm} \times 15\,\text{mm} \times 25\,\text{mm}$ and magnetization $M = 931$ A/mm suspended at a lift-off $h_0 = 1$ mm above the plate with a homogeneous conductivity $\sigma_0 = 30.61$ MS/m. The voxel size used for construction of extended regions is defined as: $\Delta x = 1$ mm, $\Delta y = 1$ mm, and $\Delta z = 1$ mm. The profiles were calculated using in (2.173) formulas (2.95), (2.176), and (2.90). For comparison, the profiles calculated using only $\mathbf{j}_D$ current ($s_{\text{ext}} = 0$) and the extended region with $s_{\text{ext}} = 5$ together with the reference solution obtained from the FEM are presented. As a result of the EAA method, a significant reduction in normalized mean square errors (NRSMEs) can be observed.

2.3 Surface charge simulation method

In experiments using the LET method, multiple scans of lines or surfaces are made that determine the profiles of forces acting on the magnetic system [80]. Force profiles are then used to identify and reconstruct defects in the tested objects. In identification procedures (inverse problems), multiple simulations are carried out on LET models requiring fast and efficient computational methods [156]. Several such semianalytical methods have been presented in previous sections. Recent optimization studies [157] showed that rather complicated magnet systems, which include highly saturating ferromagnetic materials such as iron-cobalt alloys, may be advantageous compared to standard magnet geometries. However, the implementation of nonlinear magnetic material significantly increases the computational effort needed to determine the force profiles numerically.

In this section, a numerical approach based on principles of the surface charge simulation method (SCSM) [158] is presented. Not like in the charge simulation method (CSM) [159–161] where electric charges are placed near the surface of the analyzed objects, electric charges used in SCSM are located directly on the boundaries of the object. SCSM proves its popularity in electrostatics because of its simplicity, accuracy and computational efficiency [162], but its application to the problems of electrodynamics associated with moving conductive media was only rarely analyzed [163–165].

The analysis of the LET system shown in Figure 2.17 is performed in the coordinate system attached to the permanent magnet system. The state of the system is defined by vector fields **B** and **E** under assumption that the speed of the moving conducting object is small enough such that the approach with WRA can be applied [111]. This assumption is valid, for example, for conductive objects made of aluminum and moving with a speed of less than 0.5 m/s [166]. In this case, eddy currents **J** induced in a moving conductor can be directly determined from Ohm's law

$$\mathbf{J} = [\boldsymbol{\sigma}](\mathbf{E} + \mathbf{v} \times \mathbf{B}_0) = [\boldsymbol{\sigma}](-\nabla\varphi + \mathbf{v} \times \mathbf{B}_0), \tag{2.177}$$

where $[\boldsymbol{\sigma}]$ is a diagonal tensor of electrical conductivity $[\boldsymbol{\sigma}] = \mathrm{diag}(\sigma_{xx}, \sigma_{yy}, \sigma_{zz})$, φ is scalar electric potential, and $\mathbf{B}_0$ is the primary magnetic field produced by the magnet system. The SCSM method is used to determine the φ potential.

For solid conductors with a homogeneous conductivity σ_0, the potential φ at any point $\mathbf{x} = x\mathbf{1}_x + y\mathbf{1}_y + z\mathbf{1}_z$ in the conductor can be determined according to the SCSM proposed in [164] as

$$\varphi(\mathbf{x}) = \frac{1}{4\pi\varepsilon_0}\int_S \frac{\kappa(\mathbf{x}')}{|\mathbf{x} - \mathbf{x}'|}\mathrm{d}S, \tag{2.178}$$

where $\kappa(\mathbf{x}')$ is the unknown surface density of the electric charge distributed on the surface $S = S_c \cup S_d$, and S_c, S_d are the conductor and the defect boundary surfaces, respectively. An example of SCSM mesh $\tilde{S}$ is shown in Figure 2.43.

In order to find the surface charge density $\kappa(\mathbf{x}')$ as well as eddy currents (2.177), the condition $\mathbf{n} \cdot \mathbf{J}|_S = 0$ must be applied [164]. It enables to formulate the following set of equations

$$\frac{1}{4\pi\varepsilon_0}\sum_{i=1}^{N_{2D}} \kappa_i \int_{\Delta S_i} \frac{\mathbf{n}_j \cdot (\mathbf{x}_j - \mathbf{x}_i')}{|\mathbf{x}_j - \mathbf{x}_i'|^3}\,\mathrm{d}S_i = -\mathbf{n}_j \cdot (\mathbf{v} \times \mathbf{B}_{0,j}), \quad j = 1, \ldots, N_{2D}, \tag{2.179}$$

where N_{2D} is the number of SCSM elements, $\tilde{S} = \bigcup_{i=1}^{N_{2D}} \Delta S_i$ is a mesh of rectangular elements covering the surface S, $\kappa_i = \mathrm{const.}$ on ΔS_i, and $\mathbf{B}_{0,j} = \mathbf{B}_0(\mathbf{x}_j)$.

For conductors with an anisotropic conductivity $[\boldsymbol{\sigma}] = \mathrm{diag}(\sigma_{xx}, \sigma_{yy}, 0)$, the potential φ is given by

$$\varphi(\mathbf{r}, z_k) = \frac{1}{2\pi\varepsilon_0}\int_{\Gamma_k} \tau(\mathbf{r}', z_k)\ln\frac{1}{|\mathbf{r} - \mathbf{r}'|}\mathrm{d}s + \varphi_0(\mathbf{r}, z_k), \tag{2.180}$$

where $\mathbf{r} = x\mathbf{1}_x + y\mathbf{1}_y$, $\tau(\mathbf{r}', z_k)$ is the unknown electric charge density on the edge line $\Gamma_k = \Gamma_{c,k} \cup \Gamma_{d,k}$. The line Γ_k is formed by intersecting the surface $S = S_c \cup S_d$ with the plane $z = z_k$ (Figure 2.44). In the simplest case, z_k corresponds to the z-coordinate of the defect center.

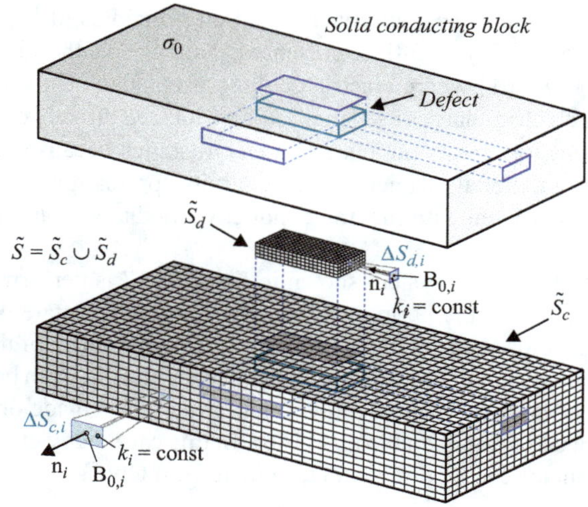

Figure 2.43 An exemplary boundary mesh of rectangular elements for the solid conductor with a single cuboidal defect [165]

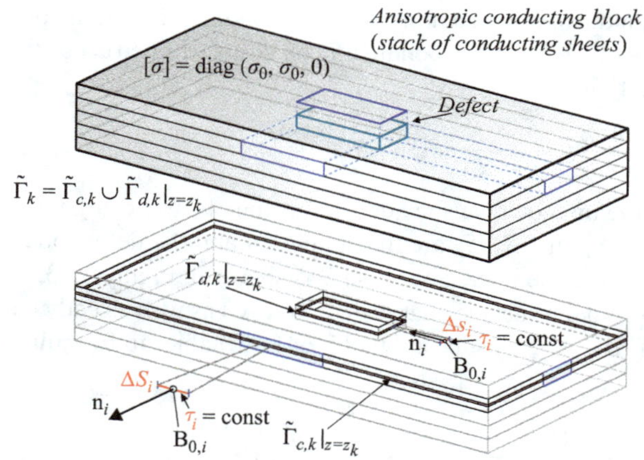

Figure 2.44 An exemplary mesh of edge elements for the anisotropic conductor with a single cuboidal defect [165]

The φ_0 potential in (2.180) is the potential determined for an infinitely wide moving plate without defect. It can be found by solving $\nabla^2\varphi_0 = 0$ with the following boundary conditions

$$\frac{\partial\varphi_0}{\partial z} = v_0 B_{0z}\,|_{z=0} \quad \text{and} \quad \frac{\partial\varphi_0}{\partial z} = v_0 B_{0z}\,|_{z=-D}\,, \tag{2.181}$$

where D is the thickness of the plate.

After applying the $\mathbf{n} \cdot \mathbf{J}|_{\Gamma_k} = 0$ condition, the following equations are obtained

$$\frac{1}{2\pi\varepsilon_0} \sum_{i=1}^{N_{1D}} \tau_i \int_{\Delta s_i} \frac{\mathbf{n}_j \cdot (\mathbf{r}_j - \mathbf{r}'_i)}{|\mathbf{r}_j - \mathbf{r}'_i|^2} \, \mathrm{d}s_i = \mathbf{n}_j \cdot (\nabla\varphi_{0,j} - \mathbf{v} \times \mathbf{B}_{0,j}) \tag{2.182}$$

with $j = 1, \ldots, N_{1D}$, $\tau_i = \text{const.}$ on Δs_i, $\mathbf{B}_{0,j} = \mathbf{B}_0(\mathbf{r}_j, z_k)$, $\nabla\varphi_{0,j} = \nabla\varphi_0(\mathbf{r}_j, z_k)$, and $\tilde{\Gamma}_k = \bigcup_{i=1}^{N_{1D}} \Delta s_i$ a mesh of 1D line elements covering the edge line Γ_k.

With the calculated electric charge density distributions (κ or τ), the density of the induced eddy currents $\mathbf{J}$ in the moving conductor can be determined (see (2.178)). Forces $\mathbf{F}$ and $\mathbf{F}_0$ acting on a magnetic system in the presence of a moving conductor with defects and no defects, and DRSs $\Delta\mathbf{F}$ can be obtained from

$$\mathbf{F} = -\int_{V-V_D} \mathbf{J} \times \mathbf{B}_0 \, \mathrm{d}V, \quad \mathbf{F}_0 = -\int_V \mathbf{J}_0 \times \mathbf{B}_0 \, \mathrm{d}V, \quad \Delta\mathbf{F} = \mathbf{F} - \mathbf{F}_0. \tag{2.183}$$

For LET systems using single permanent magnets, it is possible to use models with multiple magnetic dipoles (2.86)–(2.87) or, in special cases, analytical formulas [e.g. (2.90)], to calculate the $\mathbf{B}_0$ field. However, recent optimization studies showed that rather complicated magnet systems, which include highly saturating ferromagnetic materials such as iron-cobalt alloys, may be advantageous compared to standard magnet geometries [157]. The use of ferromagnetic material in the magnetic systems used in LET makes it impossible to apply the above methods to determine the field $\mathbf{B}_0$. For such cases, the determination of field $\mathbf{B}_0$ must be carried out using numerical methods (e.g. FEM). In the following, a method will be described that allows determining the field $\mathbf{B}_0$ in the area of the non-magnetic conductor based on $\mathbf{B}_0$ defined only on two planes parallel to the surface of the conductor. Magnetic flux density $\mathbf{B}_0$ on the planes must be determined only once using any method, which significantly reduces the calculation time and reduces the amount of data necessary to process.

It is assumed that the primary field $\mathbf{B}_0$ is given only on regular grids of points truncated by the rectangular window $W = \{(x, y) : w_x \times w_y\}$ and located on the planes $z = 0$ and $z = -D$ corresponding to the upper and lower surfaces of the object under test.

The magnetic flux density $\mathbf{B}_0$, as well as the electric potential φ_0, can be determined at any point $\mathbf{P}$ in the region $\Omega = \{(x, y, z) : W \times < -D, 0 >\}$ by solving

$$\nabla^2 \mathbf{B}_0 = \mathbf{0}, \tag{2.184}$$

$$\nabla^2 \varphi_0 = 0 \tag{2.185}$$

with the boundary conditions defined as

$$\mathbf{B}_0|_{z=0} = \mathbf{B}_0^u = \mathbf{B}_0(x, y, 0), \qquad \frac{\partial\varphi_0}{\partial z} = v_0 B_{0y}^u, \tag{2.186}$$

$$\mathbf{B}_0|_{z=-D} = \mathbf{B}_0^d = \mathbf{B}_0(x, y, -D), \qquad \frac{\partial\varphi_0}{\partial z} = v_0 B_{0y}^d. \tag{2.187}$$

The set of boundary conditions for the time and spatial domains is schematically shown in Figure 2.45.

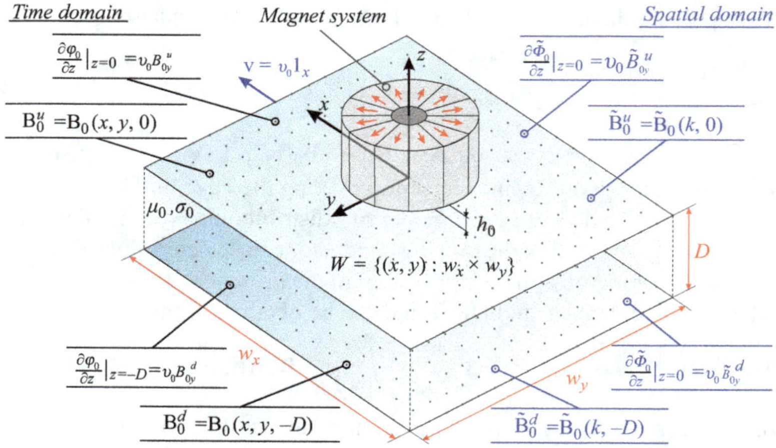

Figure 2.45 Boundary conditions in calculating the primary field $\mathbf{B}_0$ and the electric potential φ_0 in the region between the planes $z = -D$ and $z = 0$ [165]

In order to obtain solutions of (2.184) and (2.185), 2D spatial Fourier transform is applied

$$\tilde{\mathbf{B}}_0 = \tilde{\mathbf{B}}_0(k, z) = \mathscr{F}_x\mathscr{F}_y(\mathbf{B}_0), \tag{2.188}$$

$$\tilde{\Phi}_0 = \tilde{\Phi}_0(k, z) = \mathscr{F}_x\mathscr{F}_y(\varphi_0), \tag{2.189}$$

where $k^2 = k_x^2 + k_y^2$, and k_x, k_y are spatial frequencies in the x and y direction, respectively. The corresponding solutions are given by the following formulas

$$\tilde{\Phi}_0(k, z) = \frac{v_0}{k}\frac{\tilde{B}_{0y}^u \cosh k(D+z) - \tilde{B}_{0y}^d \cosh kz}{\sinh kD}, \tag{2.190}$$

$$\tilde{\mathbf{B}}_0(k, z) = \frac{\tilde{\mathbf{B}}_0^u \sinh k(D+z) - \tilde{\mathbf{B}}_0^d \sinh kz}{\sinh kD}. \tag{2.191}$$

The primary magnetic flux density $\mathbf{B}_0$ as well as the electric potential φ_0 at any plane $z \in < -D, 0 >$ can be determined using the inverse spatial Fourier transform

$$\mathbf{B}_0(\mathbf{r}, z) = \mathscr{F}_x^{-1}\mathscr{F}_y^{-1}(\tilde{\mathbf{B}}_0), \tag{2.192}$$

$$\varphi_0(\mathbf{r}, z) = \mathscr{F}_x^{-1}\mathscr{F}_y^{-1}(\tilde{\Phi}_0). \tag{2.193}$$

In practice, the presented method can be easily implemented in the MATLAB® environment by using the `fft2/ifft2` functions that return forward and inverse discrete, 2D Fourier transforms calculated using the Fast Fourier Transform algorithm.

Figure 2.46 shows the results of exemplary simulations performed under SCSM for an aluminum block (an isotropic case) and a set of aluminum sheets (anisotropic case) both moving at $v_0 = 0.2$ m/s velocity under the cylindrical Halbach structure defined in [157]. The lift-off of the magnetic system was equal to $h_0 = 1$ mm and the

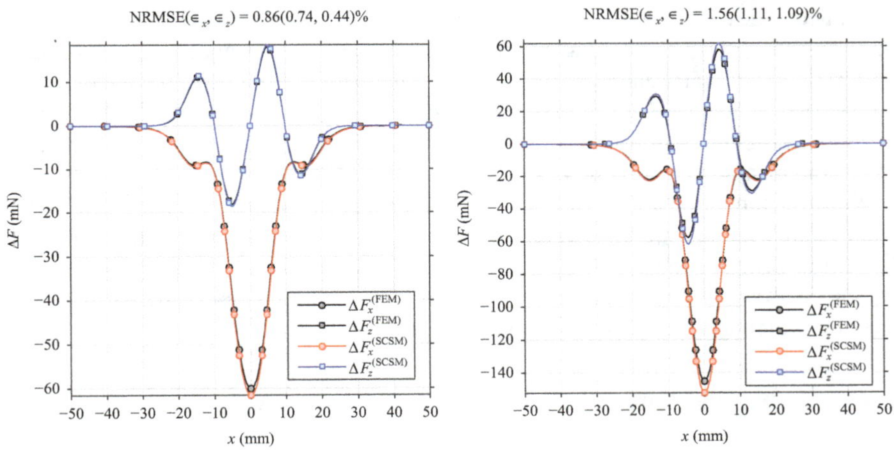

(a) Isotropic conductor (2.179). (b) Anisotropic conductor (2.182).

Figure 2.46 Simulated DRSs calculated by FEM and SCSM. The 13 mm × 2 mm × 4 mm defect was located at a depth of d = 2 mm in a solid aluminum block of 250 mm × 50 mm × 50 mm (a), and a stack of 25 aluminum sheets of 250 mm × 50 mm × 2 mm (b)

electrical conductivity $\sigma_0 = 21$ MS/m. The figure presents the DRSs for $y = 0$ from the cuboidal defect 13 mm × 12 mm × 4 mm located at a depth $d = 2$ mm calculated by means of SCSM and FEM. The normalized root mean square errors of the SCSM method in relation to FEM are small for both the isotropic and anisotropic conductor (less than 0.9% and 1.6%, respectively).

It is interesting to notice that the DRSs, i.e. ΔF_x and ΔF_z, evoked by the defect localized in the stack of sheets (anisotropic model) are more than twice the DRS from the defect in the solid conductor with isotropic conductivity. This phenomenon can be explained by analyzing the flow of distorted currents around the defect. In the case of an anisotropic conductor, distorted currents occur only in the layer that coincides with the defect, therefore their deformation is stronger than for the conductor with isotropic conductivity, where the deformation occurs in the entire volume of the conductor.

Considering the time of simulations carried out using FEM and SCSM, it should be stated that the calculation time of DRSs for the isotropic block for both methods was comparable (about 4 min per DRS point). In the case of a defect in the anisotropic block, DRS calculations using the SCSM were over 60 times faster compared to the FEM (4.4 s and 5 min per DRS point, respectively). Such a large reduction in the calculation time for SCSM results from the fact that in the SCSM formulation for anisotropic conductors, meshes of linear elements located only in the layer that coincides with the defect are used. Consequently, the number of SCSM elements and hence the size of the resultant system of algebraic equations is considerably reduced.

In summary, SCSM can only be used to simulate LET problems when the velocity of a moving non-magnetic conductive object is small enough to apply a weak reaction of induced eddy currents. The use of SCSM for conductors with isotropic electrical conductivity is time-consuming and relatively complicated because it requires a suitable mesh of boundary elements to describe a conductor with defects. To generate such a mesh, it is necessary for the user to create appropriate procedures adapted to the specific LET problem. As a result, SCSM does not compete for FEM, where it is relatively easy to adapt any commercial system to LET simulation. The simulation time for both methods is comparable and relatively long. The SCSM evaluation changes for LET problems when the object under test is a conductor with anisotropic conductivity. In this case, SCSM is an interesting alternative to FEM because, first, the generation of meshes with linear edge elements even for objects with complex shapes is relatively easy to implement, and second, simulations are much faster than in the FEM.

2.4 Numerical simulations with FEM

2.4.1 Introduction and motivation

The intrinsic phenomena associated with electric and magnetic fields affect almost all aspects of our everyday life. This is followed by continuous development and design of more sophisticated electro-mechanical devices which result in better functionality, higher efficiency, and increased safety. Depending on the particular application, this task strongly relays on the accurate modeling of the electromagnetic fields within the device, i.e. it requires solution of the governing equations described by Maxwell under well-defined conditions. Whenever it is possible, the analytical solution of the resulting system of partial differential equations (PDEs) is most wanted [142,167–171]. Apart from providing exact and fast solutions, the obtained closed-form analytical expression also helps in better understanding of the underlying physical phenomena associated with the problem under investigation. Unfortunately, these solutions are not always available and they can be obtained only for some simplified device and field configurations. Thus, in NDT and NDE applications the development and optimization of various testing techniques is performed using the numerical methods [168].

Due to its ability to handle complex geometries, anisotropic and inhomogeneous material properties, widely used numerical method in NDT and NDE applications is the FEM [170]. FEM is a numerical technique which gives approximate solutions to partial differential equations which are commonly used to describe the physical behavior of a system in engineering applications [168,172]. Unlike some other numerical techniques, e.g. finite difference method, the FEM is used to approximate the solution rather than to directly approximate the partial derivatives appearing in the governing equations. Its fundamental idea is to divide the region under investigation into small well-described elementary domains called the finite elements. In 2D investigations they represent simple geometrical forms such as triangles and quadrilaterals, whereas in 3D these domains are typically in form of tetrahedra and hexahedra. The assembly of all elements is called the finite element mesh [172].

Within each finite element the unknown solution, i.e. the unknown scalar or vector potentials, φ or $\mathbf{A}$, are usually approximated with low-order polynomial functions, which are in FEM applications widely known as shape functions. They can be defined as scalar or vector quantities. The scalar shape functions are associated with nodes of a finite element and they can be used to approximate both, scalar and vector potentials. The vector shape functions are associated with the corresponding edges of an element and they can only be used to approximate vector potentials. The resulting unknowns are also referred to as degrees of freedom (DoFs). In order to reduce the computational time it is of most interest to reduce the resulting number DoFs for the specific problem of interest. The total number of DoFs is influenced by the governing differential equation, the technique used to incorporate time-dependent effects in the model and the size of the computational domain. Further reading regarding FEM can be found in [172–176].

In general, analysis of LET systems requires accurate and time-efficient numerical approaches to allow either extensive scans of an object under test or parametric and optimization studies. For the implementation of the proposed approaches the commercial software package Comsol Multiphysics v4.4 [151] (COMSOL, Inc., Burlington, MA, USA, www.comsol.com) is used in this framework.

2.4.2 Computation of eddy current distributions including moving parts

The Lorentz force eddy current testing belongs to a special class of electromagnetic field phenomena in which various effects caused by parts set in relative motion occur. These effects, also referred to as motional effects, represent the basic operating principle over a wide range of electro-mechanical devices in different application areas such as electrical machines, magnetic levitation systems, inductive heating, eddy current brakes, NDT, NDE, etc. Due to its tremendous industrial relevance the application of FEM to this particular type of field problems, also known as the moving eddy current problems has undergone an extensive research over the past decades [177]. As a result many different techniques used for modeling of the relative motion using FEM, mostly referring to rotational electrical machines, have been developed [177–180]. However, either due to their high computational requirements or difficulties in their implementation only a few techniques are nowadays commonly used in general-purpose commercial FEM codes. The main aim here is to give brief overview of these techniques, whereas a more comprehensive study due to a large number of existing publications would be far beyond the scope of this work. Additional information concerning the particular technique can be found in the extensive reference list presented.

In principle, independent on the actual type of motion (translation or rotation) all existing techniques for simulation of general moving eddy current problems can be classified into (i) fixed grid methods and (ii) moving or time changing grid methods [178].

The fixed grid methods are usually applied to 2D/3D static or time-harmonic eddy current problems involving uniformly moving conducting parts having invariant cross-section in the motion direction [72,177,178,181–185]. The field problem is formulated in the moving frame of reference (see section 2.1) where the additional velocity term ($\mathbf{v} \times \nabla \times \mathbf{A}$) is used to describe the contribution of the induced eddy currents within the moving electrically conducting part. This approach, also referred to as the QSA [72], is very efficient in terms of computational time since only one stationary analysis needs to be performed to obtain an accurate steady state solution. For some simple device configurations several authors combine FEM with analytical solutions as well [185,186]. Considerably reduction of the simulation time and increased accuracy has been reported. However, apart from simple geometries the analysis is restricted to stationary and time-harmonic problems. The convection of the magnetic field lines introduced in the fixed grid methods can cause spurious numerical oscillations when the resulting Péclet number ($P_e = \mu\sigma v\Delta x$) is large [187–189]. This can be avoided by performing the mesh refinement within the conductor region or by the technique known as the upwinding [190,191]. The logical expression approach (LEA) presented later in this work, extends the applicability of the fixed grid methods to 2D/3D transient eddy current investigations having non-uniform parts set in relative motion.

The moving grid methods are more general and they can be applied to simulate wide variety of electromechanical devices involving linear or rotational movement. In principle, from the model topology point of view all available techniques are quite similar. The main idea is to decompose the whole computational domain into two parts associating them with the moving or with the fixed part of the assembly [177,178,192]. Within each part the governing equations are solved in their own frame of references, whereas the relative displacement and the field coupling is provided on the introduced interface [180,193]. Depending on the actual interface, which can have constant or variable lift-off distance, to achieve the coupling many different techniques have been applied, each of them having certain advantages and disadvantages.

One of the earliest techniques to model relative displacement of 2D induction machines is based on mesh deformation [194]. During the movement this technique can produce elements with large aspect ratios leading to the loss of accuracy [180,195]. Thus, its application is usually restricted for modeling 2D electromagnetic devices requiring small relative displacements [196,197].

In order to provide arbitrary large displacements techniques based on the continuous remeshing of the computational domain have been reported [198]. When re-meshing is performed only in the small air-gap region of the device, these techniques are referred to as the step-by-step finite elements or the moving band techniques [193,198–202]. Due to time changing grids the obtained solution can be numerically noisy. Additionally, for 3D complex geometries the re-meshing procedure can also be very time-demanding. This is why these techniques are mostly applied to 2D eddy current problems. In order to avoid tedious re-meshing of the model several authors combine the advantages of the FEM and analytical expressions which provide the field coupling without meshing the air-gap of the machine. The resulting technique is often referred to as the air-gap element or macro element method [195,203–206]. Whenever they can be obtained, the analytical expressions are simply coupled with the FEM formulation providing high accuracy compared to all other methods available [177]. However, as reported in [206] the coupling procedure can reduce the sparsity of the resulting stiffness matrix increasing the computational costs. Furthermore, due to difficulties in obtaining the analytical expressions, the analysis is usually restricted to 2D field problems [207]. On the same note, to avoid the re-meshing of the computational domain, hybrid methods based on the coupling of boundary element method (BEM) and FEM have been introduced. The BEM provides all the advantages to model linear unbounded air regions in which the conducting objects are free to move [189,208–211]. The conducting regions, which in general can be non-linear and inhomogeneous, are effectively modeled by means of FEM. Similarly as before, coupled BEM-FEM technique can produce partially dense matrices. To reduce the computational costs several authors propose parallelization techniques based on a domain decomposition to BEM and FEM part [212,213].

One of the widely applied moving grid methods to model relative displacements in general eddy current problems is the so-called sliding mesh technique (SMT) [214]. Similarly to all other methods, SMT also referred to as the moving mesh method [177] or slip surface method [215], requires two independent meshes to be defined. To provide the relative displacement, the meshes are simply slid relative to each other eliminating any need to alter their structure. The governing equations are solved independently in the fixed reference frame of each moving part, thereby avoiding the convection (velocity) terms. Depending on the mesh distribution along the introduced interface, which can be conforming or non-conforming, the field continuity can be preserved using several coupling techniques. In case of conforming meshes, the unknown potentials on each side of the sliding interface are made equal in the same way as Dirichlet boundary conditions are imposed [177,215–219]. However, the displacement is strictly controlled by the size of the finite elements in the motion direction and the time-step size of the transient solver. To overcome this limitation non-conforming meshes along with Lagrange multipliers have been introduced [182]. The Lagrange multiplier approach introduces an additional set of variables on the sliding interface which ensures the continuity of the field in a weak sense [220–223]. Unfortunately, the existence of additional variables considerably deteriorates the conditioning of the stiffness matrix [223–225]. Furthermore, the matrix contains zeros on the main diagonal, which means that some standard iterative solvers, such as incomplete Cholesky conjugate gradient (ICCG) or conjugate gradient (CG), either have slow convergence or fail completely [223,225]. All these issues led to development of other coupling techniques for non-conformal meshes, such as interpolation method [198,223] and the mortar element method [214,224,226–228]. Using the interpolation methods the field continuity is only globally conserved across the interface. Additionally the coupling increases the bandwidth of the resulting matrix system increasing the computational costs for the same number of unknowns [223,229,230]. The mortar element method results in a positive definite stiffness matrix, which makes it well suited for 3D moving eddy current problems. However, up to now this method has been only applied to solve 2D moving eddy current problems. Additional issues such as complex implementation and increase of non-zero elements in the resulting matrix system have also been reported as well [177,225].

In order to solve electromagnetic field problems with parts in arbitrary motion (variable air-gap sizes) the so-called overlapping or composite grid methods have been developed [177,231–234]. The method applies an iterative algorithm to couple two sets of overlapped grids. The grid of the moving part is discretized with a fine grid, whereas the fixed part is discretized with coarser grids. Due to longer simulation time and increased number of unknowns [233] the method is usually applied to 2D field problems [177]. However, some 3D implementations using nodal finite elements have been reported [233,235].

2.4.3 Numerical modeling of conductivity anomalies

The main aim of this section is to introduce new FEM-based methodology, which can be used to analyse and develop future LET systems. The particular emphasis is placed on the reduction of the overall computational requirements while maintaining the accuracy of the solution. Additional goals include development of simplified numerical models which enable fast 2D and 3D LET analysis in conjunction with the verification of assumed simplifications. The problem is simplified step-by-step, starting from time-dependent approaches, applying quasi-static approximations, and assuming a weak reaction from the conductor. For comparison and verifications of different approaches a benchmark problem which represents a typical LET configuration has been considered.

2.4.3.1 Benchmark problem definition

In this section a typical LET benchmark problem representing a generic conductor with pre-defined artificial defects, moving across the static magnetic field is described (Figure 2.47).

The conductor under test is considered to be non-magnetic with the electrical conductivity denoted by σ and magnetic permeability equal to the permeability of vacuum $\mu = \mu_0$. It has a rectangular cross-section determined by its width W_c and height H_c, whereas its length is denoted by L_c. A cylindrical permanent magnet described by magnetization **M** is used as a source of the static (primary) magnetic field. The diameter and the height of the magnet are denoted by D_m and H_m, respectively. The magnet is placed centrally above the conductor under test ($\delta y = 0$) at a lift-off distance δz.

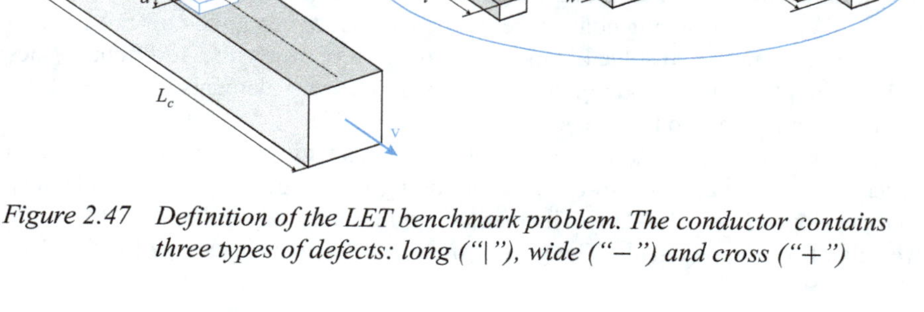

Figure 2.47 Definition of the LET benchmark problem. The conductor contains three types of defects: long ("|"), wide ("–") and cross ("+")

For the analysis, three different types of artificial defects, namely long ("|"), wide ("−") and cross ("+") have been considered. Defects are placed centrally within the conductor at depth d below its surface. They are characterised by their width w, height h, and length l. The conductivity of the defect is denoted by σ_d.

In order to reduce the number of dependent variables, the magnetic Reynolds number (R_m) has been used for the analysis (1.6). It involves the characteristic length-scale parameter L whose definition depends on the particular problem at hand and characterizes the conductor in motion. For the given benchmark problem (Figure 2.47), there are several possibilities to define L. In fluid dynamics, in order to measure the efficiency of different channel flows, all channel configurations are usually approximated with an equivalent circular pipe having the equivalent diameter D_e defined as

$$D_e = \frac{2A}{P} = \frac{2W_c \cdot H_c}{W_c + H_c}, \qquad (2.194)$$

where A is the cross-section area of the conductor, and P is its perimeter [236]. Following the same principle and choosing the radius of an equivalent cylinder ($D_e/2$) as the characteristic length-scale for the given LET benchmark problem, the value of R_m has been defined as follows

$$R_m = \mu_0 \sigma |\mathbf{v}| \frac{W_c \cdot H_c}{W_c + H_c}. \qquad (2.195)$$

2.4.3.2 Logical expression approaches

In this section the main idea of the logical expression approaches (LEAs) that allows fast computations of 2D/3D eddy current problems including parts in relative motion is presented [92]. Using the proposed methodology, the spatial coordinates of moving parts, either conducting or non-conducting, are modeled on a fixed computational grid using logical expressions (LEs). By applying the principles of Boolean algebra directly in finite element analysis (FEA) the shape of moving parts is determined on the fly by calculating the constraints given by LE and filtering the finite elements in those domains where LEs are introduced. Figure 2.48 shows three basic geometrical primitives, i.e. box, cylinder, and sphere which are modeled using LE.

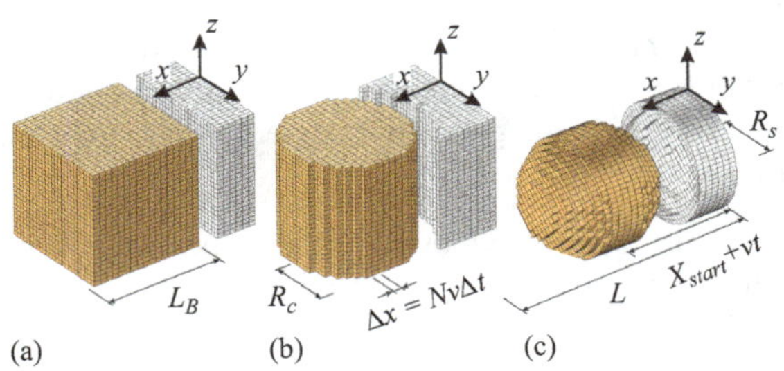

Figure 2.48 Basic geometric primitives defined by logical expressions: (a) box, (b) cylinder, and (c) sphere

Table 2.2 Logical expressions for basic geometric primitives

Shape	Logical expression (LE)
Box (B)	$[X_{start} + vt \le x \le X_{start} + L_B + vt\,]\,?\,1\,:\,0$
Cylinder (C)	$[\ (x + X_{start} + vt)^2 + y^2 \le R_C^2\,]\,?\,1\,:\,0$
Sphere (S)	$[\ (x + X_{start} + vt)^2 + y^2 + z^2 \le R_S^2\,]\,?\,1\,:\,0$

x, y, z: Cartesian coordinate system; X_{start}: starting position of the moving part; v: prescribed velocity of the displacement; t: actual simulation time.

Independent to the type of reference frame used in LET analysis, modeling of moving parts using LE requires the existence of a homogeneous zone in which these expressions are applied. This zone is referred to as the moving domain and it is determined by the shape of the moving part and its relative displacement L (Figure 2.48). In order to introduce the motion and determine the shape of moving parts, the constraints given by LE are defined as time dependent. This step represents the basic idea of the logical expression approach [92]. Table 2.2 presents some sample logical expressions for basic primitive shapes moving with velocity v along the model x-axis (Figure 2.48). Logical constructions "[*condition*]?1 : 0" in Table 2.2 should be interpreted as: if the *condition* in square brackets is valid then take 1. Otherwise if the condition is false take 0.

The implementation of LEA in the existing finite element codes can be performed easily by multiplying material properties assigned to the moving domain with the appropriate LE. This modification does not require any major changes in the FEM code and it can be applied in any commercially available FEM software. To assign material properties to the moving domain not occupied by the moving part, a term multiplying these properties with the negation of the pre-defined LE, i.e. by $1 - LE$ should be added as well.

As a summary, any implementation of LEA requires the following modeling steps:

(i) creation of the moving domain
(ii) definition of the corresponding LE
(iii) modification of material properties of the moving domain by combining the properties of the part under motion and the remaining region with appropriate LEs. This, the corresponding material properties have to be multiplied by LE_i or $1 - LE_i$, respectively.

During the discretization of the moving domain, the element size in the moving direction Δx has to satisfy the following constraint:

$$N\Delta x = v\Delta t, \tag{2.196}$$

where Δt represents the size of the time step used during the solution process, and N is any integer number greater than zero. Then the outer surface of the moving part is forced to stay unchanged during its motion. Otherwise, its volume can fluctuate in subsequent time steps introducing additional errors that are observed as non-physical Lorentz force oscillations.

One of the obvious disadvantages of the proposed LEA is its limitation to the relatively simple shapes of the moving objects (Figure 2.48). More complex shaped moving parts can be modeled by combining several moving domains and repeating the steps from (i) to (iii). Additionally, to model curved geometries using LEA the finite element mesh in the moving region should be relatively fine.

In order to test the proposed LEA approach the LET benchmark problem has been considered (cf. Figure 2.47). Depending on the definition of the global frame of reference, the LET analysis has been performed using two different implementations of LEA, (i) in case of fixed frame of reference, the LEs are used to model the motion of the permanent magnet, (ii) in case of moving frame reference, the LEs are used to model the relative motion of the defect. These two specific LEA implementations are referred to as the moving magnet approach (MMA) and the moving defect approach (MDA), respectively.

2.4.3.2.1 Moving magnet approach

In the implementation of the MMA, the global coordinate system is associated with the conducting object (fixed frame of reference) and the LEs are used to describe motion of the used cylindrical permanent magnet (Figure 2.49). The magnet is moving with constant velocity v along the model x-axis in close vicinity of the conductor containing an artificial defect below its surface. Using steps (i)–(iii) which describe the implementation of the LEA, the moving domain is defined in the surrounding air region (Figure 2.49). The cross section of the moving domain is determined by the height H_m and the diameter D_m of the magnet, while its length depends on the starting position X_{start} and its relative displacement L (Figure 2.49). In general, the starting position X_{start} is the distance of the moving object to the origin of the coordinate system at $t = 0$. It has to be large enough to avoid any influence of the used initial conditions on the resulting Lorentz force perturbations.

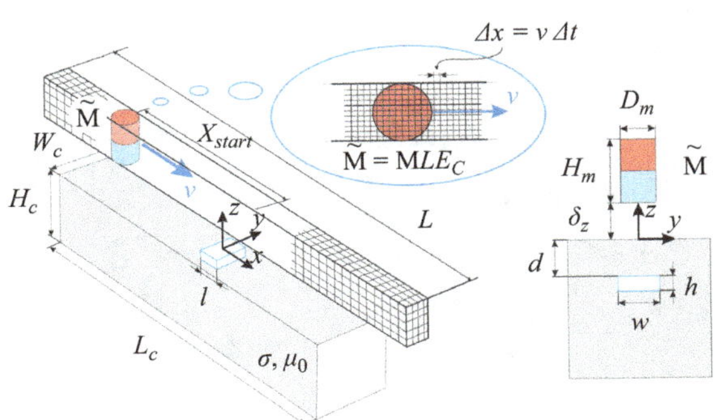

Figure 2.49 Implementation of the MMA. The moving domain is defined in the air region

To model the motion of the magnet in the moving domain, the magnetisation **M** assigned to the moving domain has been modified as $\tilde{\mathbf{M}} = \mathbf{M}LE_C$, where LE_C is the LE for a cylinder (Table 2.2). When defining the particular LE, it is important to keep the equivalent volume of the cylinder V_{LE_C} obtained by LE and the volume of the real cylinder given by $V_{cyl} = \pi(D_m/2)^2 H_m$ the same ($V_{LE_C} = V_{cyl}$). This can be achieved, e.g. by changing the diameter of the magnet D_m. In the given LET configuration, the relative velocity between the magnet and the conductor under test is constant. Thus, the size of the finite element mesh in the moving direction Δx is uniform (2.196).

Applying the optimal $\mathbf{A}^*$ formulation, the governing system of equations in MMA is given by:

$$\nabla \times \left(\frac{1}{\mu_0} \nabla \times \mathbf{A}^* - \mathbf{M} \right) = -[\boldsymbol{\sigma}] \frac{\partial \mathbf{A}^*}{\partial t}, \tag{2.197}$$

$$\nabla \cdot \left(-[\boldsymbol{\sigma}] \frac{\partial \mathbf{A}^*}{\partial t} \right) = 0. \tag{2.198}$$

This requires both nodal and edge finite element formulations in single computational domain. Due to the fact that (2.197) and (2.198) do not introduce the additional velocity term the resulting system of equations remains symmetric. Another important feature of MMA is that the stiffness matrix has to be assembled only once during the entire motion of the magnet which additionally reduces the total computational time. This is because the motion is provided by simple modification of the magnetization vector **M** which appears as a source term in the resulting FEM formulation.

2.4.3.2.2 Moving defect approach

In the MDA the global coordinate system is assigned to the permanent magnet (moving frame of reference). In this reference frame the magnet is stationary and the conductor is moving in opposite direction with velocity $-v$ along the x-axis (Figure 2.50).

For the given LET problem, the use of LE to describe the motion of the whole conducting domain would be computationally very expensive. However, the implementation of LEA in the moving reference frame can be considerably simplified

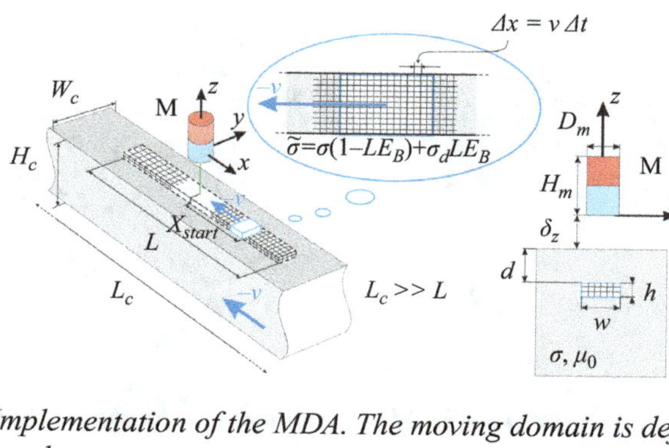

Figure 2.50 Implementation of the MDA. The moving domain is defined inside the conductor

if only perturbations of Lorentz force caused by defects are required, which is in fact a typical LET problem. Thus, it is sufficient to model only the movement of the defect relatively to the magnet, instead of modeling the motion of the whole conductor (Figure 2.50).

In MDA the moving domain is defined entirely inside the conductor where LEs are used to model the motion of the particular defect [92]. The shape and position of the moving domain is defined by the cross-section of the defect $w \times h$, its relative displacement L, and depth d (Figure 2.50). In order to simulate the defect motion, the electrical conductivity assigned to the moving domain is modified as $\tilde{\sigma} = \sigma(1 - LE_B) + \sigma_d LE_B$, where LE_B is the logical expression given in Table 2.2, and σ_d is the electrical conductivity of the defect. Due to the constant velocity, the size of the finite element mesh in the moving direction Δx is uniform (2.196).

Similarly, to the previous MMA applying the optimal $\mathbf{A}^*$ formulation in MDA results in the system of governing equations given by:

$$\nabla \times \left(\frac{1}{\mu_0} \nabla \times \mathbf{A}^* - \mathbf{M} \right) = [\sigma] \left(-\frac{\partial \mathbf{A}^*}{\partial t} + \mathbf{v} \times \nabla \times \mathbf{A}^* \right), \tag{2.199}$$

$$\nabla \cdot \left[[\sigma] \left(-\frac{\partial \mathbf{A}^*}{\partial t} + \mathbf{v} \times \nabla \times \mathbf{A}^* \right) \right] = 0. \tag{2.200}$$

This requires both nodal and edge finite element formulations to be applied in the computational domain. In contrast to MMA the formulation used in MDA involves the additional velocity term which makes the resulting system of equations non-symmetric. Additionally, the modification of the electrical conductivity by time dependent LE, introduced by MDA, modifies the resulting stiffness matrix as well. Therefore, the stiffness matrix has to be re-assembled at every time step which increases the computational time compared to MMA.

2.4.3.3 Quasi-static approach

The MMA and the MDA assume no simplifications for the given LET analysis. They offer accurate results for any relative testing velocity $\mathbf{v}$ between the magnet system and the conductor, and for any material and geometry parameters involved. Thus, they are valid for finite values of the magnetic Reynolds number (2.195), whether the conductor contains material defects or not.

If LET configuration under investigation is time independent, i.e. it involves uniformly moving conductors with a constant cross-section normal to the direction of motion (conductors free of defects) the analysis can be considerably simplified [142]. In principle, only one stationary analysis can be performed to obtain an accurate steady state solution, e.g. the Lorentz force acting on the magnet system. This assumption requires the moving frame of reference, where the additional velocity term $(\mathbf{v} \times \nabla \times \mathbf{A})$ is used as a source of the induced eddy currents inside the conductor in uniform motion

$$\nabla \times \left(\frac{1}{\mu_0} \nabla \times \mathbf{A} - \mathbf{M} \right) = [\sigma] \left(-\nabla \varphi + \mathbf{v} \times \nabla \times \mathbf{A} \right), \tag{2.201}$$

$$\nabla \cdot [[\sigma] \left(-\nabla \varphi + \mathbf{v} \times \nabla \times \mathbf{A} \right)] = 0. \tag{2.202}$$

The second equation results from the current conservation law $\nabla \cdot \mathbf{J} = 0$ and it is an additional equation for the electric scalar potential φ. This system of equations takes the deformation of the magnetic field lines correctly into account making it valid for any value of the R_m.

Although (2.201) and (2.202) provide fully correct solutions only for conductors without any material defects, their use can be still extended to NDT applications. In [72], it has been shown that for LET systems resulting in small R_m, they can be used for fast Lorentz force calculations on the moving magnet even for conductors with defects. Here, this method is referred to as a quasi-static approach (QSA). As a direct consequence of low magnetic Reynolds numbers ($R_m < 1$), the diffusion time of the magnetic field into the conducting object, estimated as $\tau \sim R_m \cdot L/v$, where L is the characteristic length-scale of the conductor (2.194), is small as well [237,238]. This basically justifies the instantaneous field reaction ($\partial \mathbf{B}/\partial t \to \mathbf{0}$) to any perturbation of induced currents, which is assumed in QSA. Nevertheless, if this is not the case, the full transient form of (2.201) and (2.202) has to be considered, which in fact represents the governing equation of already presented MDA.

In the implementation of QSA, only the change in relative position between the magnet and the defect must be provided. This is done either by moving the magnet system relative to the defect, or vice versa [72]. In any case, this requires a time consuming re-meshing procedure of the entire model geometry for each new configuration. The re-meshing of the geometry can be avoided if the basic principle of the logical expression approach (LEA) is combined with the QSA given by (2.201) and (2.202). In this LEA implementation the time variable used in different LE is just a parameter which needs to be changed from one stationary solution to another providing the displacement of the moving part (magnet or defect). Basically, this means that the same geometry, used for the implementation of the LEA, i.e. MMA and MDA (Figures 2.48 and 2.49), can be used for implementation of QSA as well. The only difference introduced by QSA is in the governing equation in the conducting region, which is now in its stationary form and contains an additional scalar potential φ. In regions free of eddy currents (surrounding air region and permanent magnet), the magnetic scalar potential formulation ψ is used.

2.4.3.4 Weak reaction approaches

The induction problem at hand can be further simplified in the case of low magnetic Reynolds numbers ($R_m \ll 1$). In this case, the induced eddy current density is so small that its magnetic field $\mathbf{B}^{(s)}$ is vanishingly small compared to the primary magnetic field $\mathbf{B}^{(p)}$ of the magnet system. By setting $\mathbf{B}^{(s)} = 0$, the magnetic and electric fields are decoupled, and therefore, can be treated independently. Hereinafter, this effect will be referred to as a *weak reaction* by the conductor to the magnetic field. Special attention must be paid to the emerging Lorentz forces when using WRAs. By neglecting the secondary magnetic field, the spatial symmetry of the electric and magnetic field is enforced. As a consequence, the lift component of the Lorentz force vanishes if the conductor is free of defects and if the magnet is far from any outer edge of the conductor, such that $F_z^{(0)} = 0$. However, in the presence of defects, the symmetry of the fields no longer holds and the DRS $\Delta\mathbf{F}$ can be determined. In the following section, two approaches are presented to illustrate the weak reaction principle, because the procedures have major advantages in terms of computational cost.

2.4.3.4.1 Extended weak reaction approach

The basic principle of the extended weak reaction approach (eWRA) is shown in Figure 2.51. The electromagnetic fields are determined in the laboratory frame of reference such that the conductor moves with a velocity $\mathbf{v}$ with respect to the magnet system. The eWRA is based on a two-step procedure. In the first step, the primary magnetic field $\mathbf{B}^{(p)}$ is determined using the (primary) scalar magnetic potential $\psi^{(p)}$, including (possibly present) ferromagnetic material and neglecting any conductor in motion:

$$\nabla \cdot \left(-\nabla \psi^{(p)} + \mathbf{M}\right) = 0. \tag{2.203}$$

In the second step, only the moving conductor is considered. The primary magnetic field, $\mathbf{B}^{(p)} = -\mu_0 \nabla \psi^{(p)}$, is imported from the first step and mapped onto the nodes of the finite elements inside the conductor. The induced eddy currents are calculated using the scalar electric potential φ. Using Ohm's law for moving conductors, the induced eddy current density is given by:

$$\mathbf{J} = [\boldsymbol{\sigma}]\left[-\nabla \varphi - \mu_0 \mathbf{v} \times \nabla \psi^{(p)}\right]. \tag{2.204}$$

Applying the law of current conservation $\nabla \cdot \mathbf{J} = 0$ yields:

$$\nabla \cdot ([\boldsymbol{\sigma}]\nabla \varphi) = -\mu_0 \nabla \cdot \left([\boldsymbol{\sigma}]\mathbf{v} \times \nabla \psi^{(p)}\right). \tag{2.205}$$

The right-hand side of (2.205) can be simplified because the velocity $\mathbf{v}$ and the primary magnetic field $\nabla \psi^{(p)}$ are curl-free inside the conductor, such that:

$$\nabla \cdot \left(\mathbf{v} \times \nabla \psi^{(p)}\right) = -\mathbf{v} \cdot \underbrace{\left(\nabla \times \nabla \psi^{(p)}\right)}_{=0} + \nabla \psi^{(p)} \cdot \underbrace{(\nabla \times \mathbf{v})}_{=0} = 0. \tag{2.206}$$

Thus, the governing equation for φ is given by the following elliptic differential equation of second order with piecewise homogeneous material properties:

$$\nabla \cdot ([\boldsymbol{\sigma}]\nabla \varphi) = 0. \tag{2.207}$$

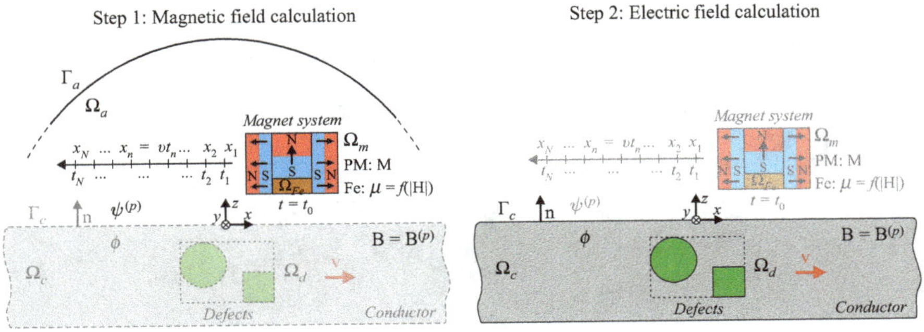

Figure 2.51 General principle of the eWRA

Defining the boundary conditions $\mathbf{n} \cdot \mathbf{J} = 0$ at the boundaries of the conducting domain allows the computation of φ:

$$\nabla\varphi \cdot \mathbf{n} = -\left(\mu_0 \mathbf{v} \times \nabla\psi^{(p)}\right) \cdot \mathbf{n}. \tag{2.208}$$

In this way, the current density is forced to flow inside the conducting domain.

Despite using a two-step procedure, eWRA has a higher computational efficiency than QSA, because only scalar potentials are involved, which leads to a decrease in the number of DoFs. The eWRA provides efficient numerical analysis, which is needed, for example, in an optimization framework.

2.4.3.4.2 Direct weak reaction approach

The direct weak reaction approach (dWRA) is similar to eWRA described in the previous section. However, in dWRA, the primary magnetic field is analytically calculated. Thus, the numerical procedure is reduced to the calculation of the electric scalar potential φ. The general principle of the approach is shown in Figure 2.52. The governing equation and the boundary conditions are given in (2.207) and (2.208), respectively. The dWRA has an even higher computational efficiency than eWRA. However, the analytical treatment of $\mathbf{B}^{(p)}$ permits the analysis of simple magnet geometries and prohibits the presence of ferromagnetic material in Ω_{Fe}.

Analytical expressions for the magnetic flux density for spherical magnets are obtained by using a single magnetic dipole with a magnetic moment of $\mathbf{m} = \frac{4}{3}\pi R^3 \mathbf{M}$ [107]. Closed-form analytical expressions for parallelepipedal or cuboidal magnets are given in [110]. However, the magnetic flux density of cylindrical magnets involves elliptic integrals, which cannot be expressed in terms of elementary functions [117]. An implementation in the framework of LET is presented by the author in [238]. The elliptic integrals are approximated using the mid-point rule [239] and the iterative arithmetic geometric mean (AGM) method [240]. AGM proves suitable because it provides fast convergence, which is needed to evaluate the force density in every node within the conductor so as to determine the total force using (1.4).

The magnetic flux density of more complex magnet geometries can be approximated by employing the principle of superposition of the field generated by multiple magnetic dipoles. This modeling approach is addressed by Mengelkamp *et al.* [100]

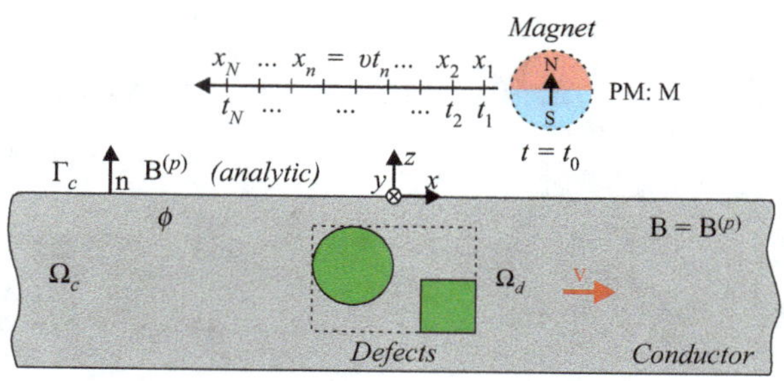

Figure 2.52 General principle of the dWRA

in a framework of Lorentz force evaluation. More information regarding dWRA can be found in [238] and [241].

2.4.3.5 Summary and overview

An overview of the presented methods is given in Table 2.3. The methods can be classified as *time-dependent*, *quasi-stationary*, or *weak-reaction* with decreasing computational complexity, as indicated by the governing equations. Additionally, the table provides the unknown quantities (DoFs) to be determined in the respective domains and the driving term of the induced eddy currents (**J**-term). The relation between the secondary magnetic field and the range of validity with respect to R_m can be readily identified. In the following section, the methods are compared for a typical LET problem so as to provide more information regarding the actual applicability for different magnetic Reynolds numbers R_m.

2.4.4 *Comparison of numerical approaches*

The approaches differ in their treatment of the secondary magnetic field $\mathbf{B}^{(s)}$. Hence, it is necessary to investigate their applicability in terms of magnetic Reynolds number R_m. The ratio between primary and secondary magnetic fields depends on the underlying geometry of the problem, as indicated by the characteristic length L in R_m in (1.6). To conduct an expressive comparison, an exemplary LET problem that corresponds to the dimensions of the available laboratory setup is defined. In this numerical experiment, a cuboidal permanent magnet, which is magnetized in the z direction, acts as the magnetic field source. Because of the simple geometry of the magnet, dWRA is used in the present analysis. Because eWRA uses the same treatment of secondary magnetic fields, the analysis is limited to dWRA (it will be simply referred to WRA in the following discussion). The direction of motion is defined such that the conductor moves with a positive velocity v_x along the x axis in MDA, QSA, and WRA; in the case of MMA, the magnet moves with a negative velocity $-v_x$ with respect to the conductor. The magnet and the defect are located symmetrically with respect to the specimen at $y = 0$ such that the object is analyzed on its centerline. Therefore, the side component of the Lorentz force F_y vanishes. The geometrical and material parameters of the exemplary problem are summarized in Table 2.4. Because the exemplary problem is strongly related to the experimental setup, all calculations are performed for isotropic specimens where $\sigma_{xx} = \sigma_{yy} = \sigma_{zz}$ (e.g. solids) and anisotropic specimens where $\sigma_{xx} = \sigma_{yy} \neq 0$ and $\sigma_{zz} = 0$ (e.g. composites or stacked sheets).

The absolute defect response signal (ADRS) $\Delta\mathbf{F}$ is defined as the force perturbation resulting from a defect. Because Lorentz forces are also present in the unperturbed case, the ADRS can be mathematically defined by the difference between the perturbed force profile $\mathbf{F}$ and unperturbed force profile $\mathbf{F}^{(0)}$:

$$\Delta\mathbf{F} = \mathbf{F} - \mathbf{F}^{(0)}. \tag{2.209}$$

The ADRSs are calculated for different magnetic Reynolds numbers R_m by varying the velocity. The force profiles in the case of isotropic and anisotropic specimens are shown in Figures 2.53 and 2.54, respectively. The ADRS is plotted over the spatial coordinate x. Positive x values are sampled first (in time), based on the direction

Table 2.3 Overview of numerical approaches to modeling the electromagnetic field problem in LET

	Approach	Governing equation	Eq.	Ω_c	$\Omega_{a\|m}$	J−term	$\mathbf{B}^{(s)}$	$\frac{\partial \mathbf{B}^{(s)}}{\partial t}$	R_m
Time-dependent	MDA	$\nabla \times \left(\frac{1}{\mu_0}\nabla \times \mathbf{A}^* - \mathbf{M}\right) = [\boldsymbol{\sigma}]\left(-\frac{\partial \mathbf{A}^*}{\partial t} + \mathbf{v} \times \nabla \times \mathbf{A}^*\right)$	(2.199)	$\mathbf{A}^*$	$\mathbf{A}^*$	$\mathbf{v} \times \nabla \times \mathbf{A}^* - \frac{\partial \mathbf{A}^*}{\partial t}$	yes	yes	high
		$\nabla \cdot \left[[\boldsymbol{\sigma}]\left(-\frac{\partial \mathbf{A}^*}{\partial t} + \mathbf{v} \times \nabla \times \mathbf{A}^*\right)\right] = 0$	(2.200)						
	MMA	$\nabla \times \left(\frac{1}{\mu_0}\nabla \times \mathbf{A}^* - \mathbf{M}\right) = -[\boldsymbol{\sigma}]\frac{\partial \mathbf{A}^*}{\partial t}$	(2.197)	$\mathbf{A}^*$	$\mathbf{A}^*$	$-\frac{\partial \mathbf{A}^*}{\partial t}$	yes	yes	high
		$\nabla \cdot \left(-[\boldsymbol{\sigma}]\frac{\partial \mathbf{A}^*}{\partial t}\right) = 0$	(2.198)						
Quasi-stationary	QSA	$\nabla \times \left(\frac{1}{\mu_0}\nabla \times \mathbf{A} - \mathbf{M}\right) = [\boldsymbol{\sigma}](-\nabla\varphi + \mathbf{v} \times \nabla \times \mathbf{A})$	(2.201)	$\mathbf{A}, \varphi$	ψ	$\mathbf{v} \times \nabla \times \mathbf{A}$	yes	no	med.
		$\nabla \cdot [[\boldsymbol{\sigma}](-\nabla\varphi + \mathbf{v} \times \nabla \times \mathbf{A})] = 0$	(2.202)						
Weak reaction	eWRA	$\nabla \cdot \left(-\nabla\psi^{(p)} + \mathbf{M}\right) = 0$	(2.203)	φ	$\psi^{(p)}$	$-\mu_0 \mathbf{v} \times \nabla\psi^{(p)}$	no	no	low
		$\nabla \cdot ([\boldsymbol{\sigma}]\nabla\varphi) = 0$	(2.207)						
	dWRA	$\mathbf{B}^{(p)}$ (analytic)	-	φ	-	$\mathbf{v} \times \mathbf{B}^{(p)}$	no	no	low
		$\nabla \cdot ([\boldsymbol{\sigma}]\nabla\varphi) = 0$	(2.207)						

Table 2.4 Parameters of the exemplary LET problem used for comparisons among different model approaches

Parameter	**Value**	**Description**
B_r	1.17 T	Remanence
X_m	10 mm	Length of the magnet
Y_m	10 mm	Width of the magnet
Z_m	10 mm	Height of the magnet
h	1 mm	Lift-off distance
X_d	12 mm	Length of the defect
Y_d	2 mm	Width of the defect
Z_d	2 mm	Height of the defect
d	2 mm	Defect depth
X_s	250 mm	Length of the specimen
Y_s	50 mm	Width of the specimen
Z_s	50 mm	Height of the specimen
σ_{Al}	30.61 MS/m	Electrical conductivity of aluminum
σ_{Cu}	59.8 MS/m	Electrical conductivity of copper

of motion defined above. Thus, the curves must be read from the right to left when considering the signal over time and not over space.

The first row shows the drag and lift components of the ADRS for low values of R_m. In this case, the secondary magnetic field is considerably smaller than the primary field from the magnet ($\mathbf{B}^{(s)} \ll \mathbf{B}^{(p)}$). The induced eddy current distribution and the total magnetic field are nearly symmetric, which results in a symmetric force profile when the magnet passes the defect. No significant differences can be identified between time-dependent approaches and WRA, which indicates that time-dependent effects are negligible. When increasing R_m, secondary fields and time-dependent effects become prevalent, resulting in non-symmetric field and force profiles. The ADRS obtained using WRA retains its symmetry because the secondary fields are neglected. As a consequence, WRA overestimates the ADRS amplitude by more than 100% compared to time-dependent approaches in the case of high R_m (see Figures 2.53(e) and (f) and 2.54(e) and (f)). The ADRS obtained using QSA is closer to ADRS values obtained using MDA and MMA, because it includes the stationary part of the secondary magnetic field ($\mathbf{B}^{(s)} \neq 0$). Specifically, in the case of high R_m, the time-dependent part of the secondary magnetic field $\partial \mathbf{B}^{(s)}/\partial t$ has an increasing influence on the ADRS. By comparing the curve of QSA to those of MDA and MMA in Figure 2.53(e) and (f), it can be seen that this results in a delayed and damped force response. As expected, the solutions from MMA and MDA are equivalent and yield very similar force profiles, because they only differ in the definition of the frame of reference.

The described effects pertain to both isotropic and anisotropic specimens. However, the ADRS has higher amplitudes in the case of anisotropic specimens than in the case of isotropic specimens. This phenomenon can be explained based on the imposed condition that $J_z = 0$ because $\sigma_{zz} = 0$. As a consequence, the current flows

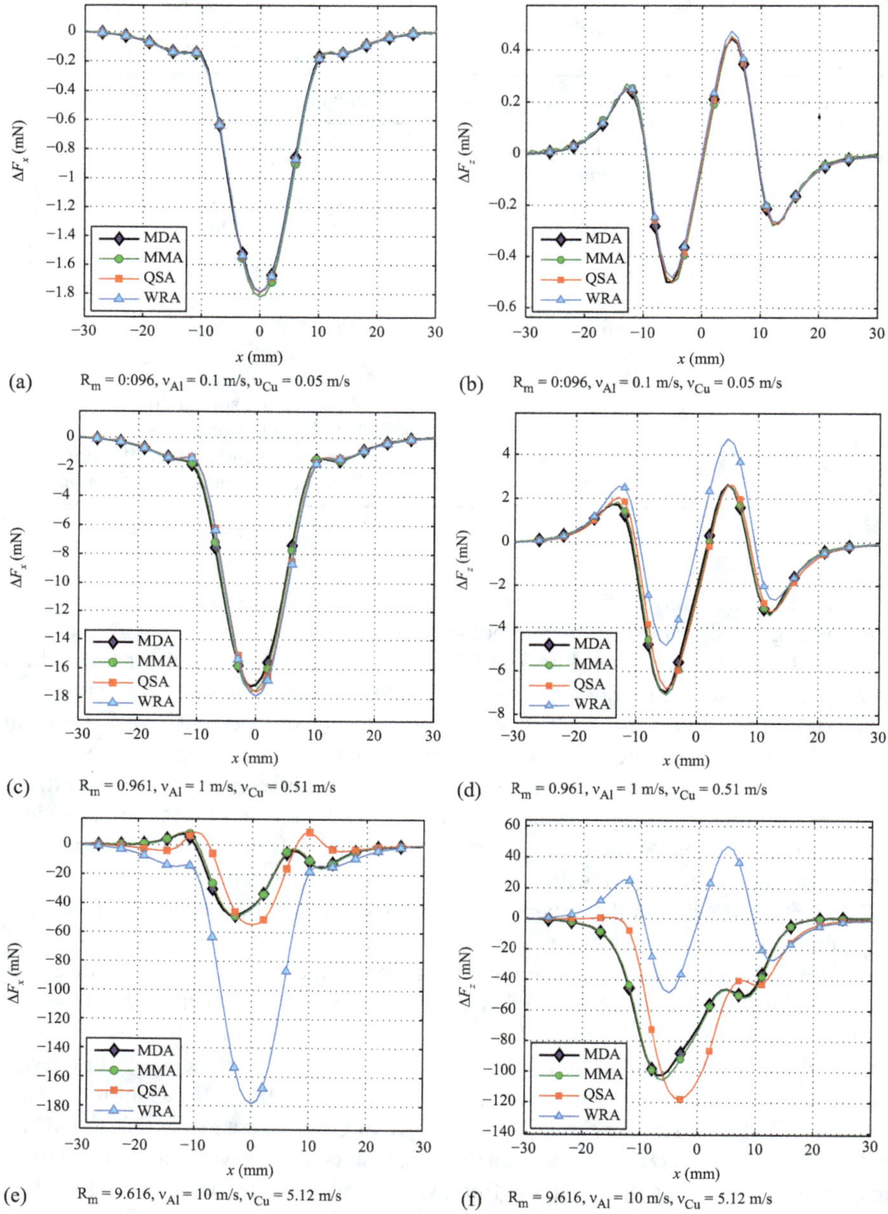

Figure 2.53 Comparison of the ADRSs of the drag force ΔF_x (left) and lift force ΔF_z (right) determined using different model approaches in the case of isotropic specimens. The magnetic Reynolds number R_m and the corresponding velocities in the case of specimens made of aluminum ($\sigma_{Al} = 30.61$ MS/m) or copper ($\sigma_{Cu} = 59.8$ MS/m) are provided. (a) and (b) low R_m, (c) and (d) medium R_m, (e) and (f) high R_m

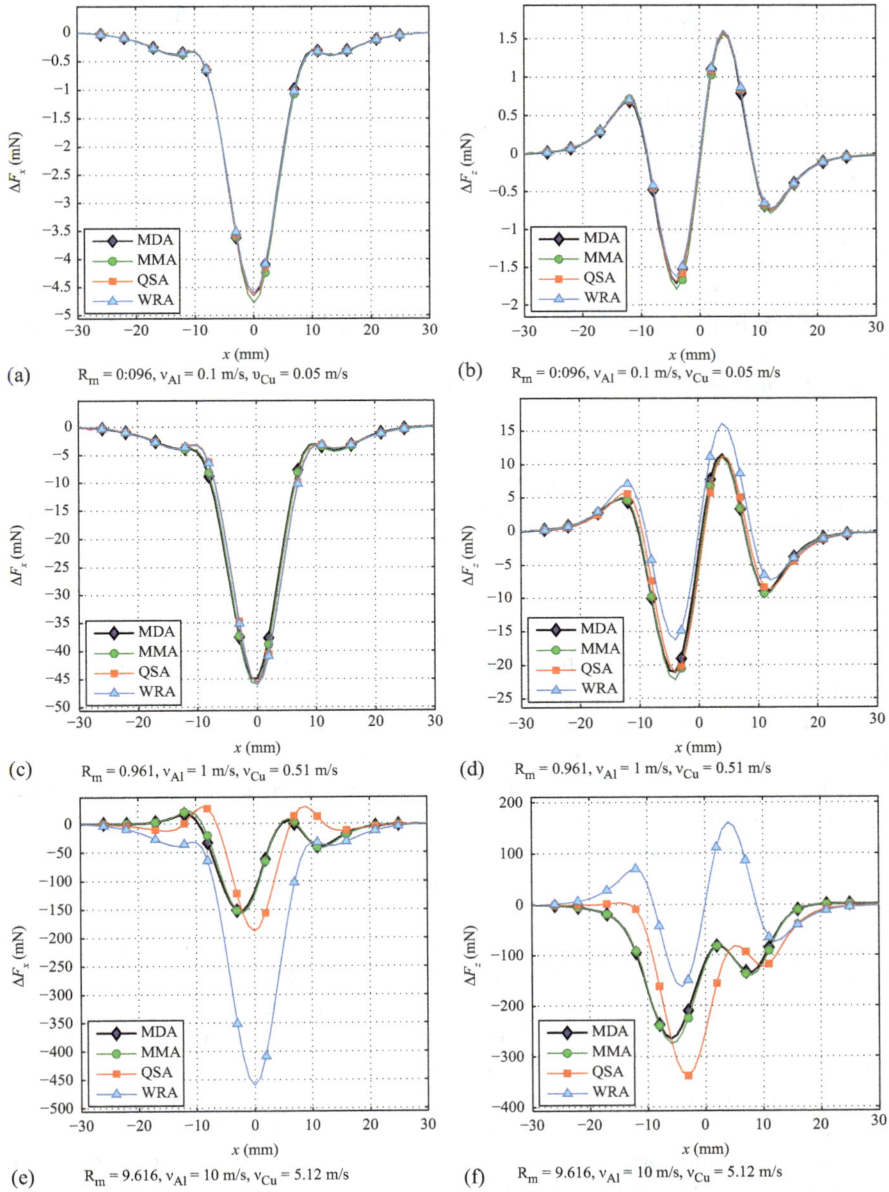

Figure 2.54 *Comparison of the ADRSs of the drag force ΔF_x (left) and lift force ΔF_z (right) determined using different model approaches in the case of anisotropic specimens. The magnetic Reynolds number R_m and the corresponding velocities in the case of specimens made of aluminum ($\sigma_{Al} = 30.61$ MS/m) or copper ($\sigma_{Cu} = 59.8$ MS/m) are provided. (a) and (b) low R_m, (c) and (d) medium R_m, (e) and (f) high R_m*

around the defect only in the xy-plane (i.e., not vertically). This phenomenon positively influences the resulting Lorentz force in terms of the ADRS amplitude. The shape of the ADRS is weakly influenced by this condition because in the unperturbed case, the induced eddy currents already flow solely in the xy-plane. However, some differences between the two cases can be identified; their anisotropic profiles show slightly sharper ADRSs, producing higher gradients. We conclude that the present anisotropy condition influences the profile but does not significantly change it, as is, for example, intended in the case of transformer sheets to prevent eddy current losses. This result confirms the applicability of layered specimens for the investigation of deep-lying defects.

To quantify the differences between the individual approaches depending on R_m, the normalized root mean square deviation (NRMSD) is calculated for the solutions obtained using MDA. The NRMSD is defined as:

$$\mathrm{NRMSD}_{x|z} = \frac{100\%}{\max\left(\Delta F_{x|z}^{(\mathrm{MDA})}\right) - \min\left(\Delta F_{x|z}^{(\mathrm{MDA})}\right)} \times \sqrt{\frac{1}{N}\sum_{i=1}^{N}\left(\Delta F_{x|z,i} - \Delta F_{x|z,i}^{(\mathrm{MDA})}\right)^2}. \tag{2.210}$$

The force components $\Delta F_{x|z,i}$ are compared at discrete points separated by $\Delta x = 1$ mm indexed by i over the plotted range, as shown in Figures 2.53 and 2.54. The defined error allows us to quantify the derivation of the shape and the amplitude between the different methods with respect to MDA. The NRMSDs are shown with double logarithmic scale in Figure 2.55. The abscissa is shown in two different velocity ranges considering specimens made of copper and aluminum to better illustrate the process in terms of potential NDT applications. During the analysis, the applied discretization (a finite element mesh) is defined such that it is as similar as possible among the individual approaches.

The error in the drag force is shown in Figure 2.55(a). WRA and QSA are at nearly the same level, up to moderate values of R_m. However, the error in WRA increases to 100% when R_m reaches values of roughly 10, which corresponds to velocities of roughly 6 m/s or 10 m/s for specimens made of copper or aluminum, respectively. The error in the lift force perturbation is shown in Figure 2.55(b). The NRMSD is significantly larger in the case of WRA, relative to QSA. This limits the applicability of WRA to the low R_m regime. For both force components, the error in MMA is at a nearly constant level, resulting from numerical inaccuracies. The conductivity anisotropy has a minor effect on the relative error in the case of the drag force. Slightly larger errors can be observed for the lift force of the isotropic specimen when applying WRA (see the dotted blue line in Figure 2.55(b)).

In addition to the drag force perturbation $\Delta\mathbf{F}$, the methods differ in the estimation of the unperturbed drag force $\mathbf{F}^{(0)}$. Because the Lorentz forces are measured in an absolute sense, a correct estimation of $\mathbf{F}^{(0)}$ is necessary when designing new systems. The absolute value of the relative difference (RD) of the unperturbed drag force is

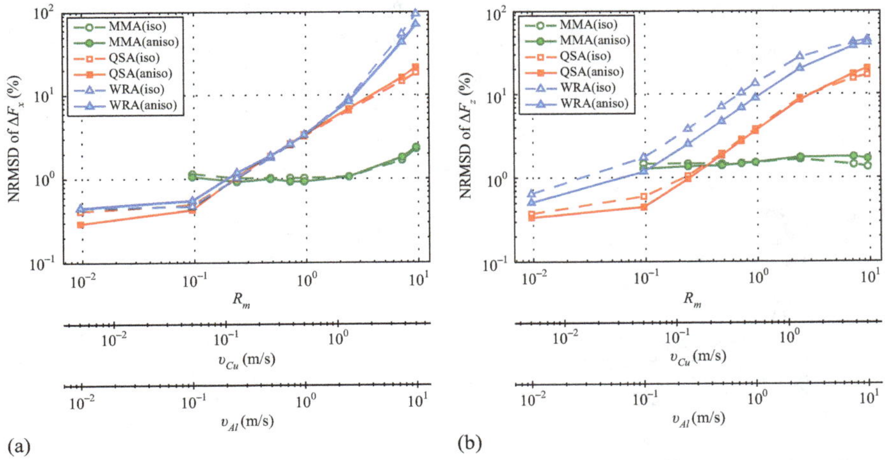

Figure 2.55 NRMSD of the DRSs of (a) the drag force ΔF_x and (b) the lift force ΔF_z for different model approaches in the case of isotropic and anisotropic conductivity profiles. Additional abscissae are provided for velocities of specimens made of copper ($\sigma_{Cu} = 59.8$ MS/m) and aluminum ($\sigma_{Al} = 30.61$ MS/m)

evaluated with respect to the reference solution obtained using MDA. The RD is defined by:

$$\mathrm{RD}_{x|z} = \left| \frac{F_{x|z}^{(0)} - F_{x|z}^{(0,\mathrm{MDA})}}{F_{x|z}^{(0,\mathrm{MDA})}} \right| 100\%. \tag{2.211}$$

Similar to the NRMSD, the individual force components indexed by c are independently compared to each other. The relative errors are shown in Figure 2.56. Regarding WRA, similar behavior can be observed concerning the drag force when increasing the magnetic Reynolds number R_m. However, the error in QSA remains at a constant level (as does that of MMA). This result is expected, because QSA yields exact results as long as the stationarity of the process is ensured (which is the case if the material is free of defects).

One major drawback of WRA is the absence of the unperturbed lift force $F_z^{(0)}$, which is an immediate result of the decoupling of electric and magnetic fields. The imposed symmetry in $\mathbf{B}^{(p)}$, and therefore also in $\mathbf{J}$, eliminates the lift force after the volume integration:

$$F_z^{(0)} = -\int_{\Omega_c} \left(J_x B_y^{(p)} - J_y B_x^{(p)} \right) \mathrm{d}\Omega = 0. \tag{2.212}$$

As a consequence, the relative difference of $F_z^{(0)}$ is 100% in Figure 2.56(b).

Apart from testing the applicability of different numerical approaches for LET problems, the present study also provides an introduction to the underlying physics of

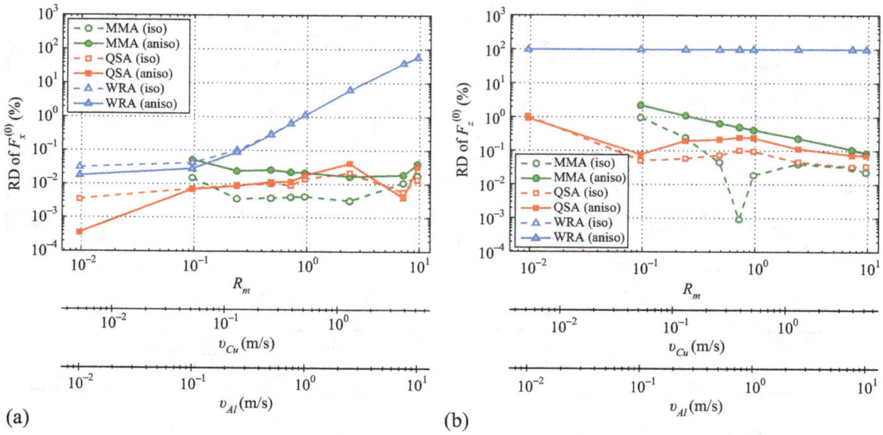

Figure 2.56 Relative difference of (a) the unperturbed drag force $F_x^{(0)}$ and (b) the unperturbed lift force $F_z^{(0)}$ for different model approaches in the case of isotropic and anisotropic conductivity profiles. Additional abscissae are provided for velocities of specimens made of copper ($\sigma_{Cu} = 59.8$ MS/m) and aluminum ($\sigma_{Al} = 30.61$ MS/m)

motion-induced eddy currents. The derived errors indicate the limits of applicability with respect to the magnetic Reynolds number R_m. The derived errors may slightly change when altering the geometrical parameters of the problem, for example, the size or location of the magnet or the defect. However, the present analysis provides valuable insight regarding how time-dependent effects are projected onto the Lorentz force profile in the case of moving conductors.

Chapter 3
Sensors for MIECT

Matthias Carlstedt[1], *Hartmut Brauer*[1] *and Konstantin Weise*[1]

3.1 Force measurement systems

This section aims to give an overview of the wide range of techniques available to measure forces. The most common measurement principles of force transducers are introduced with special focus on strain gauge load cells and piezoelectric crystal force transducers. Furthermore, the characteristics of force measurement systems are discussed, as well as the importance of calibration.

The force is a vectorial physical quantity that acts on a single point. In order to measure this vector, it has to be translated into a scalar quantity. Since real materials have limited permissible stress, a force cannot be transferred via a single point but always by a finite surface. So strictly speaking, not the force itself is measured, but the stress tensor field which is caused by the force.

A real force measurement system is therefore composed of a force transducer and the associated instrumentation, as well as perhaps mechanical installation aids. A force transducer is a device which converts the applied force into a measurable scalar quantity, e.g. change of electric resistance, through a known physical relationship. The instrumentation associated with a force transducer is used to generate an analogue or digital electrical output to represent the indicated value. Depending of the requirements of the measurement application the instrumentation may contain a number of separate elements for signal conditioning, indication, analogue-to-digital conversation, and data collection.

3.1.1 Principles of force transducers

Various measurement principles are used in force measurements today. The two principles of force transducers most widely used in industrial and laboratory environments are *piezoelectric transducers* and *elastic devices* such as strain gauge load cells. In addition, other important measurement principles are applied such as *electromagnetic force compensation*, *mechanical resonant circuits*, and *inverse magnetostrictive effect*.

[1] Advanced Electromagnetics Group, Technische Universität Ilmenau, Germany

3.1.1.1 Elastic deformation with resistance measurement

The measurement principles of most widely used force transducers are based on some kind of elastic element or a combination of such elements. When a force acts on the elastic element, it causes the deformation of that element. If the deformation is small, strictly speaking for small displacement gradients, it scales linearly with respect to the applied force as described by Hooke's law. This deformation is sensed by a secondary transducer for measuring the longitudinal and lateral strain or the displacement of a reference surface. These transducers are called elastic devices and come in different forms such as loading columns, toroidal ring, compression cylinder, shear beam, proving rings, and much more.

The materials used for the elastic element have ideally a linear relationship between stress and strain, low hysteresis, low creep, and fast creep recovery. Therefore, usually tool steel and stainless steel, aluminum or beryllium copper, are used and a special heat treatment is applied.

The most common type of force transducers that is an example of an elastic device is the strain gauge load cell, of which three types are industrially relevant today: foil strain gauge, fine wire strain gauge, and thin-film strain gauges.

The *foil strain gauge* is the most widely used type which is used in the majority of precision load cells. It consists of a metal foil pattern ($\leq 5\ \mu$m) mounted on an insulating carrier foil (10–50μm) and is produced by photo-etching processes like printed circuit boards. Thus, they can be produced in high numbers with diminishing costs per unit. Another kinds of foil strain gauges are semiconductor or piezoresistive strain gauges, which have greater sensitivity compared with a metallic strain gauge. However, both show nonlinear dependency to strain, an increased temperature dependence, and are relative fragility.

Fine wire strain gauges were the first type of bonded strain gauges. While today widely replaced by cheaper foil or thin film types, fine wire strain gauges are still used for high-temperature transducers, and are available in a wide range of materials such as nickel–chrome, copper–nickel alloys as constantan, or platinum. Common applications are strain measurement in airbreathing jet engines at very high temperatures and even measurements in cryogenic fluids near absolute zero.

Thin-film strain gauges are made by sputtering or evaporating thin films of metals or alloys directly onto the elastic element or on a small metallic carriers. They show long-term stability and higher temperature capability than other strain gauge technologies and are even more suited for use in large-volume products. Modern high-temperature thin-film strain gauges provide minimally intrusive surface strain measurements up to 1100°C directly applied onto the elastic element.

3.1.1.2 Elastic deformation with displacement measurement

Another way to utilize elastic elements is by measuring the displacement of a reference surface. The most basic type of these transducers is a *spring scale*, known as spring balances used to weigh heavy loads in road transportations and storages. Significantly more often in use are noncontact electronic proximity sensors like inductive proximity sensors, capacitive displacement sensor, photoelectric like through-beam sensors, and ultrasonic proximity sensors.

Further relevant ways to measure the displacement of the elastic element are *optical interference force sensors* which are based on laser interferometers, *atomic force microscopes*, where forces between the tip of a cantilever and a sample lead to a deflection of the cantilever, and *waveguide force microscopes* for measuring the cell adhesion forces in biological systems. In the latter case, the elastic element is substituted by the waveguide.

3.1.1.3 Inverse magnetostrictive effect

The inverse magnetostrictive effect also known as magnetoelastic effect or Villari effect refers to the interaction between the mechanical stresses, on the one hand and the magnetic quantities induction and magnetic field strength in ferromagnetic materials, including iron, nickel, and cobalt, on the other hand. In the solid state, these materials form magnetic domains, in which the magnetic moment of the atoms is rectified. Usually, these districts are differently oriented, so that macroscopically no magnetization of the material can be observed. An external magnetic field or mechanical stress influences the magnetic domains in magnitude and direction, so that a macroscopic magnetization arises.

With magnetoelastic load cells one differentiates between the *intensity transducers* and the *anisotropy transducers*. For intensity transducers, the magnetic field is guided because of the geometric conditions of the transducer. Consequently, the changes in permeability are dominant compared with the changes in the direction of the magnetic field. For anisotropy transducers, the magnetic field is not significantly guided and so it changes both in amount and direction when a mechanical load is applied. The change in the magnetic field can be detected inductively or transformatively. In practice, the inductive detection is insignificantly small. Thus, only the transforming anisotropy converters have gained importance.

A typical design of a transforming anisotropy converter consists of stacked ferromagnetic sheets that are electrically insulated from each other, and a pair of orthogonally orientated coils that are arranged at 45° with respect to the sheets. The two coils build a transforming converter. The first coil is the excitation coil which induces no voltage in the second coil while no mechanical load is applied. If a force is applied, the arising mechanical stress changes the magnetic field and voltage is induced in the secondary coil which is proportional to the force.

Magnetoelastic force transducers are relatively cost-effective compared with other measurement principles and are characterized by low creep and high creep recovery, determined only by the ferromagnetic material.

3.1.1.4 Piezoelectric effect

The piezoelectric effect is described by the change of electric charge, and therefore the generation of voltage, at the surface of some solid materials due to an internal mechanical stress caused by an applied load. Piezoelectric transducers use this effect to measure various process quantities such as pressure, force, mechanical stress, or acceleration. They are sensing elements, so no power supply is needed in order to generate an electric signal. In fact, when a force is applied to the transducer, the piezoelectric crystals generate an electrostatic charge proportional to the acting force.

This output is collected on electrodes placed between the crystals and is then amplified by a charge amplifier or converted into a low impedance voltage signal close to the transducer.

Two types of materials are used: piezoelectric ceramics and single crystal materials. Both material groups differ in sensitivity and in long-term stability. Single crystal materials, e.g. gallium orthophosphate, tourmaline, or quartz show significantly better long-term stability but are two orders of magnitude less sensitive as piezoelectric ceramics, which are manufactured in a sintering process.

Depending on the orientation of the cutting plane relative to the main axis of the piezoelectric material, four operational modes can be distinguished: longitudinal, transverse, shear, and volume (hydrostatic effect). With a *longitudinal cut* the applied force and the displaced charge share the same axis. The amount of displaced charge is directly proportional to the applied force and independent of size and shape of the piezoelectric element. In order to increase the transducer sensitivity, multiple of these elements can be put mechanically in series, while electrically in parallel. With a *transverse cut* the axis of the displaced charge and the applied load are perpendicular to each other. The amount of displaced charge is also directly proportional to the applied force but also on size and shape of the piezoelectric element. A *shear cut* results in a displaced charge which is directly proportional to the shear stress implied by the applied force. Like longitudinal cuts produce shear cuts charge outputs which are independent of shape and size of the piezoelectric element. At last, a *polystable cut* is beneficial in applications where the direction of propagation is not known like for shock wave sensors (volume effect). The charge is generated at the same two opposite surfaces dependent only on the average hydrostatic pressure.

The possible applications of piezoelectric force transducers are different from strain gauge-based sensors. While the latter show almost no drift, they are well suited for long-term monitoring tasks. For piezoelectric force transducers, small leakage of charge is inherent in the charge amplifier, which causes the force signal to drift significantly in the order of magnitude of about 1 Nmin^{-1}. Because this amount of drift is independent of the applied force, the relative measurement error is dominant when measuring particularly small forces. The greatest benefits can be achieved when using piezoelectric force transducers for dynamic force measurements. Most piezoelectric sensors show very high stiffness which results in a high resonance frequency. Combined with generally large applicable forces, the dynamic capabilities are often mainly limited by the bandwidth of the used instrumentation like charge amplifiers. Furthermore, piezoelectric force sensors are more compact when compared with other measurement principles which supports integration for many application.

3.1.1.5 Electromagnetic force compensation

Force transducers based on the principle of electromagnetic force compensation are similar to a moving coil loudspeakers and are primarily used in modern high-precision weighing cells. There the applied force is compensated by the electromagnetic force between a permanent magnet (PM) and a closed loop controlled voice coil which operates in the flux gap of the PM. The position of the movable coil is measured via a noncontact displacement sensor whose signal is used as the reference signal (input) for the closed loop control. The current through a coil is proportional to the generated force and therefore also proportional to the force to be measured in the balanced state.

Electromagnetic force transducers are in general suited for high-resolution measurements. Furthermore, they benefit also from tunable stiffness by variation of the control parameters and are therefore able to measure also dynamic forces. In order to optimally run these measurement systems care must be taken to avoid temperature changes or exposure to external magnetic forces.

3.1.2 Differential Lorentz force eddy current testing sensor

Based on the LET analysis presented previously, it might be necessary to perform the testing with high velocity and thus using PM configurations that are better adapted to the current measurement task. Apart from increasing the force perturbations this would increase the absolute values of the Lorentz force, due to Faraday's law of induction. This can be even few orders of magnitudes larger than the perturbations caused by defects.

Unfortunately, precise measurements of small force variations in relatively large range of applied forces is very difficult [91]. Thus, there is a strong demand for the usage of differential force measurements in the LET system. Similar tendency is observed in traditional ECT systems as well. In ECT, various differential pick-up probe configurations offering higher testing sensitivity have already been designed and successfully implemented [36,242].

Currently there are several possibilities to obtain the differential force signals resulting from LET systems. However, the application of commercially available differential force sensors would lead to higher spatial integration requirements and considerably higher costs. This would be even more important when designing sensor arrays for LET which could simplify and advance the defect detection and reconstruction, respectively.

The main aim in this section is to propose the simple and low-cost modification of LET setup which could be used for measurements of differential Lorentz force signals caused by material defects. The proposed modification affects the used magnet where three independent and passive pick-up coils have been winded on its surface (see Figure 3.1). The principal idea is to use voltages induced in the additional coil system and correlate the voltage signals with the corresponding differential Lorentz force signals exerted on the magnet. In fact, the resulting magnet system can be applied

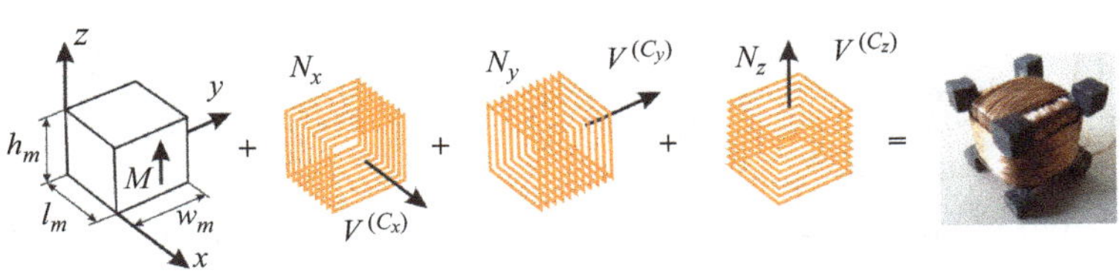

Figure 3.1 Scheme of possible differential Lorentz force sensor for DiLET applications

directly to the existing LET experimental setup, i.e. it can be used as a complete differential Lorentz force sensor. Since, this specific modification of LET system allows differential Lorentz force measurements, the proposed technique has been termed as differential Lorentz force eddy current testing (DiLET).

The concept of the corresponding sensor is based on [91,102,243], where a set of passive coils fixed to a PM is proposed to allow the detection of perturbations in the eddy current distribution caused by defect inside the specimen. In this case, the primary magnetic field $\mathbf{B}^{(p)}$ produced by a PM is constant in time, while the secondary magnetic field $\mathbf{B}^{(s)}$, connected with the eddy current distribution inside the specimen, is time dependent when a defect is present. Thus, the induced voltage $V_i (i \in \{x, y, z\})$ in a coil fixed to the PM is proportional to perturbations of the secondary magnetic field $B_i^{(s)}$ and therefore sensitive to disturbances caused by the defect. As shown in [102], the induced voltage V_i is proportional to the time derivative of the force component parallel to the respective coil axis.

The main idea of using simple coil system to obtain the differential Lorentz force signals can be traced back to the following force relation [24,244].

$$\mathbf{F} = \int_{\Omega_M} (\mathbf{M} \cdot \nabla)\mathbf{B}\, \mathrm{d}\Omega = -\int_{\Omega_C} \mathbf{j} \times \mathbf{B}\, \mathrm{d}\Omega. \tag{3.1}$$

Assuming a cubic PM uniformly magnetized along the z-axis ($\mathbf{M} = M\mathbf{e}_z$) the Lorentz force acting on the permanent magnet can be described as

$$\mathbf{F} = \int_0^{w_m} \int_0^{l_m} \int_0^{h_m} M \frac{\partial \mathbf{B}}{\partial z}\, \mathrm{d}x\, \mathrm{d}y\, \mathrm{d}z, \tag{3.2}$$

where w_m, l_m, and h_m represents the width, length, and height of the magnet, respectively (see Figure 3.1). Differentiating (3.2) with respect to time and by considering all three force components separately the differential force signals can be described as

$$\frac{\partial F_x}{\partial t} = M \frac{\partial}{\partial t} \int_0^{w_m} \int_0^{l_m} \int_0^{h_m} \frac{\partial B_x}{\partial z}\, \mathrm{d}x\, \mathrm{d}y\, \mathrm{d}z, \tag{3.3}$$

$$\frac{\partial F_y}{\partial t} = M \frac{\partial}{\partial t} \int_0^{w_m} \int_0^{l_m} \int_0^{h_m} \frac{\partial B_y}{\partial z}\, \mathrm{d}x\, \mathrm{d}y\, \mathrm{d}z, \tag{3.4}$$

$$\frac{\partial F_z}{\partial t} = M \frac{\partial}{\partial t} \int_0^{w_m} \int_0^{l_m} \int_0^{h_m} \frac{\partial B_z}{\partial z}\, \mathrm{d}x\, \mathrm{d}y\, \mathrm{d}z. \tag{3.5}$$

If dimensions of the magnet are small enough the following approximation applies

$$\frac{\partial B_i}{\partial z} \simeq \frac{B_i|_{z=h_m} - B_i|_{z=0}}{h_m}, \tag{3.6}$$

where $i \in \{x, yz\}$. By using (3.6) directly in (3.3)–(3.5) the differential Lorentz force acting on the magnet can be well approximated as

$$\frac{\partial F_x}{\partial t} \sim \frac{M}{h_m} \int_0^{w_m} \int_0^{l_m} \int_0^{h_m} \left[\left.\frac{\partial B_x}{\partial t}\right|_{z=h_m} - \left.\frac{\partial B_x}{\partial t}\right|_{z=0} \right] \mathrm{d}x\,\mathrm{d}y\,\mathrm{d}z, \tag{3.7}$$

$$\frac{\partial F_y}{\partial t} \sim \frac{M}{h_m} \int_0^{w_m} \int_0^{l_m} \int_0^{h_m} \left[\left.\frac{\partial B_y}{\partial t}\right|_{z=h_m} - \left.\frac{\partial B_y}{\partial t}\right|_{z=0} \right] \mathrm{d}x\,\mathrm{d}y\,\mathrm{d}z, \tag{3.8}$$

$$\frac{\partial F_z}{\partial t} \sim \frac{M}{h_m} \int_0^{w_m} \int_0^{l_m} \int_0^{h_m} \left[\left.\frac{\partial B_z}{\partial t}\right|_{z=h_m} - \left.\frac{\partial B_z}{\partial t}\right|_{z=0} \right] \mathrm{d}x\,\mathrm{d}y\,\mathrm{d}z. \tag{3.9}$$

In the DiLET system, the pick-up coils are moving together with the primary magnetic field $\mathbf{B}^{(p)}$, i.e.

$$\frac{\partial \mathbf{B}}{\partial t} = \frac{\partial(\mathbf{B}^{(p)} + \mathbf{B}^{(s)})}{\partial t}. \tag{3.10}$$

Using the reasonable assumption

$$\frac{\partial B|_{z=h_m}}{\partial t} \ll \frac{\partial B|_{z=0}}{\partial t}, \quad i \in \{x, y, z\} \tag{3.11}$$

(3.7)–(3.9) can be simplified as

$$\frac{\partial F_x}{\partial t} \sim -\frac{M}{h_m} \int_0^{w_m} \int_0^{l_m} \int_0^{h_m} \left.\frac{\partial B_x}{\partial t}\right|_{z=0} \mathrm{d}x\,\mathrm{d}y\,\mathrm{d}z, \tag{3.12}$$

$$\frac{\partial F_y}{\partial t} \sim -\frac{M}{h_m} \int_0^{w_m} \int_0^{l_m} \int_0^{h_m} \left.\frac{\partial B_y}{\partial t}\right|_{z=0} \mathrm{d}x\,\mathrm{d}y\,\mathrm{d}z, \tag{3.13}$$

$$\frac{\partial F_z}{\partial t} \sim -\frac{M}{h_m} \int_0^{w_m} \int_0^{l_m} \int_0^{h_m} \left.\frac{\partial B_z}{\partial t}\right|_{z=0} \mathrm{d}x\,\mathrm{d}y\,\mathrm{d}z. \tag{3.14}$$

From (3.12) to (3.14) it can be observed that to obtain the differential Lorentz force signals it is sufficient to measure the time variation of the magnetic flux density ($\partial B_i/\partial t$ at $z = 0$), i.e. at the bottom surface of the magnet. As it will be shown, this can be accomplished by a simple pick-up coil system presented in Figure 3.1. Voltages induced in each of the coils can be calculated as follows

$$V^{(C_x)} = \sum_1^{N_x} \left[\int_0^{l_m} \int_0^{h_m} \frac{\partial B_x}{\partial t} \mathrm{d}y\,\mathrm{d}z \right] \sim \frac{N_x}{h_m} \int_0^{w_m} \int_0^{l_m} \int_0^{h_m} \frac{\partial B_x}{\partial t} \mathrm{d}x\,\mathrm{d}y\,\mathrm{d}z, \tag{3.15}$$

$$V^{(C_y)} = \sum_1^{N_y} \left[\int_0^{w_m} \int_0^{h_m} \frac{\partial B_y}{\partial t} \mathrm{d}x\,\mathrm{d}z \right] \sim \frac{N_y}{h_m} \int_0^{w_m} \int_0^{l_m} \int_0^{h_m} \frac{\partial B_y}{\partial t} \mathrm{d}x\,\mathrm{d}y\,\mathrm{d}z, \tag{3.16}$$

$$V^{(C_z)} = \sum_1^{N_z} \left[\int_0^{w_m} \int_0^{l_m} \frac{\partial B_z}{\partial t} \mathrm{d}x\,\mathrm{d}y \right] \sim \frac{N_z}{h_m} \int_0^{w_m} \int_0^{l_m} \int_0^{h_m} \frac{\partial B_z}{\partial t} \mathrm{d}x\,\mathrm{d}y\,\mathrm{d}z. \tag{3.17}$$

where $V^{(C_x)}$, $V^{(C_y)}$, and $V^{(C_z)}$ are the voltages induced in the coils assigned to the x-, y-, and z-axis, respectively. The total number of windings of each coil is denoted by N_x, N_y, and N_z, respectively. By direct comparison of (3.15)–(3.17) with (3.12)–(3.14) the differential Lorentz force acting on the magnet, which is caused by the presence of material defects can be well approximated as

$$\frac{\partial F_x}{\partial t} \sim -\frac{h_m}{N_x} V^{(C_x)}, \tag{3.18}$$

$$\frac{\partial F_y}{\partial t} \sim -\frac{w_m}{N_y} V^{(C_y)}, \tag{3.19}$$

$$\frac{\partial F_z}{\partial t} \sim -\frac{l_m}{N_z} V^{(C_z)}. \tag{3.20}$$

In order to test the proposed DiLET methodology, the LET benchmark problem presented in Figure 2.49 has been solved using the moving defect approach (see page 139). Instead of the cylindrical PM, a cubic magnet with the edge length equal to $a = 1$ cm placed at a lift-off distance $\delta z = 1$ mm is used. All other geometrical and material parameters are given in Table 2.4. The DiLET sensor consists of three mutually perpendicular coils wound on a cubic PM (cf. Figure 3.1), each with a number of windings equal to $N = 300$.

Figure 3.2(a) and (c) shows the differential Lorentz force signals calculated directly by time differentiation of (3.1), whereas Figure 3.2(b) and (d) shows the corresponding voltages induced in coils oriented along x-axis and z-axis calculated according to (3.15) and (3.17). From the obtained results, a very good correlation between the differential Lorentz force signals and the corresponding voltages can be observed. The analysis confirms the validity of simplifications assumed in order to derive (3.18)–(3.20). Nevertheless, to determine differential force signals directly from voltages induced in the corresponding coils an additional calibration of the proposed differential force sensor has to be performed. The peak-to-peak perturbations of the obtained differential force signals have been compared directly with the corresponding perturbations of the induced voltages (cf. [102]). Furthermore, from (3.18)–(3.20) follows that this dependency is linear. Moreover, due to the cubic shape of the magnet and coils with the same number of windings, the calibration curves for both force components are almost identical [102]. Numerical simulations have shown that the correlation coefficient between the differential Lorentz force and voltage induced in corresponding coils was more than 0.99 for magnetic Reynolds numbers less than 5.

So far only one PM was included in the analysis. However, it was evident from the parametric studies that to increase the testing capabilities of LET systems, strongly focused magnetic flux densities are required. Therefore, future work should involve the investigation of complex magnet systems including optimized magnet/coil arrays, perhaps including ferromagnetic shields or even superconductors. Most industrial applications employ ferromagnetic materials and therefore, there is a strong need to extend LET for testing such materials as well. However, such an endeavor creates additional challenges due to the nonlinearity in magnetic permeability. The extremely

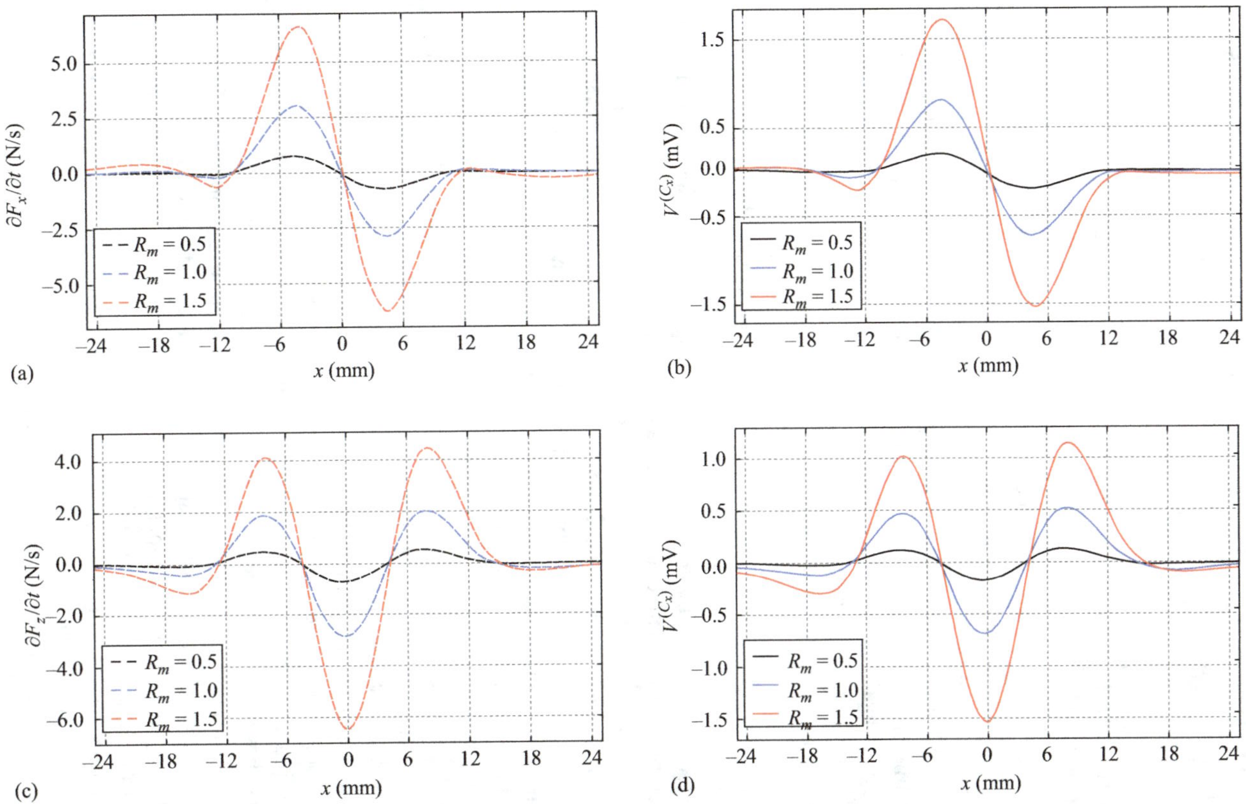

Figure 3.2 *Direct comparison of the differential Lorentz force signals acting on the cubic magnet ($a = 1$ cm) for different values of R_m. Calculations have been performed for long ("|") defect located at $d = 2$ mm using MDA. The magnet is located at lift-off distance $\delta z = 1$ mm. (a) Drag component. (b) Voltage induced in C_x. (c) Lift component. (d) Voltage induced in C_z*

high-attraction forces between the specimen and the strong PM make the application of the actual Lorentz force measurement system very difficult or even impossible. This problem can be tackled by the DiLET methodology. However, this requires further numerical and experimental investigation.

3.1.3 Characteristics and calibration of force measurement systems

The force measurement systems described earlier are based on different physical principles but can be described by a number of common characteristics. The behavior of all these systems can be expressed by plotting the *response curve* which represents the indicated output value against the applied force. An ideal response curve is a straight line from zero to the rated capacity of the force measurement system and then back again to zero. Real measurement systems differ from this ideal curve in multiple ways. These are commonly categorized by their systematic deviation with respect to the least-squares optimal line through the origin for increasing outputs.

The output at zero force is known as *offset* and is often caused by an imperfect initial state of the measurement system. A common way to encounter this error is to add the negative value to the output at the beginning of the measurement. The deviation of output from the fitted line for increasing loads is defined as *non linearity* and the largest and in some cases also the average deviation is given in the data sheet of the system. *Hysteresis* is the dependence of the state of a measurement system on its loading history. It is described by the difference in output between increasing and decreasing load. For systems with strictly progressive or degressive behavior, the largest value of hysteresis is usually at mid-range. In some cases, it is useful to describe hysteresis and nonlinearity as *combined error*. This can be done by providing two additional lines which are parallel to the least-squares optimal line and do enclose the increasing and decreasing output curves. Generally, the above-mentioned errors can result in an underestimation of the actual error of measurement.

Another important concept in measurement systems is known as *repeatability* which is described by the agreement of the outputs for repeated applications of the same load. In practice, the repeatability of transducers is also provided by many manufacturers; however, this value can only be a reference for optimal conditions during the testing procedure. In operation, it is generally an overestimation of the performance of the force measurement system due to limitations of the actual electrical measurement equipment, present temperature gradients from first to last application, and other environmental influence quantities.

Further characteristics of force measurement systems can be summarized by the imperfections of applying the force to be measured to the loading surface of the transducer. One of the most important specifications is thereby the sensitivity to off-axis forces which result in parasitic torque. Also for single-component transducers, it is important to consider the sensitivity to orthogonal forces which is equivalent to misalignment of the transducers principal axis to the force to be measured.

3.2 Optimization of PM systems

3.2.1 Introduction and motivation

Comparative studies between LET and ECT indicated the potential and competitiveness of LET [44,105]. However, the performance of an LET system can be enhanced further by applying optimization schemes to determine advanced magnet systems with improved characteristics. This involves an appropriate problem definition and associated criteria.

The optimization goal in LET is to maximize the response resulting from an inclusion surrounded by conductive material, thereby increasing the signal-to-noise ratio and, hence, improving the detection rate. However, due to the high variety of NDT problems, it is self-evident that the final details of an optimized setup strongly depend on the detection goal and external testing conditions for the particular application. The proposed methodology is developed as generally as possible, to describe and address the problem specificity. However, when considering PM systems, generality is limited by practicability, and so the geometry and associated design variables are chosen such that practical feasibility is assured. The optimization process is performed with nondimensional parameters. This provides the opportunity to determine scale-independent and generalized optimization results while decreasing the number of independent parameters. This approach can then be applied to different applications to determine optimal magnet designs for specific cases.

At first, the required parameters and the optimization problem are described in Sections 3.2.2.1 to 3.2.2.4. It is followed by the definition of the objective function and the corresponding constraints in Sections 3.2.2.5 and 3.2.2.6, respectively. The developed optimization strategies are presented in Section 3.2.2.7 and the applied optimization algorithm, sequential quadratic programming (SQP), is briefly described in Section 3.2.2.8. The numerical procedure to evaluate the objective function and the constraints is described in Section 3.2.2.9. In Section 3.2.3, the numerical optimization results are presented and discussed. The design process of prototypes of optimized magnet designs is described in Section 3.2.4. It is followed by a study regarding the current detection limit of deep-lying defects in Section 3.2.5. The chapter is summarized and conclusions are drawn in Section 3.2.6.

3.2.2 Methods

3.2.2.1 Problem definition

The optimal magnet design presented here is focused on nonmagnetic, electrically conducting specimens. However, it is also applicable to ferromagnetic specimen taking into account adaptions in the objective function evaluation. The optimization is performed under the assumptions of a smooth specimen surface and that the defect is located far from any lateral boundary, to neglect parasitic edge effects. Since the resulting Lorentz force profile depends on the shape and the depth of the inclusion, an equivalent defect of cuboidal shape is defined to represent a general flaw. The assumptions can be modified to any particular case of interest, since this would

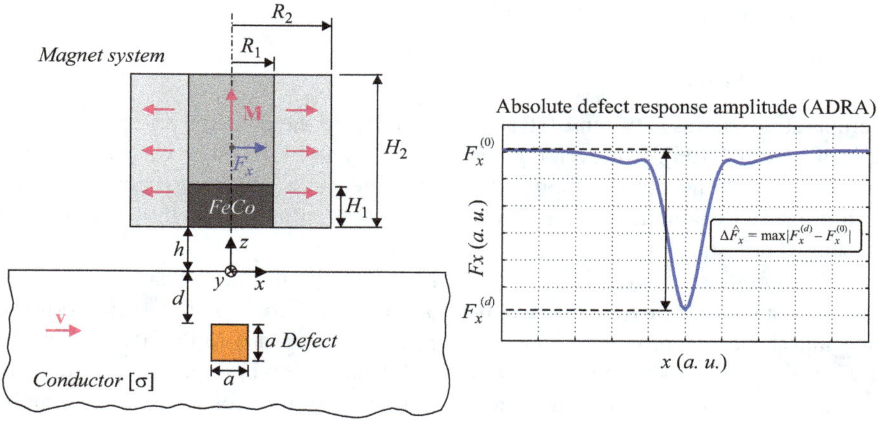

Figure 3.3 Geometrical parameters of the LET setup, design variables of the magnet system and illustration of the ADRA used as objective function [245]

involve only the geometry of the specimens defined in the forward solution, which are described in one of the following sections. The optimization is performed with respect to the drag-force F_x and the associated absolute defect response amplitude (ADRA) $\Delta\hat{F}_x$, resulting from the difference between the unperturbed drag-force $F_x^{(0)}$ and the perturbed drag-force $F_x^{(d)}$:

$$\Delta\hat{F}_x = \max\left|F_x^{(d)} - F_x^{(0)}\right|. \tag{3.21}$$

As it has been shown in the previous chapters, the force profile is symmetric, if the interaction between the primary magnetic field $\mathbf{B}^{(p)}$, generated by the PM, and the secondary magnetic field $\mathbf{B}^{(s)}$ from the induced eddy currents is negligible.

In Figure 3.3, the geometrical parameters of the problem are shown together with the ADRA. The specimen is modeled as a pseudoinfinite half-space including a defect with edge length a, located at a depth d. The magnet system is located at a lift-off distance h above the specimen.

The optimization scheme presented here covers, but is not limited to, purely isotropic specimens ($\sigma_{xx} = \sigma_{yy} = \sigma_{zz}$) and laminated structures ($\sigma_{xx} = \sigma_{yy} \neq 0, \sigma_{zz} = 0$), as they are shown in Figure 1.7.

3.2.2.2 Magnet system and design variables

Three related magnet geometries with increasing manufacturing complexity are investigated, originating from empirical preinvestigations. These are:

- Standard cylindrical PMs (C)
- Cylindrical Halbach-structures (HC)
- Cylindrical Halbach-structures supported by highly saturated soft magnetic material, such as iron–cobalt alloys (HCp)

In the general case, the nondimensional design variables $\mathbf{x}$, shown in Figure 3.3, are defined as:

$$\mathbf{x} = \left[\frac{H_2}{R_2}, \frac{R_1}{R_2}, \frac{H_1}{H_2}\right], \tag{3.22}$$

where H_2 and R_2 are the height and radius of the outer cylinder, respectively; R_1 is the inner radius of the Halbach-structure; and H_1 is the height of the ferromagnetic material. The inner cylinder is axially magnetized, whereas the surrounding cylinder is magnetized in the radial direction.

For C- and HC-systems, particular design variables become constant and the number of free variables is reduced such that for C-systems, $R_1/R_2 = 1$ and $H_1/H_2 = 0$, and for HC-systems $H_1/H_2 = 0$. Thus, the configurations are subproblems of each other; i.e. C-magnets are included in HC-configurations, which in turn are included in HCp-systems. As a consequence, by applying the presented optimization scheme considering all three design variables as free variables, a wide variety of different designs is included in the optimization. An optimal magnet system can be determined for a standard cylinder or a Halbach-structure with or without ferromagnetic material.

The construction of interchanging magnetization direction corresponds to the concept of Mallinson [246] and Halbach [247]. Changing the magnetization direction of adjacent parts of the magnet forms a semiopen magnetic circuit. Hence, the magnetic flux density is increased on one side of the magnet system and decreased on the opposite side. In the ideal case, it is possible to eliminate the magnetic flux on one side completely by employing dual magnetizations determined by means of the Hilbert transformation [246]. Practical feasibility of this form of magnetization is still a challenge. Nevertheless, geometrical approximations in the form of segments can be employed instead. These structures are typically termed as Halbach-arrays and find application in particle accelerators [247], high-speed motors/generators and servomotors [248], loudspeakers [249], magnetic bearings [250,251], and nuclear magnetic resonance spectroscopy [252]. In the framework of Lorentz force velocimetry of electrically weak conducting liquids, commonly applied linear Halbach-arrays are already used to increase the drag-force signal [253]. These systems are not optimized in terms of the defective specimen, which further motivates to investigate the optimization problem in LET.

A radially magnetized Halbach-cylinder of infinite height produces no magnetic field in the inner or outer air domain surrounding the cylindrical magnet [254]. However, in case of finite heights, this effect vanishes and strong-directed fields are present at the terminations. In this work, the concept of Halbach-arrays is extended to a novel rotationally symmetric structure including an axially magnetized cylinder, which is supported by ferromagnetic material to intensify the magnetic flux density close to the object under test.

In this work, a part of the PM is replaced by ferromagnetic material with high-saturation magnetization and hence a stronger residual field [255]. In general, ferromagnetic material can be used to focus and concentrate the flux, known as flux compression. Thus, the magnetic flux density is amplified when the ferromagnetic

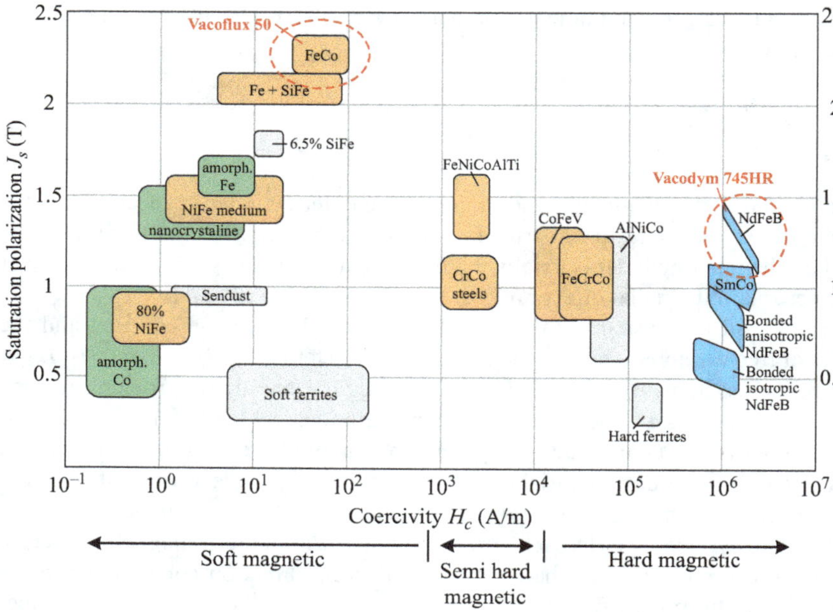

Figure 3.4 Overview of soft and hard magnetic materials. The used materials in the optimization, VACOFLUX® 50 [256] and VACODYM® 745HR [257], are highlighted

part is exposed to the external magnetic field of the permanent magnets. Consequently, the resulting Lorentz force is potentially increased. The choice of suitable materials is decisive for a successful optimization. An overview of soft and hard magnetic materials is given in Figure 3.4.

In the present study, the hard magnetic material VACODYM® 745HR [257,258] with a nominal magnetic remanence of $B_r = 1.44\,\text{T}$ is used together with the soft magnetic iron–cobalt alloy VACOFLUX® 50 [256] with a saturation polarization of 2.3 T. The $B(H)$ curves of both materials are shown in Figure 3.5, illustrating the principle to increase the magnetic flux density. A similar approach is proposed during the design process of focus lenses for linear collider accelerators [259,260], as well as for superconducting cyclotron magnets [255]. The chosen materials represent the current state of the art in magnetism and are thus most suitable for an optimal LET sensor.

3.2.2.3 Scaling parameters

To reduce the number of parameters, it is advisable to exploit *a priori* known dependencies between certain input parameters and the drag-component of the Lorentz force. A scaling factor S can be defined to scale the forces accordingly. If the ADRA, which will be used later as the objective function, can be simply scaled, the global optimum of the design variables $\tilde{\mathbf{x}}$ is scale invariant with respect to the reduced

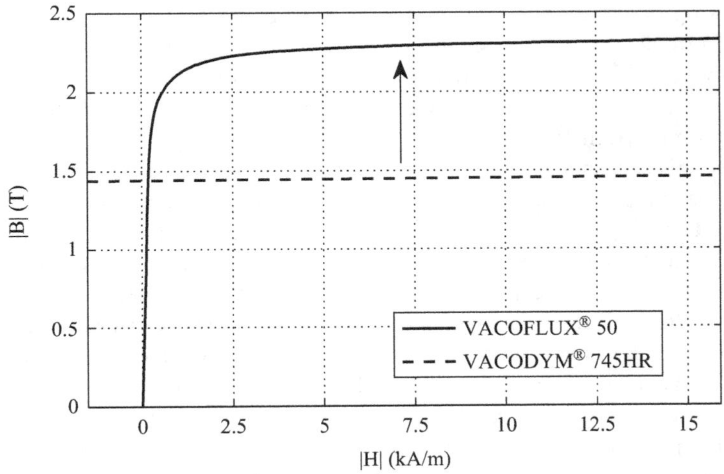

Figure 3.5 *B*(*H*) *curve of the iron–cobalt–alloy VACOFLUX® 50 used as ferromagnetic material in HCp-magnet systems in FEM simulations and a part of the hysteresis of the hard magnetic PM material NdFeB of type VACODYM® 745HR ($B_r = 1.44$ T) at room temperature [245]*

parameters. This reduces the need to rerun numerically expensive simulations and provides the desired generality. The scaling quantities are summarized in the vector of scaling parameters:

$$\mathbf{s} = [\sigma, v, B_r, h]. \tag{3.23}$$

The scaling properties of individual quantities can be determined considering the following estimate, which is valid provided secondary magnetic fields generated by the induced eddy currents are negligible [89]:

$$F_x \sim \sigma v m^2 h^{-3}, \tag{3.24}$$

where m is the equivalent magnetic dipole moment of the magnet system, which can be expressed in terms of remanence and magnet volume V_m such that $m \sim B_r V_m$. For geometric scaling, it is assumed that the whole geometry of the problem scales with the lift-off distance h so that $V_m \sim h^3$, without loss of generality. Thus, (3.24) can be expressed as:

$$F_x \sim \sigma v B_r^2 h^3. \tag{3.25}$$

Note that the drag-force F_x depends linearly on the velocity v and the electrical conductivity σ. This property can be extended to the case of possibly anisotropic specimens by defining the conductivity tensor $[\boldsymbol{\sigma}]$ in terms of a scalar conductivity σ, which defines the magnitude of the conductivity and the anisotropy vector $\mathbf{a}_\sigma$:

$$[\boldsymbol{\sigma}] = \sigma\,\mathrm{diag}(\mathbf{a}_\sigma^{\mathrm{T}}). \tag{3.26}$$

In the isotropic case, the anisotropy vector is given by $\mathbf{a}_\sigma = [1, 1, 1]^T$. If the specimen is made from metal sheets, as in the case of the laboratory setup (see Figure 1.7), the anisotropy vector is $\mathbf{a}_\sigma = [1, 1, 0]^T$. The prescribed scaling property of the electrical conductivity addresses the scalar conductivity σ, and, thus, holds for both isotropic and anisotropic specimens.

Closer inspection of (3.25) shows that the Lorentz force increases with the square of the remanence B_r. However, this scaling property can be applied if and only if nonlinear ferromagnetic material is omitted or considered as linear in the whole domain. Since the nonlinearity between **H** and **B** is accounted for, this factor can only be modified for C- and HC-systems, and must be fixed at a predefined remanence in the case of HCp-systems. The magnetic remanence can be used as a scaling factor if the whole magnet system is made from the same magnetic material. If this is not the case, the remanence of each compartment has to be scaled in the same way.

The system parameters, described later in detail, define the geometrical relationships of the problem. However, it is of interest how $F_x^{(0)}$ and $\Delta\hat{F}_x$ scale with respect to the geometrical size of the problem. Reformulating (3.24) to (3.25) provides the scaling property of the drag-force directly as a cubical relationship with respect to h, serving as the characteristic length and thus defining the geometrical size of the problem. In this sense, h is eliminated from the set of independent parameters and only the geometrical ratios with respect to h define the problem. According to the chosen h, the actual geometry can be stretched or clinched to any scale of interest.

Finally, by combining the individual scaling parameters into one, the scalar scaling factor S contains linear, quadratic, and cubic terms:

$$S = \left(\frac{\sigma v}{\sigma_0 v_0}\right)\left(\frac{B_r}{B_{r0}}\right)^2\left(\frac{h}{h_0}\right)^3. \tag{3.27}$$

As reference values for the simulations, the velocity is set to $v_0 = 1$ m/s, the conductivity is $\sigma_0 = 1$ MS/m, the magnetic remanence is $B_{r0} = 1$ T, and the lift-off distance is $h_0 = 1$ mm. Consequently, the scaling factor can be used to convert the forces $F_x^{(0,s)} = SF_x^{(0)}$ and $\Delta\hat{F}_x^{(s)} = S\Delta\hat{F}_x$, according to the previously mentioned conditions. Thus, the total number of independent parameters is reduced by the number of scaling parameters.

3.2.2.4 System parameters

Given the design variables **x** of the magnet system and the scaling parameters **s**, the LET setup is defined by the set of system parameters:

$$p = \left\{\frac{V_m}{V_d}, \frac{d}{h}, \frac{a}{h}, \mathbf{a}_\sigma, B_r, B(H)\right\}, \tag{3.28}$$

where $V_m = \pi R_2^2 H_2$ is the total volume of the magnet system and $V_d = a^3$ is the volume of the equivalent cuboidal defect. The magnet to defect volume ratio V_m/V_d defines the weight of the magnet system at constant defect volumes. The depth-to-lift-off ratio d/h defines the defect depth measured from the surface of the conductor to the upper surface of the defect. The system is geometrically completely described by the third ratio a/h between the edge length of the defect a and the lift-off distance of the magnet h. Since the material anisotropy of the specimen affects the eddy current distribution, the anisotropy vector $\mathbf{a}_\sigma$ is included in the set of system parameters. The remanence B_r only appears as a system parameter if nonlinear ferromagnetic material is included in the magnet design. Otherwise, B_r is a scaling parameter. Finally, $B(H)$ is the initial magnetization curve of the ferromagnetic material.

3.2.2.5 Objective function

The optimization goal is to maximize the ADRA $\Delta\hat{F}_x(\mathbf{x},p)$. The scaling factor S from (3.27) allows generalizing the calculated ADRA independently for conductivity, velocity, and geometric scale (and in the linear case also for magnetic remanence). Given the defined set of system parameters $\boldsymbol{p}$, the optimal design variables $\tilde{\mathbf{x}}$ of the magnet system can be determined by applying distinct optimization schemes. Subsequently, the optimal ADRA $\Delta\tilde{F}_x(\tilde{\mathbf{x}},p)$, together with its corresponding unperturbed drag-force $\tilde{F}_x^{(0)}(\tilde{\mathbf{x}},p)$ is provided and the limits of the system are determined. To this end, the objective function $f(\mathbf{x},p)$ to be minimized is defined by the negative ADRA from (3.21):

$$\min_{\mathbf{x}\in\mathscr{F}} f(\mathbf{x},\boldsymbol{p}) = -\Delta\hat{F}_x(\mathbf{x},\boldsymbol{p}). \tag{3.29}$$

The objective function is nonlinear and depends on both the design variables and the system parameters. Thus, the optimal solutions $\tilde{\mathbf{x}}(\boldsymbol{p})$ depend on the predefined system parameters p, which can be understood as a mathematical description of problem specificity with respect to external conditions. However, the optimal design variables have to be part of the feasible set of solutions $\mathscr{F}$. The feasibility of a solution is defined by constraints, which are described next.

3.2.2.6 Definition of constraints

The feasible set of solutions $\mathscr{F}$ is defined by two types of constraints. The first are linear inequality constraints, also known as bound constraints, resulting from the limits of the design variables $\mathbf{x}$, defined by the geometry of the magnet system. They are covered by the linear inequality constraints $\mathbf{c}_\mathrm{l}$, which must not be violated:

$$\mathbf{c}_\mathrm{l}(\mathbf{x},\boldsymbol{p}) \leq \mathbf{0} \tag{3.30}$$

$$\mathbf{c}_\mathrm{l}(\mathbf{x},\boldsymbol{p}) = [\mathbf{A}_\mathrm{HCp}]\,\mathbf{x}^\mathrm{T} - \mathbf{b}_\mathrm{HCp}. \tag{3.31}$$

The linear inequality constraint matrix $[\mathbf{A}_\mathrm{HCp}]$ and constraint vector $\mathbf{b}_\mathrm{HCp}$ are determined considering the limits: $H_2/R_2 > 0$, $0 \leq R_1/R_2 \leq 1$, and $0 \leq H_1/H_2 \leq 1$ (Figure 3.3):

$$[\mathbf{A}_\mathrm{HCp}] = \begin{bmatrix} -1 & 0 & 0 \\ 0 & -1 & 0 \\ 0 & 1 & 0 \\ 0 & 0 & -1 \\ 0 & 0 & 1 \end{bmatrix}, \quad \mathbf{b}_\mathrm{HCp} = \begin{bmatrix} \varepsilon \\ 0 \\ 1 \\ 0 \\ 1 \end{bmatrix}, \tag{3.32}$$

with $\varepsilon \to 0$ to ensure magnet volumes greater than zero ($H_2/R_2 > 0$).

The second type of constraint is a nonlinear inequality c_{nl}. This constraint is defined by the maximum force $F_x^{(\max)}$ measurable by the applied force sensor, which corresponds to the maximum drag-force in the unperturbed case $F_x^{(0)}$:

$$c_{\text{nl}}(\mathbf{x}, \boldsymbol{p}, F_x^{(c)}) \leq 0, \tag{3.33}$$

where

$$c_{\text{nl}}(\mathbf{x}, \boldsymbol{p}, F_x^{(c)}) = F_x^{(0)}(\mathbf{x}, \boldsymbol{p}) - F_x^{(c)}, \tag{3.34}$$

with

$$F_x^{(c)} = \frac{F_x^{(\max)}}{S}\left(\frac{B_r}{B_{r0}}\right)^2. \tag{3.35}$$

By this definition, optimization is performed in the reference space, such that the forces are determined by the defined reference values v_0, σ_0, and h_0. However, the maximum drag-force $F_x^{(\max)}$ is defined in the unscaled space and has to be scaled accordingly using the scaling factor $S\,(B_r/B_{r0})^{-2}$. If nonlinear magnetic material is omitted in the magnet design, $F_x^{(c)}$ is simplified and the ratio $(B_r/B_{r0})^{-2}$ may be excluded from (3.35). Depending on external conditions, S can be calculated by (3.27) and acts as a weighting factor in the nonlinear constraint function c_{nl}. The normalized maximum drag-force $F_x^{(c)}$ from (3.35), directly affects the feasible set of solutions and plays a central role in the optimization. By scaling the constraint, it is possible to identify similarities between different LET setups. For example, consider two configurations with the same system parameters $\boldsymbol{p}$. The first system which obeys a scaling factor $S_1 = 10$, resulting from a velocity $v_1 = 0.5$ m/s together with a maximum drag-force of $F_x^{(\max)} = 3$ N is equivalent to the second with $S_2 = 20$, $v_2 = 1$ m/s, and $F_x^{(\max)} = 6$ N. In the same way, similarities between the optimal designs can be identified considering the geometric scale of the whole problem defined by h. A more detailed and vivid explanation is given in the results part in Section 3.2.3. The nonlinear constraint is optional, since it strongly depends on the system parameters and the force sensor technology employed. As already mentioned, a force sensor based on the strain gauge technology is used in the experimental setup at hand. Linear behavior of this sensor type is guaranteed until a nominal force $F_x^{(\max)}$ is reached. Plastic deformation of the deflection body will occur if the applied force exceeds the safe load, which is approximately in the range of $2F_x^{(\max)}$ [104]. Consequently, an optimized magnet system which operates at the global optimum probably could not be applied and must be replaced by a system which considers the drag-force limit. The constrained optimization problem can now be classified as a parametric multivariate nonlinear optimization problem with linear and nonlinear inequality constraints [261].

3.2.2.7 Optimization strategies

To address this problem, two different approaches are presented. Depending on the needs and external conditions, one or the other approach can be applied. A combination of both principles is also possible to improve the performance of the magnet system further.

3.2.2.7.1 *Volume and force constraint optimization*

The volume and force constraint optimization (VcFc) approach is based on the definition of a fixed magnet volume as well as a maximum drag-force given by the applied force sensor. Thus, the V_m/V_d ratio is fixed besides all other system parameters, and the constraints are satisfied by adjusting the design variables $\mathbf{x}$ in an optimal way. Depending on previously defined external conditions, it is possible that the nonlinear constraint is active and the constrained solution $\tilde{\mathbf{x}}_c(\boldsymbol{p})$ is located at the constraint boundary. Consequently, the nonlinear constrained optimum differs from the unconstrained solution ($\tilde{\mathbf{x}}_c(\boldsymbol{p}) \neq \tilde{\mathbf{x}}(\boldsymbol{p})$). Thus, the feasible set for the case of a HCp-magnet system is:

$$\mathscr{F} = \left\{ \mathbf{x} \in \mathbb{R}^3 \mid \mathbf{c}_{\mathrm{l}}(\mathbf{x}, \boldsymbol{p}) \leq \mathbf{0},\ c_{\mathrm{nl}}(\mathbf{x}, \boldsymbol{p}, F_x^{(c)}) \leq 0 \right\}. \tag{3.36}$$

3.2.2.7.2 *Volume adaptive force constraint optimization*

In general, the system parameters are defined by the given detection goal and the particular application of interest. However, the volume ratio V_m/V_d can be used as a free parameter by the system designer. The volume adaptive force constraint optimization (VaFc) approach is motivated by considerations related to mechanical dynamics. To improve the dynamic range of the sensor system, it is desirable to increase its eigenfrequency. This offers the possibility to perform measurements at higher testing velocities and can be achieved by reducing the mass of the magnet system as much as possible, while retaining the maximum ADRA as the primary optimization goal. The VaFc approach is proposed to determine a magnet system with a V_m/V_d ratio, which operates at the transition before the nonlinear constraint becomes active. The role of the V_m/V_d ratio is changed from fixed to a variable system parameter, while all other system parameters are kept constant. The corresponding optimal solution is then $\tilde{\mathbf{x}}(\boldsymbol{p}_c)$, and the general procedure is illustrated in Figure 3.6. The procedure starts by defining the maximum drag-force $F_x^{(\max)}$ and the fixed system parameters d/h, a/h, and $\mathbf{a}_\sigma$. The starting point $\mathbf{x}^{(0)}$ and the initial volume ratio $(V_m/V_d)^{(0)}$ are defined to initialize the VaFc procedure. The method iterates by determining the unconstrained optimal design $\tilde{\mathbf{x}}(\boldsymbol{p}^{(n)})$ by a distinct optimization method, where the superscript (n) is an iteration counter. The SQP algorithm is used, as described in the following section. After determining the optimal design by neglecting the nonlinear inequality constraint, the associated drag-force $\tilde{F}_x(\tilde{\mathbf{x}}(\boldsymbol{p}^{(n)}), \boldsymbol{p}^{(n)})$ can be calculated. The relative difference to the defined maximal drag-force is evaluated and compared with the predefined relative tolerance ε_F. If the criterion is fulfilled, the algorithm stops and provides the optimal design variables together with the corresponding volume ratio $(V_m/V_d)^{(n)}$. If not, the volume ratio is updated in a correction step. During the first iteration, the next volume is approximated proportional to the force ratio:

$$(V_m/V_d)^{(1)} = \frac{F_x^{(\max)}(V_m/V_d)^{(0)}}{SF_x(\tilde{\mathbf{x}}(\boldsymbol{p}^{(0)}), \boldsymbol{p}^{(0)})}. \tag{3.37}$$

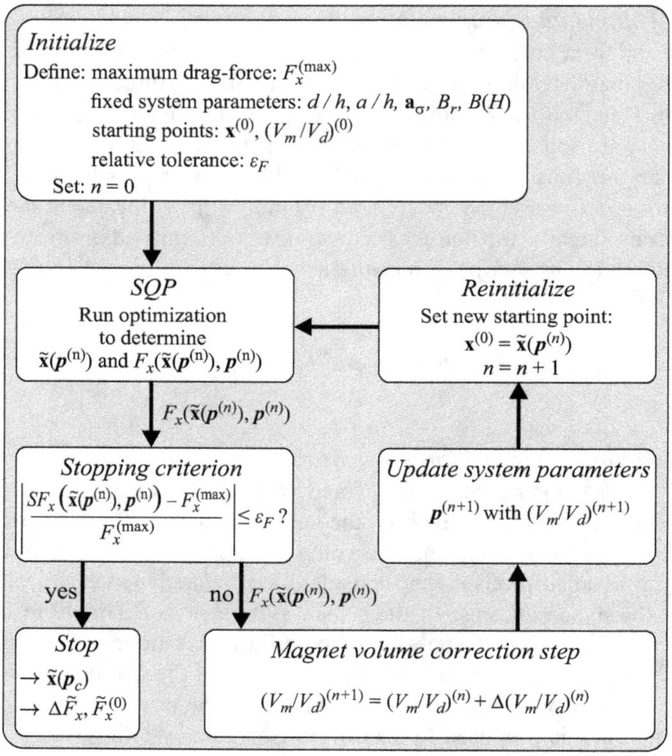

Figure 3.6 Optimization procedure of the VaFc approach to determine the optimal design variables $\tilde{\mathbf{x}}(p_c)$ of a magnet system operating at the unconstrained global optimum, while retaining the nonlinear constraint given by the limiting drag-force $F_x^{(\max)}$ [245]

For all following iterations ($n > 1$), a more robust Newton–Raphson step is used, which considers the gradient information of $\tilde{F}_x^{(0)}$ with respect to V_m/V_d:

$$\Delta(V_m/V_d)^{(n)} = -\frac{SF_x(\tilde{\mathbf{x}}(p^{(n)}), p^{(n)}) - F_x^{(\max)}}{S\partial F_x(\tilde{\mathbf{x}}(p^{(n)}), p^{(n)})/\partial(V_m/V_d)}. \tag{3.38}$$

A backward approximation is applied to determine the partial derivative of the drag-force. After determining $(V_m/V_d)^{(n+1)}$, the system parameters $p^{(n+1)}$ are updated. Before starting a new iteration, the starting point of the design variables is set to the optimal solution of the previous optimization step $\mathbf{x}^{(0)} = \tilde{\mathbf{x}}(\boldsymbol{p}^{(n)})$, since the optimal design between two iterations may be assumed to be similar. Thus, the process is accelerated. The proposed approach can be used to determine the critical V_m/V_d ratio when the optimal magnet system still operates at the unconstrained global optimum while still considering the previously defined constraint of maximum drag-force.

3.2.2.8 SQP algorithm

To solve the optimization problem, the SQP algorithm is used. This method is an extension of the Broyden–Fletcher–Goldfarb–Shanno–Quasi–Newton method by introducing Lagrangian multipliers. The first quasi-Newton-based method was introduced by Davidon [262]. The advantage of Newton-type methods is in the use of the gradient and curvature information provided by the Jacobian and Hessian matrix of the objective function. Quasi-Newton methods avoid the computationally intensive evaluation of the Hessian, which is beneficial when comparatively expensive numerical solvers are involved to evaluate the objective function. The Hessian is approximated during the iterative process successively.

Early work on SQP was done by Biggs [263], Han [264], and Powell [265]. However, the method has been continuously improved. The method is also referred to as a quadratic programming-based projected Lagrangian method [261]. SQP is one of the most powerful methods in the framework of nonlinear constrained optimization. In the following, the mathematical basics of the optimization problem and the SQP algorithm are described to provide a general overview about the applied methodology in the present context.

In general, a parametric multivariate nonlinear optimization problem with nonlinear constraints is mathematically described by:

$$\begin{aligned} &\min_{\mathbf{x}\in\mathbb{R}^n} f(\mathbf{x},\boldsymbol{p}) \\ \text{such that}\quad &\mathbf{h}(\mathbf{x},\boldsymbol{p}) = \mathbf{0} \\ &\mathbf{c}(\mathbf{x},\boldsymbol{p}) \leq \mathbf{0}. \end{aligned} \tag{3.39}$$

In the present case, the objective function $f(\mathbf{x},\boldsymbol{p})$ is given by (3.29) and the constrained function vector $\mathbf{c}(\mathbf{x},\boldsymbol{p})$ of length m contains the inequality constraints $\mathbf{c}_l(\mathbf{x})$ from (3.31) and $c_{nl}(\mathbf{x})$ from (3.34). To provide a general overview about the method, the (possibly nonlinear) equality constraints, concentrated in the vector $\mathbf{h}(\mathbf{x},\boldsymbol{p})$ of length w, are included in the description.

In constrained optimization, the Lagrangian $\mathscr{L}$ plays a central role. It combines the objective function with the given constraints by means of Lagrangian multipliers summarized in the vectors $\mathbf{u}$ and $\mathbf{v}$:

$$\mathscr{L}(\mathbf{x},\boldsymbol{p},\mathbf{u},\mathbf{v}) = f(\mathbf{x},\boldsymbol{p}) + \mathbf{u}^{\mathrm{T}}\mathbf{h}(\mathbf{x},\boldsymbol{p}) + \mathbf{v}^{\mathrm{T}}\mathbf{c}(\mathbf{x},\boldsymbol{p}). \tag{3.40}$$

For every constraint, there exists one Lagrangian multiplier. It can be proven that the local optimum $\tilde{\mathbf{x}}$ of the objective function with respect to the given constraints is a stationary point of the Lagrangian with the corresponding optimal multipliers $\tilde{\mathbf{u}}$ and $\tilde{\mathbf{v}}$, such that the gradient of the objective function is a linear combination of the gradients of the constraints [261]. The name of the SQP algorithm originates from the approach to solve the nonlinear problem from (3.39). This is done sequentially by approximating the Lagrangian using a quadratic function around the point $\mathbf{x}^{(k)}$, where the superscript (k) is the iteration counter. The constraint functions are linearized so that the quadratic subproblem of (3.39) is given by a Taylor expansion of the Lagrangian from (3.40). The problem is reformulated to determine the optimal search direction $\mathbf{d}^{(k)}$ from the point $\mathbf{x}^{(k)}$, such that the constraints are fulfilled. The vector ∇f is the Jacobian, i.e. the gradient of the objective function with respect to $\mathbf{x}$. In the SQP algorithm, the quadratic subproblem of (3.40) is solved iteratively. In every iteration, two additional subproblems have to be solved. First, the search direction $\mathbf{d}^{(k)}$ and the Lagrangian multipliers $\mathbf{u}^{(k+1)}$ and $\mathbf{v}^{(k+1)}$ are evaluated. Then, the step size α is determined to provide a certain decrease in the objective function while still satisfying the given constraints. The next iteration is performed at the updated point $\mathbf{x}^{(k+1)} = \mathbf{x}^{(k)} + \alpha\mathbf{d}^{(k)}$. In order to determine the first step, the following linear system of equations is solved for $\mathbf{d}^{(k)}$, $\mathbf{u}^{(k+1)}$, and $\mathbf{v}^{(k+1)}$:

$$\begin{bmatrix} [\mathbf{L}^{(k)}] & \nabla\mathbf{h}^{(k)} & \nabla\mathbf{c}^{(k)} \\ \nabla\mathbf{h}^{(k)\mathrm{T}} & 0 & 0 \\ \nabla\mathbf{c}^{(k)\mathrm{T}} & 0 & 0 \end{bmatrix} \begin{bmatrix} \mathbf{d}^{(k)} \\ \mathbf{u}^{(k+1)} \\ \mathbf{v}^{(k+1)} \end{bmatrix} = \begin{bmatrix} -\nabla f^{(k)} \\ -\mathbf{h}^{(k)} \\ -\mathbf{c}^{(k)} \end{bmatrix}. \tag{3.41}$$

The symmetric matrix $[\mathbf{L}^{(k)}] = \nabla^2\mathscr{L}$ is an approximation of the second-order partial derivatives of the Lagrangian from (3.40), which is also called the Hessian of the Lagrangian. Since the Hessian is not given analytically, the matrix $[\mathbf{L}^{(k)}]$ is successively approximated rather than explicitly calculated. In the present case, a very robust and efficient approximation, the Broyden-Fletcher-Goldfarb-Shanno (BFGS) update formula, derived by Broyden [266,267], Fletcher [268], Goldfarb [269], and Shanno [270], is used. The first row of the system matrix in (3.41) corresponds to the partial derivative of the quadratic approximation of the Lagrangian from (3.40) with respect to $\mathbf{d}^{(k)}$. The remaining rows correspond to the linearly approximated constraint functions.

The step size α is determined by means of a merit function. This function is incorporated into an SQP algorithm for the purpose of robustness. In constrained optimization, a merit function balances the drive to decrease the objective function while satisfying the defined constraints, and measures the progress of convergence toward $\tilde{\mathbf{x}}$ as a function of α. The solution of the quadratic subproblem has a unit step size $\alpha = 1$. If the constraints are not violated, this step size is taken and a new iteration is started. However, if the constraints are violated, the step-length is reduced to the nearest constraint. For the proposed approach, the merit function developed by Han [264] and Powell [265] is employed. The procedure is repeated until the Karush–Kuhn–Tucker conditions [271] are satisfied up to a certain limit of defined tolerances. The SQP converges to a local optimum and returns the corresponding solution. Further literature about nonlinear constrained optimization can be found in, for example, [261,272,273]. SQP theory is covered in detail by Han [264,274] and Powell [265,275,276], to note some of the first but still frequently applied concepts of this method. A more general overview about SQP is given by Boggs and Tolle [277] and by Gill and Wong [278].

The performance of the proposed approach strongly depends on the implementation due to nontrivial technical and algorithmic issues. For that reason, it is highly recommended to use professional and well-tested software. Some of the first successful implementations are reported by Schittkowski [279–281]. These references also cover the proof of convergence of the SQP. More advanced but commercial code is available, for example by the MATLAB® function *fmincon* [282] or FORTRAN-based routines NPSOL [283] and SNOPT [284,285] from Gill *et al.* Partially free software with the restriction to apply the routines in an academic framework is available

(NLPQLP) [286]. This software has been tested on 306 optimization problems with a reported success rate of better than 90%, which underlines the efforts for implementation of SQP-based algorithms [286]. In this work, the MATLAB® implementation *fmincon* is used to couple the FEM Solver of COMSOL Multiphysics® and the SQP algorithm by means of the MATLAB Livelink™ [151]. The implementation includes the BFGS-update formula as well as an adaptive step size using the merit function discussed earlier. In the subsequent sections, the methodology to evaluate the objective and nonlinear constraint function with low computational cost is described.

3.2.2.9 Objective and constraint function evaluation

To apply the SQP algorithm, a time-efficient approach is essential to evaluate both the objective function $f(\mathbf{x},\boldsymbol{p})$ and the nonlinear constraint function $c_{\text{nl}}(\mathbf{x},\boldsymbol{p})$. The governing equations are simplified such that secondary magnetic fields are neglected. Hence, it is possible to analyze the problem in the stationary case, which takes into account the symmetry of the field and force profiles. This significantly reduces the computational cost and offers the possibility for efficient numerical analysis. The FEM in combination with eWRA, described in Section 2.4.3.4, is applied as the numerical method. The nonlinearity of the ferromagnetic material significantly influences the profile and magnitude of the magnetic field. Consequently, the resulting Lorentz force is also affected. This necessitates the use of nonlinear models. Linear models are too inaccurate and falsify the optimization results, especially if the drag-force must not exceed $F_x^{(\max)}$. In the case of HCp-magnet systems, the iron–cobalt alloy VACOFLUX® 50, together with the corresponding $B(H)$ curve from Figure 3.5, is used. To minimize computational cost, the field problem is subdivided into three successive steps, illustrated in Figure 3.7.

3.2.2.9.1 Step 1: Primary magnetic flux density (2D)

In the first step, the primary magnetic flux density $\mathbf{B}^{(p)}$ of the PM is calculated with a scalar magnetic potential formulation ψ, neglecting any moving conductor. Thus, the primary magnetic field is given by (2.203). Given the axisymmetry of the magnet geometry, the magnetic field is determined in 2D employing a cylindrical coordinate system, which significantly accelerates the solving process of the nonlinear problem.

3.2.2.9.2 Step 2: Induced eddy currents in the conductor free of defects (3D)

In the second step, only the conductor in motion is considered. It is modeled as a large cylindrical domain with finite radius $10R_2$ proportional to the outer radius of the magnet system, assuming that the edges are far enough away to prevent any parasitic disturbance of the induced eddy currents. The height H_s of the moving conductor is defined according to the following conditional relation:

$$H_s = \begin{cases} 10R_2, & \text{if } 10R_2 \geq 2d + a \\ 10R_2 + 2d + a, & \text{if } 10R_2 < 2d + a. \end{cases} \tag{3.42}$$

This implies that the size of the conducting domain is adjusted according to the magnet system, while ensuring geometrical models with defects deeper than $10R_2$. Thus, ensuring that the conducting domain is sufficiently large for all sets of $\boldsymbol{p}$ and $\mathbf{x}$.

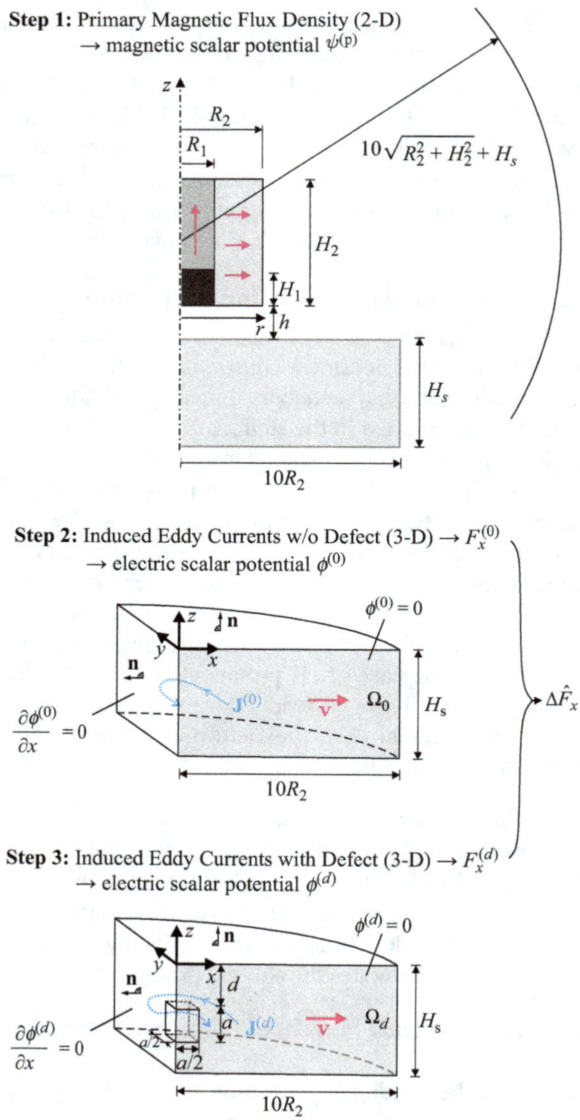

Figure 3.7 Procedure to compute the nonlinear objective and constraint function [245]

The primary magnetic field $\mathbf{B}^{(p)}$ is imported from the first step and transformed from cylindrical coordinates into the three-dimensional Cartesian coordinate system. The induced eddy currents inside the conductor in motion are calculated by (2.207) and (2.208) using a scalar electric potential formulation φ. The 3D model contains symmetry with respect to the xz- and yz-plane, when secondary magnetic fields are neglected. This reduces the modeled geometry of the specimen to only one quarter. On the xz-plane, the tangential components of the current density vanish:

$$\mathbf{J}_{\mathrm{t}} = \mathbf{J} \times \mathbf{n} = [\sigma]\left(-\nabla\varphi + \mathbf{v} \times \mathbf{B}^{(p)}\right) \times \mathbf{n} = \mathbf{0}. \tag{3.43}$$

As a result of the axisymmetry of the magnet system, the magnetic field is zero in the direction of the boundary normal at this plane ($B_y^{(p)} = 0$). Considering the linear motion of the conductor in the x-direction $\mathbf{v} = [v_x, 0, 0]^{\mathrm{T}}$ leads to $\mathbf{v} \times \mathbf{B}^{(p)} = -v_x B_z^{(p)} \mathbf{e}_y$. Substituting this into (3.43) shows that $\frac{\partial\varphi}{\partial x} = 0$ and $\frac{\partial\varphi}{\partial z} = 0$, which can be achieved by defining the Dirichlet boundary condition $\varphi = \text{const.}$ at the xz-plane.

In contrast, the normal component of the induced current density J_{n} vanishes at the yz-plane, such that:

$$J_{\mathrm{n}} = \mathbf{J} \cdot \mathbf{n} = [\sigma]\left(-\nabla\varphi + \mathbf{v} \times \mathbf{B}^{(p)}\right) \cdot \mathbf{n} = 0. \tag{3.44}$$

On this plane, the magnetic field is zero in the x-direction $B_x^{(p)} = 0$, so:

$$\mathbf{v} \times \mathbf{B}^{(p)} = -v_x B_z^{(p)} \mathbf{e}_y + v_x B_y^{(p)} \mathbf{e}_z. \tag{3.45}$$

Since this expression has no component normal to the boundary, a Neumann boundary condition for the electric scalar potential $\partial\varphi/\partial x = 0$ must be defined. All other boundaries of the moving conductor share the boundary condition $\mathbf{J} \cdot \mathbf{n} = 0$ to prevent any current leaving the conducting domain.

3.2.2.9.3 Step 3: Induced eddy currents in the conductor with defect (3D)

To determine the ADRA, a third simulation is performed. The primary magnetic field $\mathbf{B}^{(p)}$ from the first step is incorporated in the same way as in the previous step. However, in this step, the conductor contains an equivalent cuboidal defect with volume V_d located at a depth d. The governing equations from the previous step remain valid and in consequence, the boundary condition $\mathbf{J} \cdot \mathbf{n} = 0$ now also applies for the defect boundaries, preventing any current flow into the defect region. To further decrease the computational cost, the same 3D FEM mesh can be used in Steps 2 and 3. As a result, the 3D mesh needs only to be built once and then transferred between Steps 2 and 3 as necessary. As a positive side effect, using the same mesh decreases numerical noise when computing the ADRA. Finally, the resulting Lorentz force is calculated by spatial integration over the conductor:

$$\begin{aligned} F_x^{(i)} &= -4 \int_{\Omega_i} \mathbf{J}^{(i)} \times \mathbf{B}^{(p)} \, \mathrm{d}\Omega \\ &= -4 \int_{\Omega_i} \left[[\sigma]\left(-\nabla\varphi^{(i)} + \mathbf{v} \times \mathbf{B}^{(p)}\right)\right] \times \mathbf{B}^{(p)} \, \mathrm{d}\Omega, \end{aligned} \tag{3.46}$$

where the superscript $i \in \{0, d\}$ indicates the quantities obtained in the defect-free and defective case, respectively.

The computation of one objective function evaluation takes 20–25 s on a common desktop PC (i7-3770, 4 GHz). This includes building the geometry, meshing, assembling the systems of equations, solving, and postprocessing of all three steps. The memory consumption is moderate and does not exceed 1 GB since geometrical symmetry is taken into account in all individual steps.

3.2.3 Optimization results and discussion

The optimization results of an HCp-system are presented using VACODYM® 745HR [257] as PM material and the iron–cobalt alloy VACOFLUX® 50 [256] as ferromagnetic material. The scaling factor S is chosen such that it corresponds to the currently available laboratory setup. The specimen is constructed from stacked aluminum sheets ($\mathbf{a}_\sigma = [1, 1, 0]^\mathrm{T}$) with electrical conductivity $\sigma = 30.66$ MS/m, moving with velocity $v = 0.5$ m/s. The lift-off distance of $h = 1$ mm is assumed, together with a magnetic remanence $B_r = 1.44$ T (VACODYM® 745HR). This results in a scaling factor $S^{(\mathrm{lab})} = 31.79$. In the experimental setup, a force sensor K3D40 [104] based on the strain gauge technology is employed. According to the manufacturer, the applied load is limited to $F_x^{(\max)} = 3$ N.

Two particular setups of system parameters are investigated. The first represented the case of medium-sized defects located deep inside the specimen. The corresponding system parameters are $V_m/V_d = 56$, $d/h = 10$, and $a/h = 5$. The optimization results are shown in Figure 3.8. The insets in the upper left corner of each figure show cross sections of the optimal magnet systems colored according to Figure 3.3. The ADRA is shown in Figure 3.8(a) over the space of design variables $H_2/R_2 = (0 \ldots 2.5]$, $R_1/R_2 = (0 \ldots 1]$, and $H_1/H_2 = [0 \ldots 1]$. The planes intersect in the global unconstrained optimum at $\tilde{\mathbf{x}} = [H_2/R_2, R_1/R_2, H_1/H_2] = [0.5, 0.6, 0]$, with a denormalized ADRA of $\Delta\tilde{F}_x^{(\mathrm{lab})} = 21.9$ mN. The continuous objective function does not contain local optima.

The unperturbed drag-force is shown in Figure 3.8(b). The ADRA-optimal magnet generates an unperturbed drag-force of $\tilde{F}_x^{(0,\mathrm{lab})} = 5.2$ N, which is too high for the experimental force sensor. To fulfill the constraint, the VcFc approach is applied. To illustrate the impact of nonlinear constraints, Figure 3.8(c) and (d) shows the equi-force surfaces for different values of $F_x^{(\max)}$ over the same space of design variables. Figure 3.8(c) shows the surface for $F_x^{(\max)} = \tilde{F}_x^{(0,\mathrm{lab})} = 3$ N. This plane is colored according to the value of the objective function (ADRA). The points inside the equi-force surface correspond to solutions which violate the predefined nonlinear constraint ($c_{\mathrm{nl}} > 0$) and hence do not belong to the feasible set of solutions. Since the global optimum is not part of the feasible set, the constrained optimum is located at the constraint boundary. The SQP algorithm converged to the constrained optimal solution, which describes the magnet system as a standard cylinder $\tilde{\mathbf{x}}_c = [1.6, 1, 0]$. The solution is located at the boundary of the design space and the linear inequality constraints (bounds) are active ($H_1/H_2 = 0$). Thus, the number of free design variables is reduced from three to two at the optimal solution. Compared with the unconstrained optimum, the ADRA decreased to $\Delta\tilde{F}_x^{(\mathrm{lab})} = 15.6$ mN. Considering the ADRA projected on the constraint hyperplane, the proposed problem is nonconvex. Consequently, if initial values $\mathbf{x}^{(0)}$ are chosen such that $H_2/R_2 < 0.5$ and $R_1/R_2 < 0.4$, the derivative-based SQP algorithm converges to a local optimum located in the region of $H_2/R_2 \approx 0.2$ and $R_1/R_2 \approx 0.1$. To avoid local convergence, the use of a multistart approach using three to five different starting points is recommended. In the multistart approach, the first starting point is defined by the user, whereas the following are chosen randomly.

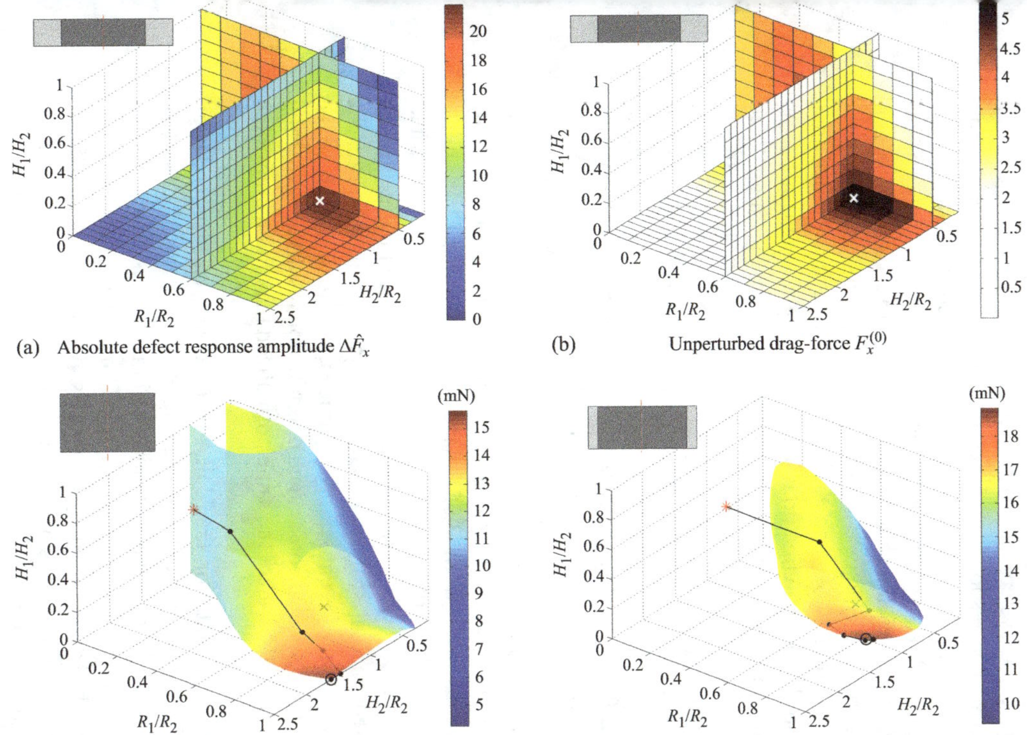

(a) Absolute defect response amplitude $\Delta\hat{F}_x$

(b) Unperturbed drag-force $F_x^{(0)}$

(c) $F_x^{(\max)} = 3$ N: $\tilde{x}_c = [1.6, 1.0]$, $\Delta\tilde{F}_x^{(\mathrm{lab})} = 15.6$ mN

(d) $F_x^{(\max)} = 4$ N: $\tilde{x}_c = [1, 0.8, 0]$, $\Delta\tilde{F}_x^{(\mathrm{lab})} = 18.8$ mN

Figure 3.8 *HCp-magnet system made of VACODYM® 745HR and VACOFLUX® 50 for the case of system parameters $V_m/V_d = 56$, $d/h = 10$, and $a/h = 5$ (anisotropic specimen $\boldsymbol{a}_\sigma = [1,1,0]^T$). The scaling factor of the laboratory setup is $S^{(lab)} = 31.79$. Crosses and circles indicate the unconstrained and constrained optima, $\tilde{\mathbf{x}}$, and $\tilde{\mathbf{x}}_c$, respectively. The insets are sketches of the individual optimized magnet systems. The data are shown as a function of the design variables $\mathbf{x} = [H_2/R_2, R_1/R_2, H_1/H_2]$. (a) Cutplanes of the denormalized ADRA $\Delta F_x(\mathbf{x},\boldsymbol{p})$ intersecting at the unconstrained global optimum $\tilde{\mathbf{x}} = [0.5, 0.6, 0]$. (b) Cutplanes of the corresponding denormalized unperturbed drag-force $F_x^{(0)}(\mathbf{x},\boldsymbol{p})$. (c) and (d) Semitransparent isosurfaces of maximum drag-forces $F_x^{(\max)} = \{3\,N, 4\,N\}$ colored by the according ADRA. The solid lines indicate the optimization paths for the initial starting point, $\mathbf{x}^{(0)} = [1.6, 0.3, 0.8]$ (red star) [245]*

The constraint hyperplane for $F_x^{(\max)} = 4$ N is shown in Figure 3.8(d). As expected, the enclosed domain shrinks and the feasible set increases. The constrained optimal solution changes to $\tilde{\mathbf{x}}_c = [1, 0.8, 0]$ and the ADRA increases to $\Delta\tilde{F}_x^{(\mathrm{lab})} = 18.8$ mN. The optimization path of the initial starting point is shown in Figures 3.8(c) and (d) with a solid line. Each dot corresponds to one iteration taken by the SQP algorithm.

In a second example, the system parameters are changed to $V_m/V_d = 875$, $d/h = 2$, and $a/h = 2$, which corresponds to the case of small defects located close to the surface of the specimen. The associated results are presented in a similar way as in the former case in Figure 3.9. The unconstrained optimum, shown in Figure 3.9(a), is relocated compared with the previous case because of the variation in the system parameters. A Halbach-structure emerges, which includes ferromagnetic material ($\tilde{\mathbf{x}} = [0.8, 0.2, 0.5]$). Considering the behavior of the ADRA and the unperturbed drag-force in Figure 3.9(a) and (b), the maxima do not correlate. Thus, magnet systems, which generate high unperturbed drag-forces, do not inherently produce high defect responses. This particular magnet system generates an ADRA of $\Delta\tilde{F}_x^{(\mathrm{lab})} = 36$ mN, while generating an unperturbed drag-force of $\tilde{F}_x^{(0,\mathrm{lab})} = 3.6$ N. Since the constraint of maximum drag-force is again not fulfilled, the VcFc approach must be applied. The constraint hyperplane of $F_x^{(\max)} = 3$ N is shown in Figure 3.9(c). The constraint optimum is $\tilde{\mathbf{x}}_c = [1.1, 0.2, 0.6]$ and is located close to the unconstrained solution. The modified magnet has an ADRA of $\Delta\tilde{F}_x^{(\mathrm{lab})} = 34.5$ mN and satisfies the 3 N constraint. Assuming a maximum drag-force $F_x^{(\max)} = 4$ N, the unconstrained global optimum is located in the feasible set of solutions. Consequently, the nonlinear constraint is inactive and the SQP algorithm converges to the global optimum as shown in Figure 3.9(d).

Both examples show that if $F_x^{(c)}$ from (3.35) changes by modifying $F_x^{(\max)}$ or the scaling factor S, the constraint hyperplane defined by c_{nl} from (3.36) grows or shrinks accordingly. In a similar sense, the constraint function is influenced by the system parameters $\boldsymbol{p}$. As a consequence of scaling, the unconstrained optimization results can be adopted to setups with a different scaling parameter, provided they share the same system parameters. However, the constrained optimization results are generally valid as long as the different setups share the same normalized maximum drag-force from (3.35).

The influence of the V_m/V_d ratio on the ADRA and the optimal magnet design is investigated further. The study is performed assuming the same two sets of system parameters defined previously. To compare the individual magnet systems with each other, the investigation is performed for HCp-, HC-, and C-magnets separately. Since the HCp-configuration covers HC- and C-magnet systems, the corresponding ADRA must be equal or larger than the other cases, which are geometrically restricted beforehand. The optimizations are performed as a function of V_m/V_d in the unconstrained case as well the VaFc and VcFc approaches.

The results for deep defects ($d/h = 10$, $a/h = 5$) are shown in Figure 3.10(a). Each point on the curves represents an optimal magnet system. The dashed lines indicate the ADRA of the unconstrained optimal solutions (uc). In this parameter range, the ADRA increases almost linearly as a function of the V_m/V_d ratio. The critical V_m/V_d ratios are determined for C-, HC-, and HCp-magnet systems using

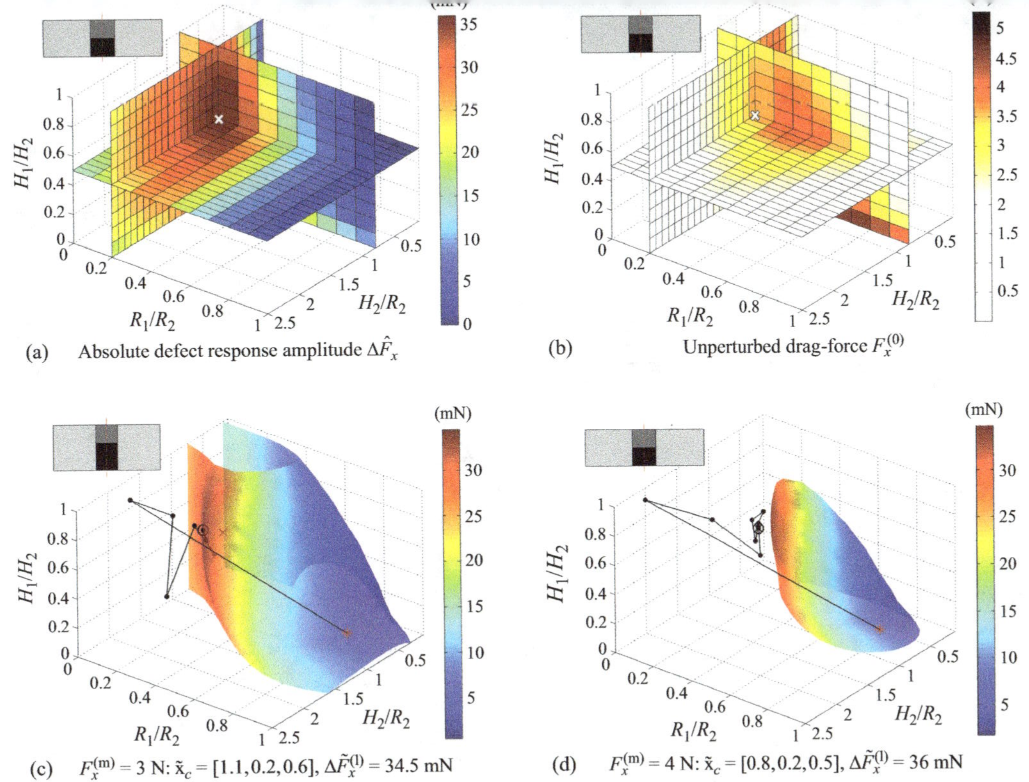

(a) Absolute defect response amplitude $\Delta\hat{F}_x$

(b) Unperturbed drag-force $F_x^{(0)}$

(c) $F_x^{(m)} = 3$ N: $\tilde{x}_c = [1.1, 0.2, 0.6]$, $\Delta\tilde{F}_x^{(l)} = 34.5$ mN

(d) $F_x^{(m)} = 4$ N: $\tilde{x}_c = [0.8, 0.2, 0.5]$, $\Delta\tilde{F}_x^{(l)} = 36$ mN

Figure 3.9 *HCp-magnet system made of VACODYM® 745HR and VACOFLUX® 50 for the case of system parameters $V_m/V_d = 875$, $d/h = 2$, and $a/h = 2$ (anisotropic specimen $\boldsymbol{a}_\sigma = [1,1,0]^T$). The scaling factor of the laboratory setup is $S^{(lab)} = 31.79$. Crosses and circles indicate the unconstrained and constrained optima, $\tilde{\mathbf{x}}$, and $\tilde{\mathbf{x}}_c$, respectively. The insets are sketches of the individual optimized magnet systems. The data are shown as a function of the design variables $\boldsymbol{x} = [H_2/R_2, R_1/R_2, H_1/H_2]$. (a) Cutplanes of the denormalized ADRA $\Delta F_x(\boldsymbol{x}, \boldsymbol{p})$ intersecting at the unconstrained global optimum $\tilde{\mathbf{x}} = [0.8, 0.2, 0.5]$. (b) Cutplanes of the corresponding denormalized unperturbed drag-force $F_x^{(0)}(\boldsymbol{x}, \boldsymbol{p})$. (c) and (d) Semitransparent isosurfaces of maximum drag-forces $F_x^{(m)} = \{3\,N, 4\,N\}$ colored by the according ADRA. The solid lines indicate the optimization paths for the initial starting point $\boldsymbol{x}^{(0)} = [0.9, 1, 0.2]$ (red star) [245] [$(m) \equiv (\max)$, $(l) \equiv (lab)$]*

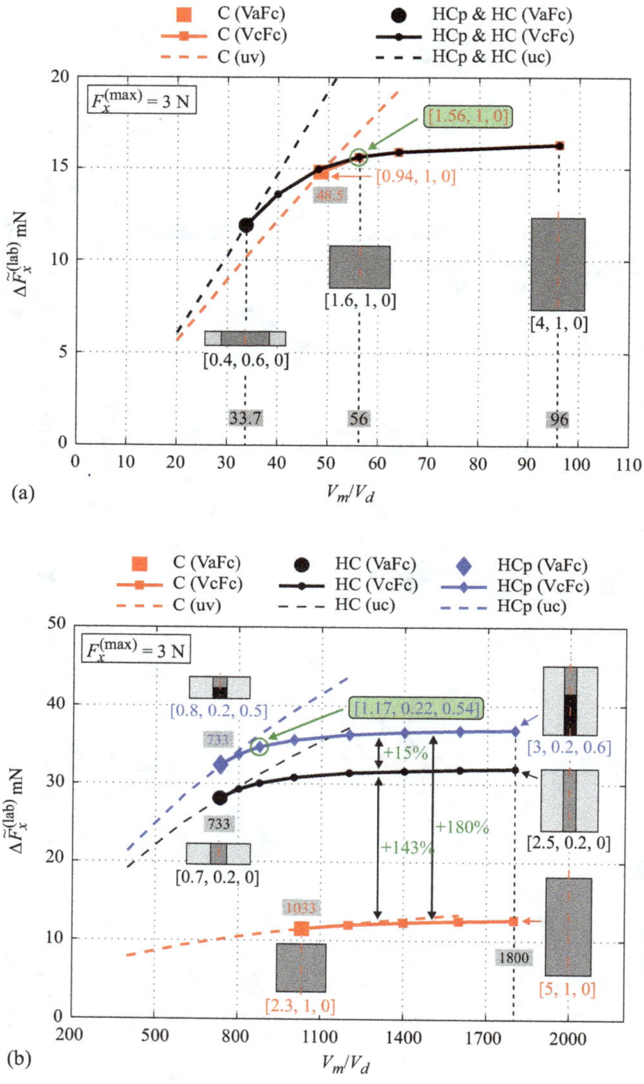

Figure 3.10 Comparison of the denormalized ADRA between C-, HC- and HCp-magnet systems as a function of the system parameter V_m/V_d in the case of anisotropic specimens, $\boldsymbol{a}_\sigma = [1,1,0]^T$, for two different sets of system parameters. The optimizations are performed in the unconstrained case (dashed lines) as well using the VaFc (single marker) and VcFc (solid lines) approaches. A maximum drag-force of $F_x^{(\max)} = 3$ N is considered. The numbers in brackets correspond to the optimal design variables $\tilde{\mathbf{x}}_c = [H_2/R_2,\ R_1/R_2,\ H_1/H_2]$ [245]. (a) Medium-sized deep defects ($d/h = 10$, $a/h = 5$). (b) Small subsurface defects ($d/h = 2$, $a/h = 2$)

the VaFc approach, and are shown with a single marker each. The magnet system is defined by the unconstrained global optimum but still satisfies the constraint. The magnet designs which obey these parameters are lying on the constraint hyperplane $c_{\rm nl}$ intersecting with the global optimum $\tilde{\mathbf{x}}$. Hence, the critical V_m/V_d ratio is where the constraint becomes active and the curve of the constrained solution diverges from unconstrained solution. However, magnet systems with higher V_m/V_d ratios have to be restricted by applying the VcFc approach such that all presented configurations indicated with solid lines fulfill the defined force constraint $F_x^{(\max)} = 3$ N. The insets show cross sections of particular magnet systems together with the corresponding optimal design variables. As expected, optimal magnet systems do not include ferromagnetic material ($H_1/H_2 = 0$) if the defect is located deep within the specimen, which was also shown in Figure 3.8. Thus, the curves of HCp- and HC-magnet systems are the same. The critical volume ratio obtained by the VaFc approach is 33.7 and 48.5 for HC- and C-magnets, respectively. Hence, VaFc-optimal HC-magnets are smaller but have a lower ADRA with 11.9 mN compared with 14.8 mN for cylindrical magnets. Increasing the V_m/V_d ratio beyond the critical point and applying the VcFc, saturates the ADRA while keeping the maximum drag-force at the defined limit. The further gain in ADRA results in consequential increased magnet volume and weight. For this set of system parameters, that the optimal design variables of the HC-magnet system converge to a C-magnet, which was observed in context of Figure 3.8(c). Hence, a further increase in the magnet volume ratio beyond $V_m/V_d > 56$ leads to an overlap of the VcFc-optimal ADRA curves (see solid lines in Figure 3.10(a)). Meanwhile, the optimal H_2/R_2 ratio increases with V_m/V_d in a specific way such that the magnet gets higher to fulfill the given constraint. Considering the present case for practical reasons, it can be concluded that an increase beyond $V_m/V_d \approx 55\ldots60$ does not lead to further significant gain in ADRA. Regular cylinders are favorable in the case of deep defects if the unperturbed drag-force is limited. However, Halbach-structures generate higher ADRAs compared to standard cylinders when comparing both at a given V_m/V_d ratio omitting any constraint (see dashed lines in Figure 3.10(a)).

Figure 3.10(b) shows the second set of system parameters which covers the case of small defects located close to the surface of the specimen ($d/h = 2$, $a/h = 2$). The optimal magnet designs are presented in a similar way as in the former case. There are distinct performance differences between C-, HC-, and HCp-magnet systems. Cylindrical magnets show a critical volume ratio of 1033 together with an ADRA of 11.5 mN. In contrast, HC- and HCp-configurations are approximately 30% smaller and generate defect responses of 28.1 and 32.4 mN which correspond to a gain of approximately 140% and 180%, respectively. Considering the VcFc solutions, the ADRA increases slightly further by increasing the V_m/V_d ratio beyond the critical point. However, this effect is comparatively smaller than in the case of deep defects (see Figure 3.10(a)). Comparing HC- and HCp-systems, the presence of ferromagnetic material increases the ADRA by about 15%. Regarding the unconstrained solutions, the ADRA is saturating slowly when increasing the V_m/V_d ratio. This has also been observed but not explicitly shown here for high V_m/V_d ratios of the case from Figure 3.10(a).

In the following, two distinct magnet systems of both scenarios are compared considering the same magnet volume V_m. For medium-sized deep defects (green marker in Figure 3.10(a)), the optimal C-magnet with a volume ratio of $V_m/V_d = 56$ is chosen. The corresponding HCp-magnet system optimized for small subsurface defects (green marker in Figure 3.10(b)) has a volume ratio $V_m/V_d = 875$. The geometrical parameters are obtained by denormalizing both systems assuming a lift-off distance and hence geometric scale of $h = 1$ mm. The spatial distributions of the magnitude of the magnetic flux density **B** and the induced eddy current density **J** are shown in Figure 3.11. The eddy currents for regular C-magnets (Figure 3.11(a)) are less concentrated than HCp-systems (Figure 3.11(b)). The Halbach-structure leads to a considerably more focused magnetic flux and eddy current distribution under the inner part of the magnet system. The flux density is increased to 1.6 T on the surface of the specimen, which is significantly larger than standard magnet systems.

All investigations are also performed assuming a specimen with an isotropic conductivity profile ($\mathbf{a}_\sigma = [1,1,1]^{\mathrm{T}}$). In the range of system parameters considered, the optimal design variables differed only by approximately ±5% compared with the anisotropic cases. For deep defects, the ADRA decreased by 23% compared with the anisotropic case. However, for $d/h \leq 2$, the ADRA only decreased by approximately 15%, which can be described by the circumstance that the isotropic profile becomes gradually anisotropic from the top side if the defect gets closer to the surface of the specimen. The nonlinear constraint function is not affected when comparing the two cases of $\mathbf{a}_\sigma = [1,1,1]^{\mathrm{T}}$ and $\mathbf{a}_\sigma = [1,1,0]^{\mathrm{T}}$. This is due to the unperturbed drag-force $F_x^{(0)}$ being the same for both conductivity profiles, since the unperturbed eddy currents only flow in the xy-plane. Hence, anisotropy in the z-direction does not influence $F_x^{(0)}$, and so the nonlinear constraint function c_{nl} is unaltered.

3.2.4 Prototypes of optimized LET magnet systems

Prototypes of the proposed optimized magnet systems from Figure 3.11 are designed and manufactured. The optimal C-magnet is a custom design ordered from the company HKCM engineering e.K. (www.hkcm.de). As assumed during the optimization, it is made of NdFeB with a material grade of N52 with a nominal remanence of 1.43 T. The diameter D and height H are $[D,H] = [22.5\text{ mm}, 17.6\text{ mm}]$.

The realization of the HCp-magnet system is more intricate. The radially magnetized outer part of the magnet system is constructed by diametrically magnetized segments. The influence of the segmented structure on the ADRA, compared with the ideal case with a continuous radial magnetization, is shown in Figure 3.12. As expected, the ADRA converges with an increasing number of segments. For the prototype of the HCp-magnet system, a structure made of 12 segments is chosen, which results in an ADRA of 98% compared with the ideal case. The final geometry and the manufactured prototype are shown in Figure 3.13. It is made of the hard magnetic material VACODYM® 745HR and the iron-cobalt alloy VACOFLUX® 50 as it is assumed during the optimization. The prototype is manufactured in collaboration with the company Vacuumschmelze Hanau GmbH & Co. KG (www.vacuumschmelze.de). An experimental validation of the proposed magnet

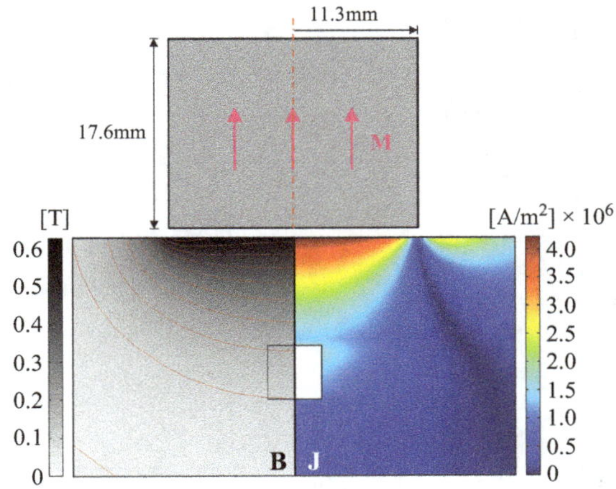

(a) Deep defect: $V_m/V_d = 56$, $d/h = 10$, $a/h = 5$.

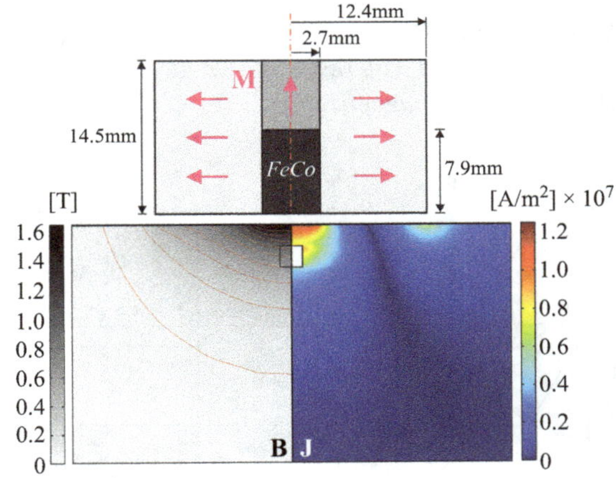

(b) Subsurface defect: $V_m/V_d = 875$, $d/h = 2$, $a/h = 2$.

Figure 3.11 Magnitude of the magnetic flux density **B** *and induced eddy current density* **J** *of VcFc optimized magnet systems for the case of anisotropic specimens. The magnet systems are denormalized assuming a lift-off distance of h = 1 mm. Both generate an unperturbed drag-force of* $F_x^{(lab)} = 3\,N$ *considering a scaling factor of* $S^{0,(lab)} = 31.79$*. Cross section of (a) a cylindrical magnet with* $\tilde{\mathbf{x}}_c = [1.56, 1, 0]$ *and (b) a Halbach-cylinder with iron-cobalt* $\tilde{\mathbf{x}}_c = [1.17, 0.22, 0.54]$ *[245]*

systems considering the investigated defect scenarios is presented in [103]. Moreover, both the optimized C- and HCp-magnet are used to detect defects in glass laminate aluminum reinforced epoxy [156]. The presented results demonstrate the expected performance of the proposed designs. In the next section, the depth-optimized C-magnet is used for the detection of deep-lying slits to point out the current state of the art and further perspectives in LET.

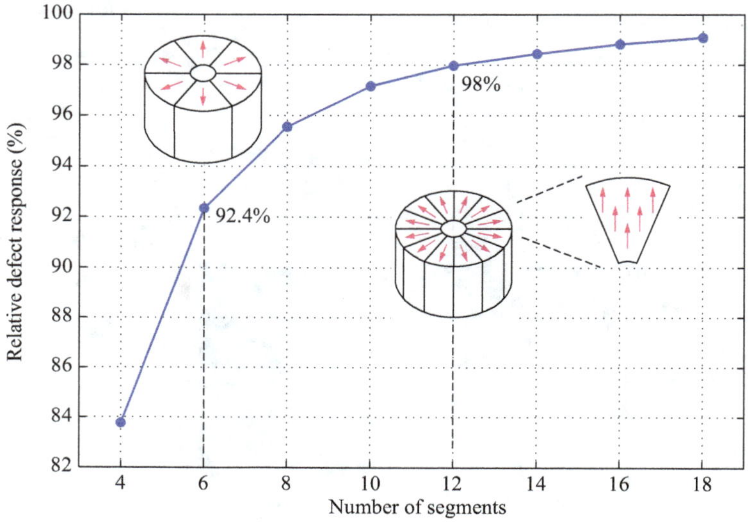

Figure 3.12 Relative defect response of the optimized segmented HCp-magnet. The ADRA is calculated with respect to an ideal HCp-magnet assuming a continuous radial magnetization

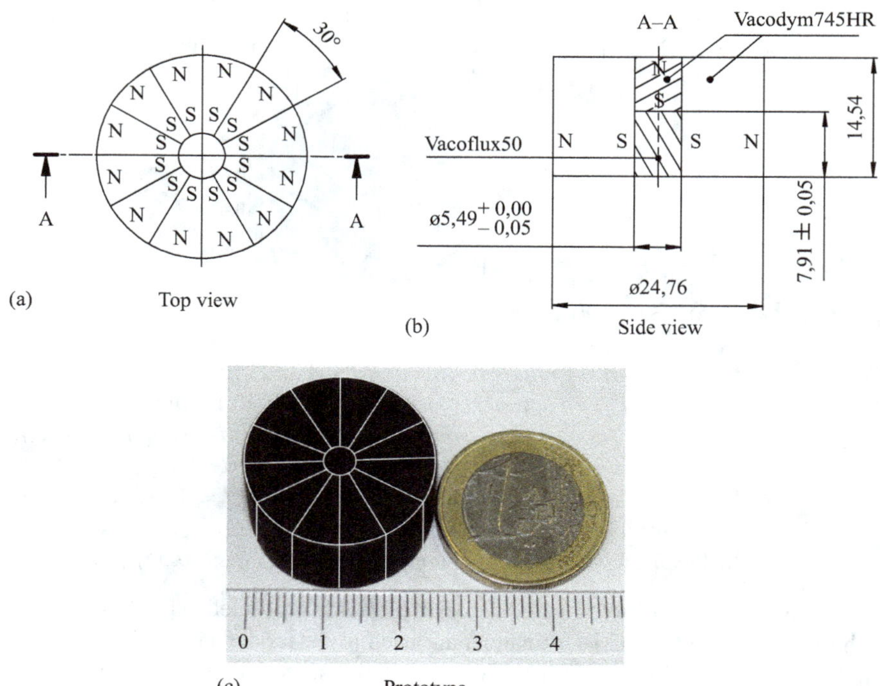

Figure 3.13 Geometry of the HCp-prototype manufactured by the company Vacuumschmelze Hanau GmbH & Co. KG (www.vacuumschmelze.de). (a) Top view. (b) Side view. (c) Prototype

3.2.5 Defect depth study

Previous ECT studies often assumed quasi-infinite cracks to evaluate the detection limit (see Table 1.1). In this case, the defect is a slit obeying a pronounced length compared with the sensor system. In order to provide comparability to the results reported in the literature, the following benchmark problem is defined [287]. The problem geometry is inspired by the study from Mook *et al.* [42] and is shown in Figure 3.14. The specimen consists of a solid block of size [250 mm, 50 mm, 24 mm] made of aluminum, which contains a slit of size $[X_d, Y_d, Z_d] = [75\,\text{mm}, 1.5\,\text{mm}, 24\,\text{mm}]$. The artificial crack is oriented in parallel to the direction of motion. On top of this structure, a variable number of aluminum sheets is situated. Each sheet has a thickness of 2 mm. The defect depth d is varied from 0 to 36 mm using 18 sheets. The sheets which are not on top of the specimen are situated on the bottom in order to ensure a constant height of the total assembly and to not alter the outer dimensions of the problem. By doing so, the magnetic Reynolds number is also kept constant. The overall dimensions of the specimen are then $[X_s, Y_s, Z_s] = [250\,\text{mm}, 50\,\text{mm}, 60\,\text{mm}]$. During the measurements, the magnet is located on the centerline with respect to the y-axis in a height of $h = 1$ mm. All measurements are performed at a velocity of $v = 0.5$ m/s. The picture from Figure 1.7 shows the particular laboratory setup with 18 aluminum sheets on top of the slotted bar.

The measurement data are postprocessed with a 10th order Butterworth low-pass filter having a cutoff frequency of 100 Hz. The force profiles are normalized with respect to the stationary values which occur far away from the defect. The results of the normalized drag- and lift-force over the whole specimen are shown in Figure 3.15(a) and (b), respectively. The area, where the slit is located, is shown on the right-hand

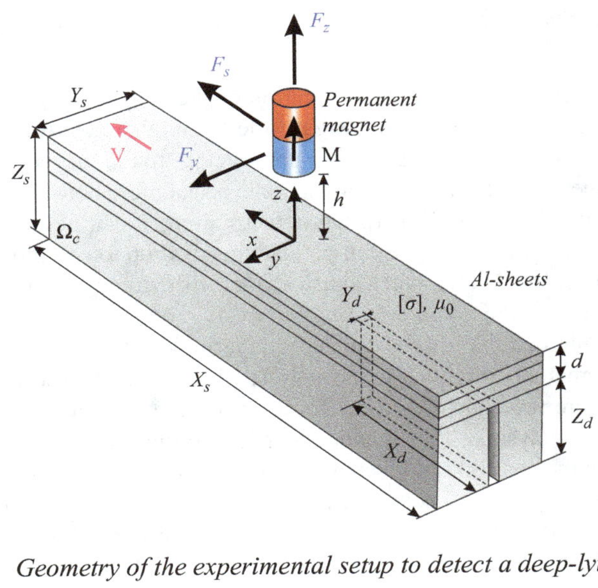

Figure 3.14 Geometry of the experimental setup to detect a deep-lying slit defect

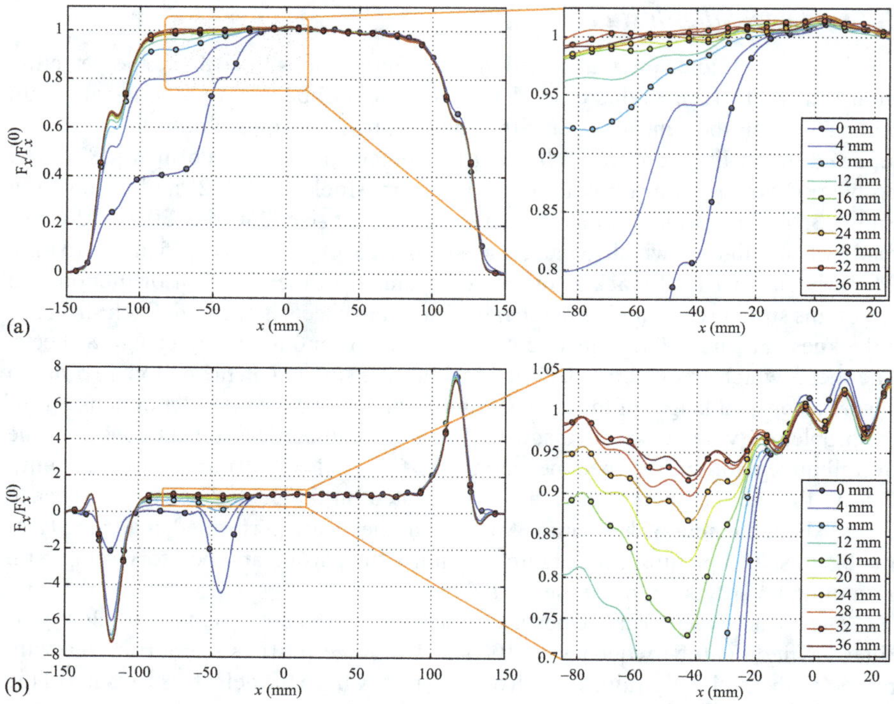

Figure 3.15 Measured profiles of the drag- and lift-force during the investigation on the maximum defect depth. The specimen contains a slit of size $[X_d, Y_d, Z_d] =$ [75 mm, 1.5 mm, 24 mm], which is located at different depths d. (a) Normalized drag-force. (b) Normalized lift-force

side in enlarged form. Based on the definition of the velocity of the specimen, the data are recorded over time from the right to left such that positive x-positions are sampled first in time. When the specimen comes close to the magnet, the drag-force ramps up and the lift-force shows a characteristic peak before both components reach their steady state ($F_x^{(0)}$ and $F_z^{(0)}$). In the defect region, it can be observed that the perturbations of the Lorentz force can be distinguished up to a depth of $d = 12$ mm for the drag-force. In contrast, when considering the lift-force, the slit can be clearly observed up to a depth of 24 mm.

It can be seen that the lift-component is superimposed by parasitic oscillations. These are systematic nature and do partially result from the measurement frame of the system, where the magnet and the force sensor are mounted. A more detailed explanation and analysis of this phenomenon can be found in [103]. The presented results chart out the possibility to increase the reported depth limit in the future when the disturbances in both force components can be reduced. An exclusive construction of an optimized magnet system for this particular defect scenario is not necessary since it has less practical relevance.

By means of this analysis, it is possible to classify LET in the framework of electromagnetic eddy current methods in a qualitative sense. Considering the values obtained by Mook *et al.* [42] from Table 1.1, it can be stated that the LET method within its present realization is highly comparable in terms of defect depth. It is worth to mention that depth-optimized ECT probes were employed in the comparative study from Mook *et al.* (Leotest MDF 1701 and MDF 3301, Leotest-Medium Center). However, it is emphasized that in LET the object is tested when it is in motion and within fractions of a second while avoiding any contact to its surface, which is the decisive difference to traditional ECT methods.

3.2.6 Conclusions

For the first time, the optimal magnet design in the LET framework is addressed. Parameters were classified as design, system, and scaling parameters, and the number of free variables was reduced, which simplified the optimization procedure. It was possible to clearly define the optimization problem while preserving universal applicability, motivated by the high specificity of NDT problems. The definition of a scaling factor offers the possibility to identify and convert similarities between different setups. The proposed methodology considers the strong interrelation to the applied force sensor. Therefore, three different optimization approaches were presented, the unconstrained optimization scheme (uc), the VaFc, and the VcFc approach. The VaFc approach can be employed to determine the critical V_m/V_d ratio. Then, the ADRA can be further increased while still satisfying the given force limit by applying the VcFc approach.

Unconstrained optimization schemes are applicable for force sensors with high force limits, e.g. piezoelectric sensors [288]. On the other hand, the VaFc approach in combination with the VcFc approach is especially suited for systems with a limited force range, e.g. strain gauge sensors. The different locations of the unconstrained global optima demonstrate that the detection goal, expressed in form of the system parameters, strongly influences the optimal magnet design. In the case of constrained optimization for subsurface defects, a Halbach-structure in combination with soft magnetic material clearly outperforms cylindrical magnets of the same geometrical dimension. In contrast, for deep defects, the optimal magnet design converged to a regular cylindrical magnet when force constraints have to be considered.

The proposed optimization strategy is highly flexible, i.e., the magnet system can be replaced by a different system in the first evaluation step of the forward solver. Moreover, the model of the specimen and/or the defect in the second and third evaluation step can be adjusted to particular scenarios of interest. The latter has been applied during the optimization of magnet systems, which are used for the inspection of small metal injection molding specimens [289]. In this study, the proposed procedure is extended to cuboidal specimens of finite size and defects of different edge lengths.

When considering high-speed applications, secondary magnetic fields become prevalent and cannot be neglected as it is shown in Section 2.4.4. In this case, the magnetic field formulation used in the forward solver can be adjusted, e.g. to an

$\mathbf{A}, \varphi - \psi$ formulation, as it is described in Section 2.4, albeit at the expense of computational cost.

In ECT, there exists a trade-off between penetration depth and resolution, which can be controlled by means of the diameter of the exciting coil [42]. This fact can be also observed in LET and is inherently a limiting factor of both methods. Our results (particularly Figures 3.9 and 3.10(b)) demonstrate the advantage of combining active and passive magnetic materials in form of a Halbach-structure in an LET sensor for selected applications. The results of the unconstrained optimization demonstrate that the use of those structures is counteracting the trade-off between penetration depth and resolution, revealing additional potential of future sensor systems.

In the present case, the optimization procedure is applied considering two different defect scenarios taking into account a force constraint of $F_x^{(0)} = 3$ N. The associated optimal magnet designs are manufactured and made available for experimental studies presented in [103,156].

A defect depth study, adopting the concept of a quasi-infinite crack, revealed a current detection limit of 24 mm when considering the lift-component of the Lorentz force.

Chapter 4
Experiments and LET measurements

Matthias Carlstedt[1] *and Konstantin Weise*[1]

This chapter describes experimental investigations, which are the basis for the advancement of the LET method. Furthermore, the experiments provide objective data for the validation of numerical approaches used for theoretical determination of the exerted Lorentz force.

At first, the measurement procedure of LET is described. The underlying measurement principle is analyzed, and all operations to perform a LET measurement are explained in the measurement method. Secondly, the final state of the experimental setup that was used for laboratory experiments is described. The main components of the setup and their functional relationship are explained and selected elements are described in detail. The digital signal processing (DSP) developed for LET is explained afterwards. DSP is the final step of the description of the measurement method and allows to determine the value of the measurand by means of the arithmetic mean and the experimental standard deviation.

In the next section, a representative overview of the measurement performance of the described experimental setup is given. One application is presented with an artificial defect in stacked aluminum sheets.

4.1 Measurement procedure

According to the international vocabulary of metrology (VIM) [290], a *measurement procedure* is defined as a detailed description of a *measurement* according to the *measurement principles* and the given *measurement method*. The measurement method is the description of the logical organization of operations used in a measurement. It is based on a *measurement model* and includes any calculation to obtain a *measurement result*. The measurement result is a set of quantity values being attributed to a *measurand* and is expressed as a *measured quantity value* and a *measurement uncertainty*.

According to these definitions, the measurement procedure for LET is described in the following sections.

[1] Advanced Electromagnetics Group,Technische Universität Ilmenau, Germany

4.1.1 Measurement principle

The measurand in LET is the Lorentz force $\mathbf{F}(t)$ acting on the PM during an LET experiment. The basic *measurement principle* of LET is based on the electromagnetic interaction of the *PM* and the *unit under test* (UUT). For a better understanding, the interaction can be decomposed into two causally connected physical phenomena: (i) the electromagnetic induction of *eddy currents* (ECs) inside the UUT and (ii) the Lorentz force acting on the UUT and the PM.

The decomposition into a causal sequence, in terms of cause and effect, further simplifies the understanding of the LET measurement principle (cf. Figure 4.1). The initial cause in LET is the relative motion of the PM and the UUT that leads to electromagnetic induction of ECs inside the UUT. This effect causes the generation of the Lorentz force due to the interaction of the induced ECs with the magnetic field of the PM. The second causal relationship can also be triggered by other current flows, which are independent of a relative motion. In general, no other current flows are present inside the UUT; therefore, they are neglected in the following.

To measure the Lorentz force, one or more measurement principles are necessary to convert the force into an electrical signal. These force measurement principles complete the overall measurement principle of LET. The selected force measurement principles have a significant influence on the result of the measurement, but they do not alter the physical phenomena of interest.

4.1.2 Measurement method

The measurement method describes the logical organization of all operations used to measure the Lorentz force. It involves all information necessary to describe an experiment and to ensure its repeatability. In other words, the description should allow

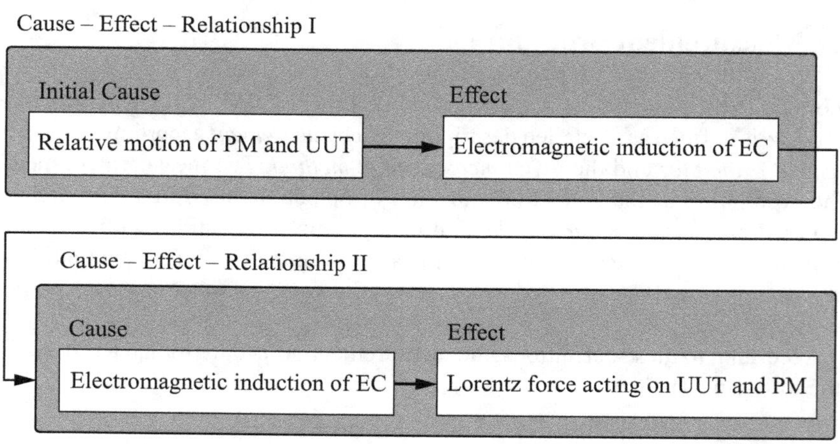

Figure 4.1 Causal sequence of the basic measurement principle in LET

a well-trained person to perform an experiment and to do all calculations necessary to obtain a complete measurement result.

Since the basic measurement principle of LET neither defines the type of relative motion, nor the concrete realization of the force measurement, the following description can only cover a particular realization of the method. The realization discussed next assumes the rectilinear motion of the UUT relative to the stationary PM and describes the attached sensor system. Figure 4.2 shows the schematic view of the experimental setup. In addition to the UUT (1) and the PM (2), four main components are necessary: the linear drive (3), the 2D-positioning stage (4), the multicomponent force sensor (5), and the measurement frame (6).

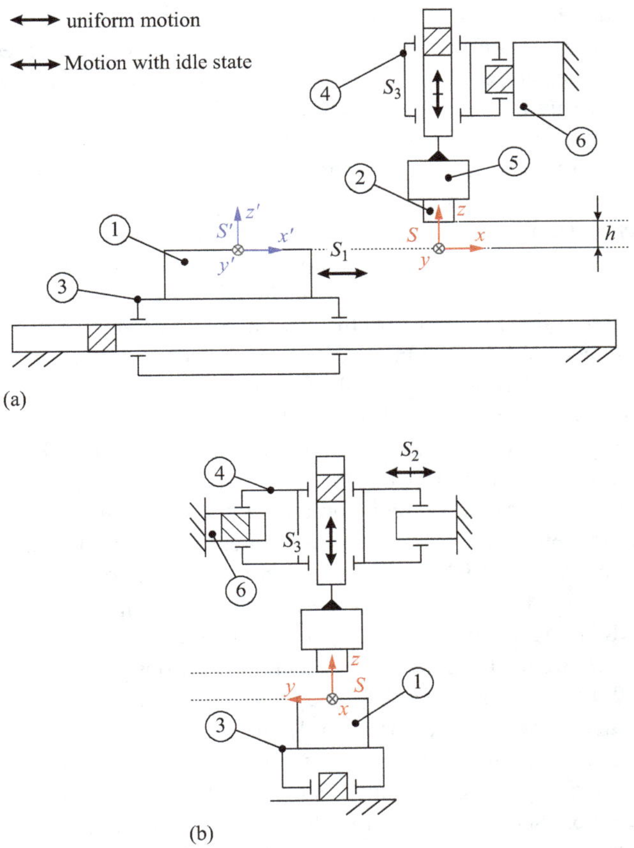

Figure 4.2 *Schematic view of experimental setup for LET measurements. The UUT (1) moves along the linear guide of a linear drive (3) relative to a stationary PM (2). The PM is mounted on the multicomponent force sensor (5). The force sensor is attached to the 2D-positioning stage (4), which is mounted on the measurement frame (6). (a) Side view. (b) Front view*

The UUT is mounted on the slide of a linear drive that provides the controlled motion and guidance of the UUT. The linear drive is fixed on a heavy load bench, which is assumed to be mechanically insulated from ground vibrations. The PM is indirectly attached to a 2D-positioning stage via a multicomponent force sensor. The 2D-positioning stage provides the positioning of the PM relative to the linear guide of the linear drive. The 2D-positioning stage is fixed on a measurement frame (6), which is assumed to be mechanically insulated from ground vibrations. Due to the placement of the multicomponent force sensor, all forces occurring between the PM and the 2D-positioning stage can be measured in the experiment.

All additional components such as power supply, hardware controller, and data acquisition (DAQ) are assumed to be mechanically decoupled from the measurement apparatus, so that they do not influence the measurement process.

The logical operations needed to perform an experiment are equal for all experimental studies presented in this book. They are explained in the following.

The first step in all experiments is to define the design of experiment (DOE), including the purpose of the investigation. The parameters to be defined in the DOE are the position of the PM during operation (y- and z-position), the desired velocity of the UUT (v), and the number of repetitions of the respective experiment. The DOE also has to include the assignment and ordering of the successive experiments in order to allow further statistical evaluation, e.g. correlation analysis or hypothesis testing.

The next step is the preparation of the UUT according to the DOE and the subsequent fixation on the slide of the linear drive. The positioning of the UUT relative to the slide is part of the definition of the laboratory frame of reference and is crucial for the repeatability of the experimental study. After the fixation of the UUT, the sensor system including PM and force sensor is mounted on the measurement frame and has to be aligned according to the surface of the UUT and the guide of the linear drive. This step completes the definition of the laboratory frame of reference. In the particular realization of the method, the procedure of alignment can be supported by measuring the contact forces between UUT and PM in order to enhance the repeatability and precision. This method is referred to as force feedback [291].

After the functional check of the sensor system and DAQ, the subsequent operations of the measurement process are sequentially motions performed by the linear drive and the 2D-positioning stage. These operations need to be monitored and controlled by separate devices (actuator control) and can therefore be fully automated. Such a process ensures a high level of repeatability and reproducibility.

The *measurement process* is defined as a sequence of single experiments according to the DOE. For each experiment, the UUT moves rectilinearly relative to the PM along the linear guide. During the controlled motion, all sensor data are sampled and temporally stored by the DAQ. At the end of the test track, the slide stops and the recorded data are stored permanently on the hard disc for later processing. Depending on the intended purpose of the investigation, the next experiment can be started from this position or the slide is moved back to the desired starting position.

The stored data are referred to as *sensor data*. Each sensor delivers an electrical signal that needs to be multiplied by the specific calibration factor in order to calculate the corresponding physical quantity which is part of the DSP and is discussed in a separate section (cf. Section 4.2.1).

4.1.3 Experimental setup

The particular realization of the laboratory setup for LET measurements, shown in Figure 4.3, consists of six main components as shown in Section 4.1.2. Both UUT and PM are regarded as problem-specific components; thus, they will be described in Section 4.2.4 for the respective investigations.

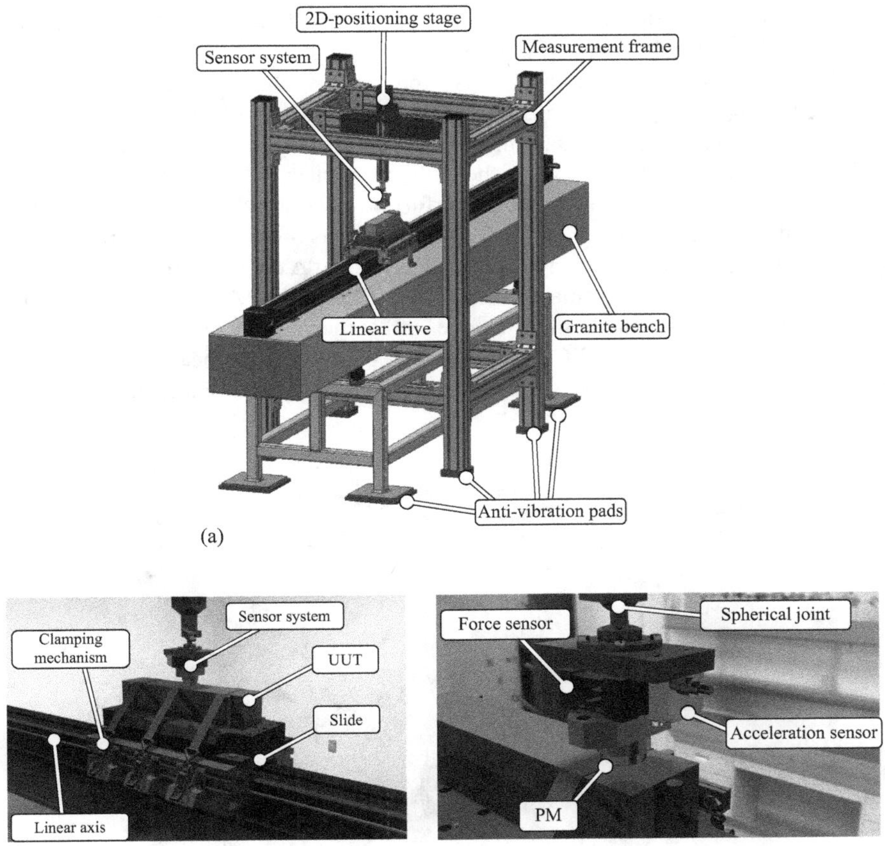

Figure 4.3 Experimental setup for LET measurements: (a) Overview shows the measurement frame and the 2D-positioning stage separated from positioning unit which is mounted on a heavy granite bench. (b) Detail view of the positioning unit for motion of the UUT. (c) View of the sensor system including a PM mounted on the force sensor and separate acceleration sensor

4.1.3.1 Linear drive

The positioning unit consists of a belt driven linear drive designed by *Jenaer Antriebstechnik GmbH* mounted on a linear guide by *Bahr Modultechnik GmbH*. A planetary gear of gear ratio $i = 3$ transmits the torque from the servomotor (type 110B) to the pulley of the belt drive. The belt is directly connected to the slide of the linear guide, which realizes the translational motion of the UUT. In the framework of design improvements, the original tooth belt drive was replaced by poly-v belt drive to eliminate a source of noise emission, which was introduced by tooth meshing [91]. The maximum velocity of the slide is about $v = 2$ m/s using a third-order motion profile for reduced jerk, with quadratic ramping and deramping phases in the velocity. If the jerk is not considered, a maximum velocity of about $v = 3.75$ m/s is possible but leads to increased structural damage (fatigue) of the linear drive.

The positioning deviations as well as pitch and yaw angles of the linear guide are qualified using a long-range multibeam interferometer by *SIOS Meßtechnik GmbH*. As shown in Figure 4.4, a measurement reflector based on three individual prismatic reflectors is placed on the slide while the multi beam interferometer is fixed at the end of the linear guide. The experiments are performed for static slide positions as well as for dynamic measurements of up to $v = 0.3$ m/s. It was observed that the positioning deviation along the full travel range of $l_G = 2$ m is about $\Delta x = \pm 0.5$ mm and the lateral displacement of the slide is smaller than $\Delta y = \pm 25\ \mu$m and $\Delta z = \pm 15\ \mu$m [292].

The velocity deviation of the slide does affect the velocity of the UUT and has a direct impact on the induced EC distribution (cf. Figure 4.2). In the specified velocity range of up to $v = 4$ m/s, the manufacturer guarantees a deviation less than 5% of the

Figure 4.4 Experimental setup for experimental characterization of the linear guide using a long-range multibeam interferometer by SIOS Meßtechnik GmbH. The measurement reflector is placed on the slide for static and dynamic measurements of positioning deviation, pitch and yaw angle [292]

predetermined velocity for the predefined control parameters [91]. In the framework of an uncertainty analysis [293], this statement is verified by means of an incremental position encoder TONiC T1000 (*Renishaw plc*) implemented into the experimental setup (cf. Section 4.1.3.3).

4.1.3.2 2D-positioning stage

The 2D-positioning stage is used for the positioning of the sensor system relative to the linear guide of the linear drive. It is an assembly of two orthogonally oriented precision linear stages of the NLS4 series [294] by *Newmark Systems Inc.* The travel range in y- and z-direction is 300 and 100 mm, respectively.

The design of this series of linear stages is improved for high stiffness and repeatability. Prestressed linear guide bearings and an internally lubricated plastic drive nut provide zero backlash operation and enable a specified accuracy of 0.6 μm/mm. The used lead screw has a pitch of 1.58 mm/rev (1/16″/rev) and the stepper motor encoder allows 4000 counts/rev, which results in a resolution of the drive nut position of about 0.4 μm, on each axis.

The orientation of the assembly enables a maximum load of the sensor system and fixture devices up to 6.0 kg (max. lifting capacity). The applied static load is given by the weight of the attached cantilever and mounted sensor system. The assembly has an estimated mass of about 1.4 kg, and the estimate absolute value of the maximum forces is up to 3 N. Due to these facts, a disturbing influence of the process forces on the specified precision during operation is not expected.

The motion controller used in this setup is an NSC-G3-E series controller [295] with up to three individual axes with encoder feedback for stepper motors by *Newmark Systems Inc.* The configured communication between the sensor system and the host computer uses an Ethernet interface.

4.1.3.3 Sensor system

The sensor system contains all transducers used to observe an experiment (Figure 4.5). It consists of four components: (1) the multicomponent force sensor, (2) an additional acceleration sensor, (3) a coil with multiple turns (to measure induced voltage), and (4) an incremental position encoder (not shown). The PM is intentionally excluded from the description since it does not provide additional information during an experiment.

As a consequence of the definition of reference frames S' and S, the recorded signals of force $\mathbf{F}(t)$ and acceleration $\mathbf{A}(t)$ are each equal in direction and magnitude for both frames of reference. However, the relative position of both coordinate systems to each other is measured with different signs in the respective coordinate system. Thus, they have to be distinguished carefully. In fact, the position of the slide of the linear drive measured by the incremental position encoder is recorded in the frame of reference S, whereas the lateral position of the PM y' given by the position of the 2D-positioning stage is only defined in the frames of reference S'. To reduce possible confusion with the use of two coordinate systems, in the following the relative position of UUT and PM is always given in frame of reference S' of the UUT, but *without* the primed notation.

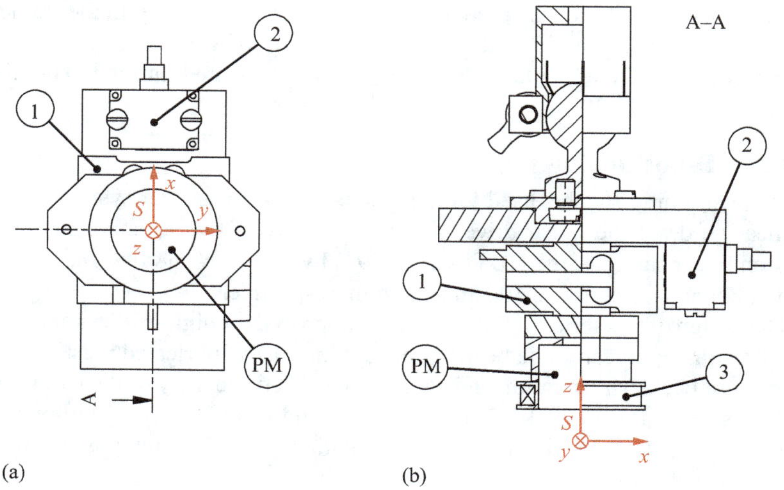

Figure 4.5 Technical drawing of the assembled sensor system of the experimental setup: (1) multicomponent force sensor, (2) acceleration sensor, and (3) coil with multiple turns. (a) Bottom view. (b) Side view.

4.1.3.3.1 3-Axes force sensor

The essential element of the sensor system is the custom build 3-axes force sensor K3D40 [104] by *ME-Messsysteme GmbH* for nominal loads of 3 N×3 N×10 N in the *x*-, *y*-, and *z*-direction, respectively. According to the data sheet, the sensor accuracy class is 0.5 % resulting in a nominal measurement error of about 15 mN × 15 mN × 50 mN for the respective measurement axes.

The measurement principle of the sensor is based on strain measurements on the surface of a compliant mechanical structure which is deflected due to the force of interest. The compliant structure is made of a single aluminum part designed as a series connection of three orthogonal-oriented single-axis flexure hinges. Each single-axis mechanism is equipped with four strain gauges in a balanced Wheatstone bridge. The underlying measurement principle is the piezoresistive effect. Due to the limited bandwidth of the deflection body, this sensor concept is primarily designed for static force measurements.

Since the force sensor has already been used in previous work [91], its reliability was tested multiple times in between experimental studies. Due to the limited dynamic information about the sensor provided by the manufacturer, a comprehensive analysis has also been carried out to determine the dynamic sensor characteristics [296].

The sensor is connected to the analogue measurement amplifier is a GSV-1A4 [297] by *ME-Messsysteme GmbH*. According to the data sheet, the measurement amplifier provides a bridge supply voltage of $V_S = 5\,V$ and an output voltage of $V_A = \pm 10\,V$. During operation and an examination of the amplifier's characteristics, both malfunctions and contradictory measurement results were observed. Thus, an overhaul of the device was carried out. The disassembly of the measurement amplifier revealed significant differences of the used hardware components in comparison to the data sheet. Actually, the amplifier consists of four individual strain gage measurement amplifiers of type GSV-1M [298] by *ME-Messsysteme GmbH.* A characterization of the four amplifiers, analogously to [296], showed a gain of $g_F = 2 \times 10^3$ with fixed input sensitivity of $s_N = 1.0 \times 10^{-}3$ V/V. Multiple contact faults have been rectified and the postrepair tests confirmed a fully functional device.

The sensor and the amplifier are connected via 37-pin Sub-D socket and have been together calibrated by the manufacturer. According to the calibration certificate, nominal output voltages at nominal loads are 3.1969 V × 4.8835 V × 7.8030 V in the x-, y-, and z-direction, respectively. Regarding the identified amplifier parameters, this corresponds to nominal sensor sensitivities for the corresponding channels are given by:

$$s_{Fx} = 1.0656 \times 10^{-4} \frac{V}{VN}, \tag{4.1}$$

$$s_{Fy} = 1.6278 \times 10^{-4} \frac{V}{VN}, \tag{4.2}$$

$$s_{Fz} = 7.8030 \times 10^{-5} \frac{V}{VN} \tag{4.3}$$

4.1.3.3.2 3-Axes acceleration sensor

The second component of the sensor system is the 3-axes accelerometer ASC 5511LN [299] by *Advanced Sensors Calibration – ASC GmbH.* The measurement principle of the accelerometer is based on a capacitive microelectromechanical system (MEMS), which is specially designed for low-frequency responses from constant value up to 5 kHz in a range of ±2 g[1]. The sensitivity of the accelerometer is specified to be invariant to the supply voltage in the range of $V_S = +8\,\text{V} \cdots + 30\,\text{V}$.

The sensor has been calibrated by the manufacturer for all three axes separately. According to the calibration certificate, the nominal accelerometer sensitivities are:

$$s_{Ax} = 98.848 \times 10^{-3} \frac{Vs^2}{m}, \tag{4.4}$$

$$s_{Ay} = 98.535 \times 10^{-3} \frac{Vs^2}{m}, \tag{4.5}$$

$$s_{Az} = 98.955 \times 10^{-3} \frac{Vs^2}{m}, \tag{4.6}$$

at a nominal acceleration of 5 m/s^2 for an excitation frequency of 16 Hz.

In the experimental setup, the 3-axes accelerometer is mounted on the same plate as the 3-axes force sensor (cf. Figure 4.5). This allows to observe the translational components of the motion of the sensor system during an experiment and to estimate the effects of inertial forces on the PM.

[1] In the context of accelerometers, the term *g-force* refers to the acceleration relative to free-fall. It is described by the standard acceleration due to gravity g_n which is defined to 9.80665 m/s^2.

4.1.3.3.3 1-Axis DiLET sensor

The third component of the sensor system is a differential Lorentz force eddy current testing (DiLET) sensor. This 1-axis sensor [243], which is a passive pick-up coil with multiple turns winded coaxially on the outer surface of the PM (cf. Figure 4.5). It is designed as an optional upgrade to measure additionally the time variations of the secondary magnetic flux density ($\partial B_z/\partial t$) at the bottom surface of the PM.

The concept of this sensor is based on [91,102], where a set of passive coils fixed to a PM is proposed to allow the detection of perturbations in the EC distribution caused by defect inside the UUT. It is based on the fact that the primary magnetic field $\mathbf{B}^{(p)}$ produced by a PM is constant in time, while the secondary magnetic field $\mathbf{B}^{(s)}$, connected with the EC distribution inside the UUT, is time dependent when a defect is present. Thus, the induced voltage V_i ($i \in \{x, y, z\}$) in a coil fixed to the PM is only proportional to perturbations of the secondary magnetic field $B_i^{(s)}$ and by that sensitive to disturbances caused by defects.

As shown by [102], the induced voltage V_i is proportional to the first time derivative of the force component parallel to the respective coil axis. Therefore, the proposed technique has been termed DiLET.

In this chapter, measurement results of the induced voltage $V_z(t)$ are shown using one particular 1-axis DiLET sensor mounted on a PM. It is mentioned that index z denotes the orientation of the coil axis, not a component of a vector. The custom-built DiLET sensors are equal with respect to the number of turns N_z but of different size due to the different outer diameters of the PMs.

Because the different 1-axis DiLET sensors are not calibrated, the respective nominal output voltage V_z and the sensor sensitivity s_{Vz} of the respective sensors are unknown. However, it is possible to provide an estimate of both factors for an *effective* variation of the magnetic flux density $\partial B_{\mathrm{eff},z}(t)/\partial t$ by the following approximations. Assuming that the coil is substituted by a single circle at the bottom of the PM, then the time variation of the magnetic flux density $\partial B_z/\partial t$ can be given by

$$\frac{\partial}{\partial t}\Phi_z(t) = \frac{\partial}{\partial t}\int_S B_z(t)\,\mathrm{d}S \approx \frac{\partial}{\partial t}B_{\mathrm{eff},z}(t)\,S, \tag{4.7}$$

with $\Phi_z(t)$ the magnetic flux through the enclosed surface S of the coil. For a circular coil with diameter D_{eff}, which can be approximated by the mean of inner and outer diameter of the real coil, (4.7) becomes

$$\frac{\partial B_{\mathrm{eff},z}(t)}{\partial t} \approx \frac{4V_z(t)}{\pi D_{\mathrm{eff}}^2 N_z} = \frac{1}{s_{Vz}}\,V_z(t), \tag{4.8}$$

with the number of turns of the coil N_z and the sensor sensitivity s_{Vz}. For the two used 1-axis DiLET sensors, one with $D_{\mathrm{eff}} = 27$ mm and the other with $D_{\mathrm{eff}} = 30$ mm, and both $N_z = 5000$ turns the sensor sensitivities are estimated by:

$$s_{Vz}|_{D_{\mathrm{eff}}=27\,\mathrm{mm}} \approx 2.86\frac{Vs}{T}, \tag{4.9}$$

$$s_{Vz}|_{D_{\mathrm{eff}}=30\,\mathrm{mm}} \approx 3.53\frac{Vs}{T}. \tag{4.10}$$

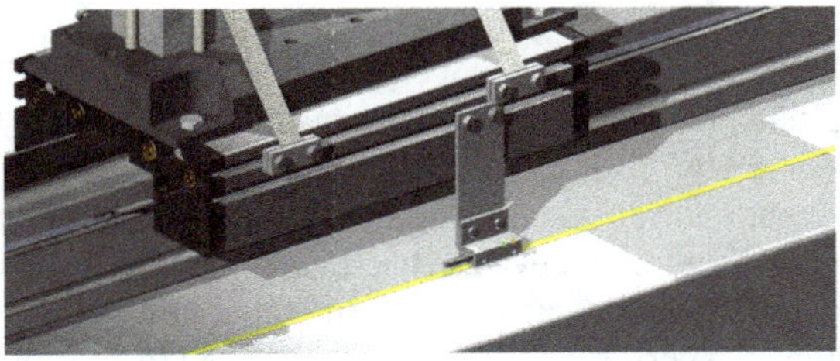

Figure 4.6 Incremental position encoder T1000-50A

The induced voltage $V_z(t)$ is amplified by a self made analogue measurement amplifier. It is based on a precision instrumentation amplifier AD624 [300] by *Analog Devices Inc.* and provides a configurable gain $g_{DF} = [1, 5, 10, 50, 100, 200, 500, 1000]$ of the DiLET signal. A response test confirmed the measurement amplifier to be fully operational.

The amplified output voltage of the analogue measurement amplifier at a nominal variation of the magnetic flux density $\partial B_{eff,z}(t)/\partial t$ is calculated by

$$V_z(t) = g_{DF}\, s_{Vz}\, \frac{\partial B_{eff,z}(t)}{\partial t}. \tag{4.11}$$

4.1.3.3.4 Incremental position encoder

The incremental position encoder system TONiC by *Renishaw plc.* serves as an external position measurement system of the UUT in the x-direction (Figure 4.6). It is composed of three components: (i) T1000-50A read head [301], (ii) RGSZ20-S gold plated steel scale [302], and (iii) encoder interface T0100 A40A [303].

This encoder system measures the position of the read head relative to reference marks at both ends of the linear scale. The read head is mounted on the side of the slide of the linear drive. The linear scale, of 1 m length, is fixed on the heavy granite bench below the PM. The combination of optical detector (read head), linear scale (scale pitch 20 μm), and encoder interface results in a specified resolution of 50 nm (interpolated) up to a maximum velocity of 5.4 m/s.

In the framework of the analysis of the linear drive, the incremental position encoder was used to validate the controller estimated slide velocity. The investigation resulted in a correction of the specified pinion diameter and adjusted controller parameters for improved velocity constancy at the velocity operating point of $v = 0.5$ m/s.

4.1.3.4 DAQ and measurement control system

The DAQ device is used for signal conditioning and digitization of incoming analogue signals from the sensor system. The analog-to-digital converter (ADC) is the main component of the DAQ. In this particular realization of the experimental setup, the NI PXI-4472 [304] by *National Instruments* (NI) is used. It provides eight simultaneously sampled analogue inputs at a sampling rate of up to 102.4 kHz each. The resolution of the ADC is specified with 24-bit per channel and a dynamic range of 110 dB. The ADC is connected to the Host-PC via the *PCI eXtensions for Instrumentation* (PXI) computer bus. A six-slot chassis NI PXI-1036 [305] by NI serves as a Host-PC. All results presented in this chapter were sampled at $f_s = 10\,\text{kHz}$ at full input range of $V_{IN} = \pm 10\,\text{V}$.

The measurement control system (MCS) is used to monitor and to control the experimental setup and its main devices. It provides a graphical user interface that allows the operator to monitor current system states and sensor inputs. Furthermore, the MCS allows to perform single experiments and to control the initialization sequence for the definition of the laboratory frame of reference. The most important task of the MCS is the batch processing of DOE without manual intervention of the operator. This enables comprehensive experimental studies including multiple repetitions without disturbing influences of the operator, while ensuring a high level of repeatability and reproducibility. The MCS has access to an specially established external database for storage of the acquired raw data at the local computing center for improved data security and accessibility. The programming language of the MCS is C++.

4.2 Validation procedure

4.2.1 DSP and basic statistics

The basic idea of DSP for LET is to provide an estimate of the expected value of the measurand $\mathbf{F}(t)$ and the corresponding experimental standard deviation $\sigma_F(t)$ for a specific experiment. The statistical analysis is necessary because even if the ideal measurement principle of LET (cf. Figure 4.1) can be considered as deterministic, where no randomness is involved, then at least the realization in a real laboratory setup introduces a vast amount of process noise and measurement errors. Thus, each physical quantity x obtained in the measurement process is considered as a *random variable.*

4.2.1.1 Concepts of signal ensembles

To provide statistical information about the measurement process, it is necessary to introduce the concepts of the *system ensemble* and *signal ensemble.* In the following, the theoretical concept of the *ideal signal ensemble* is presented as well as the deduced concept of an *artificial signal ensemble.*

4.2.1.1.1 Ideal signal ensemble

A system ensemble describes a set of H equally realizations of the measurement process (cf. Figure 4.7). Each member of that system ensemble is referred to as system realization or just member of the system ensemble. In theory, each of these H realizations generates simultaneously H individual signals $x_h(t)$ of the measurand $x(t)$. The set of H signals builds the ideal signal ensemble $\{x_h(t)\}$.

The signal ensemble is evaluated for each time of observation t_{obs} along the ensemble plane (cf. Figure 4.7). Every continuous signal ensemble $\{x_h(t)\}$ is represented by an infinite number of random variables $x_h(t_{\text{obs}})$, which are described by their statistical properties. The presumed sampling process leads to a finite number of

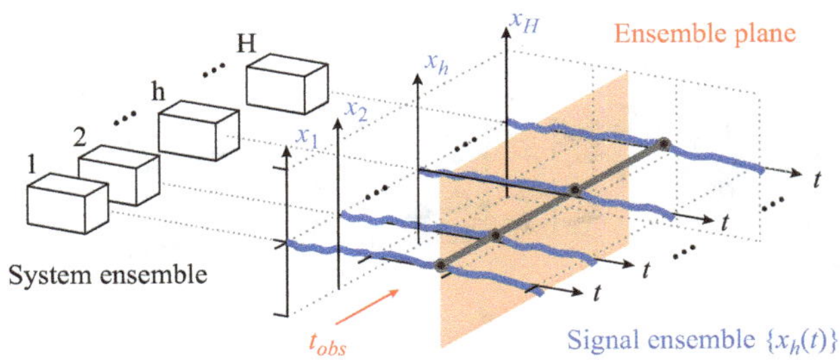

Figure 4.7 Concept of the ideal signal ensemble: a system ensemble of H equally realizations of the measurement process produces H individual signals $x_h(t)$ of the measurand x(t). This set builds the ideal signal ensemble {x(t)} that allows to estimate the statistical properties of x(t) in the ensemble plane

observations $\{x_h[n]\}$ at isochronal times $t = nT_s$, with sampling period $T_s = 1/f_s$. The sampling process is assumed to be ideal, i.e. the acquired samples $x_h[n]$ are equivalent to the instantaneous value $x_h(nT_s)$ of the continuous signal $x_h(t)$ at the desired point in time $t = nT_s$.

In the special case where all statistical properties of the signal ensemble are independent of time (constant), the signal ensemble, is called stationary. For a stationary signal ensemble, the result of the statistical evaluation along the ensemble plane is identically to the result of each individual signal. Thus, a single signal $x_h(t)$ can be used to describe the complete signal ensemble.

For experimental studies in LET, it has to be assumed that due to the finite size of the UUT or the presence of defects the statistical properties, e.g. mean value of the force, will vary over time. Because of that, it is evident that a single signal $x_h(t)$ of any component of the Lorentz force cannot provide a complete measurement result, i.e. expected value of the measurand $\mathbf{F}(t)$ and corresponding experimental standard deviation $\sigma_F(t)$.

4.2.1.1.2 Artificial signal ensemble

In practical applications, it is not economically reasonable to build multiple realizations of the same measurement process. Nevertheless, to be able to obtain a good estimate of the statistical properties of the nonstationary signal ensemble, a series of multiple experiments is executed to build an artificial signal ensemble.

Figure 4.8 illustrates the concept of building an artificial signal ensemble. In contrast to the ideal signal ensemble $\{x_h(t)\}$, a set of $K \geq H$ experiments is performed on a single realization of the experimental setup. Afterward, the individual signals $x_k(t_k)$ have to be aligned according to a virtual trigger time t_{tr} which is a representative point in time for the nonstationary signal ensemble $\{x_k(t)\}$. The resulting set of serially recorded signals $x_k(t_k)$ is merged to an artificial signal ensemble $\{x_k(t)\}$ on which all statistical evaluations can be carried out. Additionally to the ideal signal ensemble $\{x_h(t)\}$, one important requirement has to be fulfilled to obtain representative statistical

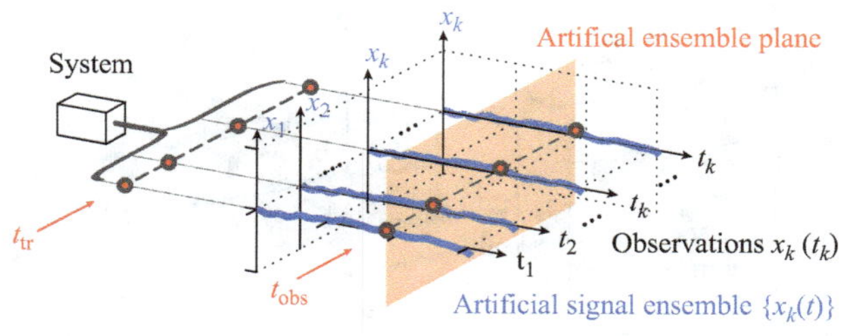

Figure 4.8 Concept of the artificial signal ensemble: a single system realization of the measurement process produces $K \geq H$ serially recorded signals $x_k(t_k)$. The artificial signal ensemble $\{x_k(t)\}$ is created by aligning the individual signals $x_k(t_k)$ according to a trigger time t_{tr}, which allows to estimate the statistical properties of $x(t)$ in the artificial ensemble plane

information from the artificial signal ensemble $\{x_k(t)\}$. The boundary conditions for each signal $x_k(t_k)$ need to be equal, i.e. all influence parameters for the random process need to be *statistically independent*. Especially, it is important that every single experiment is not influenced by any previously performed experiment.

The task of DSP for LET is to create such an artificial signal ensemble and to provide an estimate of statistical properties of nonstationary signals. The assumption of statistically independence can only be a hypothesis.

4.2.1.2 Basics of signal alignment

For the alignment of recorded signals $x_k(t_k)$, a trigger time t_{tr} corresponding to a relative position of UUT and PM has to be defined. The reliability of trigger time determination is crucial for signal alignment since it has direct impact on the expectation value and other statistical properties of the ensemble.

4.2.1.2.1 Signal alignment based on external trigger signals

One way to define a trigger time is to detect the presence of the UUT by means of an proximity sensor. This sensor would preferably be based on a measurement principle which is invariant to the investigated electromagnetic properties of the UUT.

In the experimental setup described earlier, the incremental position encoder could provide this functionality under the following two assumptions. First, the position deviation between UUT and the slide of the linear drive is small, relative to the measurement uncertainty of the incremental position encoder. Second, the relative position of linear drive and measurement frame is fixed, i.e. it does not vary between experiments.

While the first assumption is ensured by the redesigned clamping mechanism (cf. Figure 4.3), the latter is not fulfilled since the position of the measurement frame can vary between experiments in the case of maintenance and preparation work, e.g. exchange of the PM. Furthermore, the linear scale used is too short and therefore had to be equipped with magnetic triggers as reference marks for the encoder. Reliability tests revealed clear weaknesses in the reproducibility of such a triggering function, resulting in varying absolute position measurements.

An additional external trigger, mounted on the measurement frame has not been implemented. Thus, a signal alignment based on external trigger signals is insufficient for the used experimental setup.

4.2.1.2.2 Correlation of time-continuous signals

Another possibility to define a trigger time t_{tr} is to search for repeating patterns in the signals and to align the signals according to the closest match. The most common mathematical tool to find the best match for time-lagged signals is the cross-correlation function $\rho_{XY}(\tau)$.

Let (X_t, Y_t) be two stochastic processes that are jointly wide-sense stationary [306], then

$$\rho_{XY}(\tau) = \frac{E[(X_t - \mu_X)(Y_{t+\tau} - \mu_Y)]}{\sigma_X \sigma_Y}, \tag{4.12}$$

where E is the expected value operator, $\mu_{X,Y}$ are the mean values, $\sigma^2_{X,Y}$ are the variances of the random processes, and τ is the time-lag between the signals. The value of $\rho_{XY}(\tau)$ lies in the range $[-1, +1]$, with -1 indicating complete anticorrelation and $+1$ indicating complete correlation of analyzed signals. The estimated time delay between signals is determined by

$$\tau_{\text{est}} = \arg\max_{\tau} \rho_{XY}(\tau). \tag{4.13}$$

The necessary condition to apply this method is the presence of significant patterns in these signals. Since the used ADC guarantees a quasi simultaneous sampling of all recorded signals, a single signal can serve for the time delay estimation. To test signals for repeating patterns, it is useful to have a look on typical signals and their corresponding autocorrelation function $\rho_{XX}(\tau)$. Since all measured signals are real valued $X_t \in \mathbb{R}$, the following discussion is restricted only to the autocorrelation function of real valued signals.

The autocorrelation function $\rho_{XX}(\tau)$ of a wide-sense stationary random process X_t is defined as

$$\rho_{XX}(\tau) = \frac{E[(X_t - \mu)(X_{t+\tau} - \mu)]}{\sigma^2}, \tag{4.14}$$

with μ is the mean value, σ^2 is the variances of the random process, and τ is the time-lag between the signals.

In most practical cases, as well as in the discussed problem, neither μ nor σ^2 of the respective signal ensemble are known. Thus, the above definition is often used without standardization, i.e. without subtracting the mean and dividing by the variance

$$R_{XX}(\tau) = E[X_t X_{t+\tau}], \tag{4.15}$$

which to be more exact is the autocovariance without mean centering. In the following, the term autocorrelation function refers to definition (4.15) which is most common in physics and engineering. The advantage of this definition is its validity also for nonstationary processes, as in the discussed case. However, it lacks the possibility to compare different physical quantities since the resulting product of the involved units is without any physical meaning.

4.2.1.2.3 Correlation of time-discrete signals

So far, only time-continuous processes are considered. To deal with digitized signals, it is necessary to extend the definition previous definitions to time-discrete signals. The calculation of the time-discrete autocorrelation function is based on the *Wiener–Khinchin theorem* [307]. It describes how the autocorrelation function of a wide-sense-stationary random process has a spectral decomposition given by the *power spectral density* (PSD) of that process.

The autocorrelation of a time-discrete function $x(n)$ of length N is computed in two steps. Firstly, the *discrete Fourier transform* (DFT) of $x(n)$ is computed as

$$X(k) = \sum_{n=1}^{N} x(n) \exp\left(\frac{-\mathrm{j}2\pi(k-1)(n-1)}{N}\right), \tag{4.16}$$

with $1 \leq k \leq N$. Secondly, by making use of *Parseval's theorem* the inverse DFT of the PSD, i.e. squared absolute value $|X(k)|^2$, is calculated to

$$c(n) = \frac{1}{N} \sum_{k=1}^{N} X(k) \exp\left(\frac{\mathrm{j}2\pi(k-1)(n-1)}{N}\right), \tag{4.17}$$

with $1 \leq n \leq N$.

$$\tau = \{-1 \cdots +1\}\frac{N-1}{f_s} \quad \text{and} \quad R_{ii} = \frac{c_i}{N}. \tag{4.18}$$

The computation via Fourier transform and the inverse transform is implemented by using the MATLAB™ function `xcorr` [282]. In the representation shown, the time-lag τ is divided by the sampling frequency f_s of the signal to be invariant on this quantity.

4.2.2 Autocorrelation on typical force signals

To test if the measured signal does fulfill the described requirements, a typical result of an experiment is discussed. Figure 4.9 shows representative signals of the three components of the measured force $F_i(t)$ and the corresponding autocorrelation $R_{ii}(\tau)$ with $i \in \{x, y, z\}$. The respective signals are part of the experimental study presented in Section 4.2.4.

Figure 4.9(a) shows three components of the measured force $F_i(t)$ which are normalized to the mean value of the force plateau in the x-component $F_{x0}(t)$ plotted against a dimensionless time $\tilde{t}$ with

$$\tilde{t} = \frac{t}{t_c} = \frac{t\bar{v}}{L_x}, \tag{4.19}$$

with $\bar{v}$ the mean value of the velocity and L_x the length of the UUT in x-direction. The chosen characteristic time t_c for this representation is the time it takes for the UUT to pass the center of the PM. The time count starts when the first edge of the UUT passes the center of the PM.

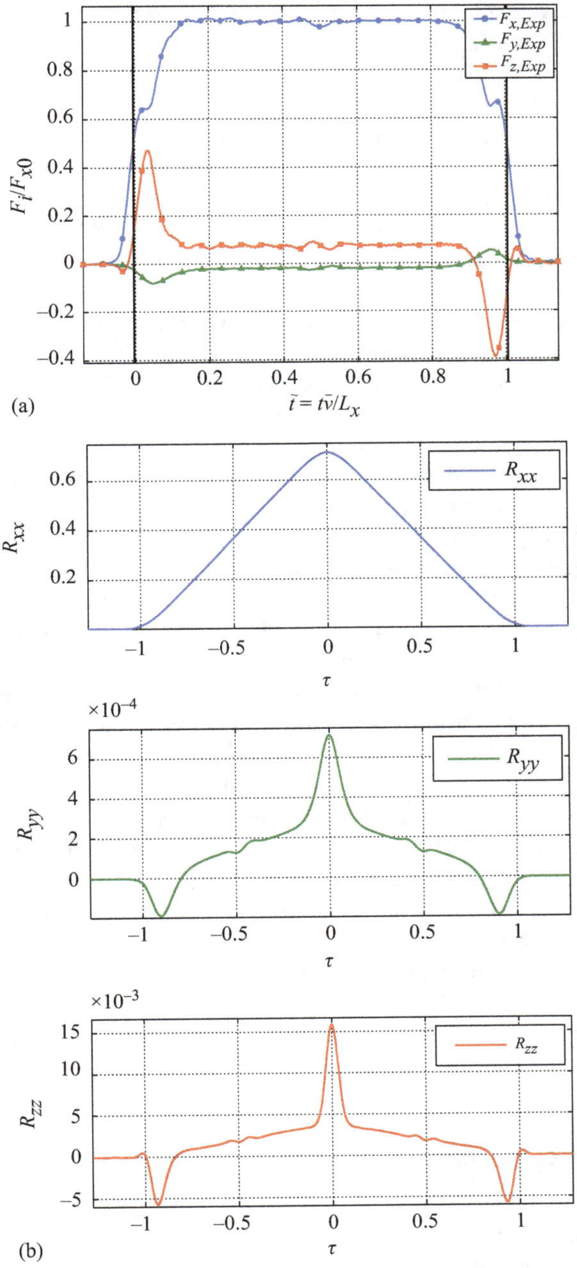

Figure 4.9 Representative signals of the three components of the measured force $F_i(t)$ normalized to max $(F_x(t))$: (a) force signals and (b) autocorrelation function $R_{ii}(\tau)$ of each component

In this normalized representation, important features of the typical force signals can be observed. The x-component of the measured force $F_x(t)$ is the dominant component and has square-wave-like shape. At $\tilde{t} = 0$ and $\tilde{t} = 1$, the force component has reached approximately 50% of its maximum which could be observed in all experimental studies. The y- and z-component of the measured force $F_y(t)$ and $F_z(t)$ show significant peaks near the edges of the UUT and a decreased, nearly constant, plateau in between. When the PM is almost completely covered by the UUT (near the UUT edges), then the absolute value of the peak of $F_y(t)$ and $F_z(t)$ becomes maximal. The force at $\tilde{t} \approx 0.1$ is in the opposite direction in comparison to $\tilde{t} \approx 1 - 0.1$, for both components.

Figure 4.9(b) shows the dimensionless autocorrelation function corresponding to the normalized force components $F_i(t)$ plotted against the dimensionless time-lag τ. As expected from the convolution of a rectangular function, the autocorrelation of the x-component $R_{xx}(\tau)$ is very similar to the triangular function with its maximum at zero lag $\tau = 0$. The autocorrelation function of the y- and z-component, $R_{yy}(\tau)$ and $R_{zz}(\tau)$, is very similar in shape. Due to similar peaks but in opposite directions, the autocorrelation shows negative correlation at $\tau \approx -0.9$ and $\tau \approx +0.9$, and a very sharp peak at $\tau = 0$ for both components.

It is important to mention that due to the performed normalization of the measured signals and the use of a characteristic time scale, the presented autocorrelation plots are representative in shape and magnitude for most studied experiments. All three components of the measured force are real valued, thus the autocorrelation function must be symmetric. According to Parseval's theorem, the autocorrelation function $R_{ii}(0)$ at zero lag $\tau = 0$ corresponds to the total power of the respective signal.

It is shown that all components do fulfill the described requirements and are therefore suitable for estimating the individual time-lags, but not all signals are equally appropriate. $R_{yy}(\tau)$ and $R_{zz}(\tau)$ clearly show a narrow peak at $\tau \approx 0$, but the total signal power of $R_{xx}(\tau)$ is two to three orders of magnitude higher. Under the assumption of a similar noise power for the nondeal measurement process, it is evident that the force component with the largest signal power provides the most reliable estimate for the time delay τ_{est} between two members of the artificial signal ensemble. In conclusion, it becomes clear that the x-component of the force is the most reliable source of information for the assembling of an artificial signal ensemble of similar experiments from multiple instances. Therefore, the cross-correlation of the x-components is used in order to estimate the time-lag between the signals and to allow statistical analysis on the nonideal measurement process.

4.2.3 Program flowchart for DSP

In the previous sections, the basic concept of DSP in LET is defined and the mathematical methods for the efficient computation are explained. In the following, an overview of all computations necessary to obtain a complete measurement result is given. This completes the description of the measurement method (cf. Section 4.1.2) and allows to repeat all experimental studies presented in this chapter. Figure 4.10 shows the flowchart of DSP used in LET for the assembling of an artificial signal ensemble of sequential measurements.

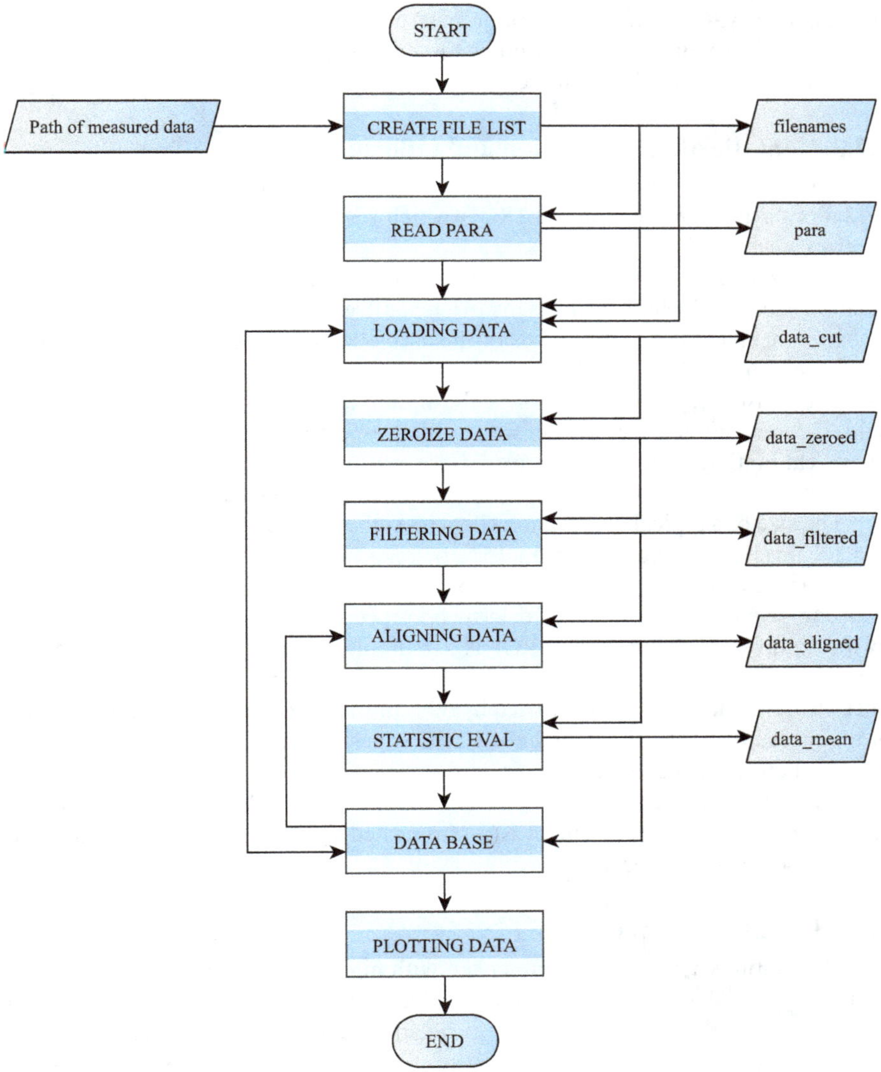

Figure 4.10 Flowchart of DSP applied in LET

4.2.3.1 Loading data

In the first step, a file list with observations is created that are members of the artificial signal ensemble $\{x_k(t)\}$. Next, the database is analyzed for experimental parameters (*ensemble parameters*) of each chosen observation, including all known information about geometry, relative position of the PM, and calibration data of the used sensors (cf. Section 4.1.3). Having ensemble parameters, the respective data files are loaded and tested for consistence, e.g. equal number of recorded channels. The observation files contain ASCII-delimited numeric data of the simultaneously sampled sensor output given in Volt, with exception of the data output of the incremental position encoder, which is stored in millimeters.

4.2.3.2 Static offset correction and tailoring

The next step combines (a) static offset correction, (b) sensor output conversion, and (c) tailoring to region of interest (ROI) which results in the *uncorrected result of measurement*.

The *static offset correction* is performed for each signal, except the incremental position encoder signal. It corrects insufficient offset nulling that is performed before an experiment and long-term trends of the used electronic devices as they may arise by temperature drifts.

The *sensor output conversion* is the multiplication of the sensor output signals of force and acceleration sensors by the respective sensor calibration factors. These calibration factors are derived from sensor-specific gain and sensitivities (cf. Section 4.1.3.3) and are stored in the ensemble parameters.

The size of a typical observation file with signals of 5 s duration is about 5.5 MB and includes on average less than 15% of relevant data where the UUT leads to a significant change in the measured force. To reduce computing time, the *signal tailoring* reduces the amount of data to be processed to an ROI that corresponds to the significant changes. The ROI is a multiple of the UUT's length L_x and is defined relative to the leading and trailing edge of the UUT. Considering findings from Section 4.2.2, the edges are estimated by 50% of the maximum of the absolute value of the measured force. Having identified the edges, the ROI is defined by from N_{ext} samples before the leading edge to N_{ext} after the trailing edge, with $N_{\mathrm{ext}} = f_s k_{\mathrm{ext}} L_x / \bar{v}$, where k_{ext} is the extension factor. The next steps of the DSP are all executed on the uncorrected result of measurement $x_k(t_k)$ (truncated signals) from which the artificial signal ensemble $\{x_k(t)\}$ is built.

4.2.3.3 Filtering data

The observations $x_k(t_k)$ are superimposed with high-frequency noise from multiple sources, e.g. analogue measurement amplifier of the force sensor (Section 4.1.3.3). To improve the estimation of the cross-correlation function, the high frequency noise is reduced by means of a digital low-pass filter. Except for the incremental position encoder, a zero-phase digital filter [308] is applied using the MATLAB™ function `filtfilt`. The designed digital filter is a Butterworth low-pass filter of fourth order with a cutoff frequency $f_c = 500$ Hz for the 3 dB point below the passband value. The filter is designed to have a maximally flat frequency response in the passband to minimize the distortion of the signals of interest.

4.2.3.4 Aligning data

The next step of DSP in LET is the alignment procedure which uses the x-component of the measured force $F_x(t)$ (Figure 4.11). Two alignment procedures associated with different assumptions can be applied.

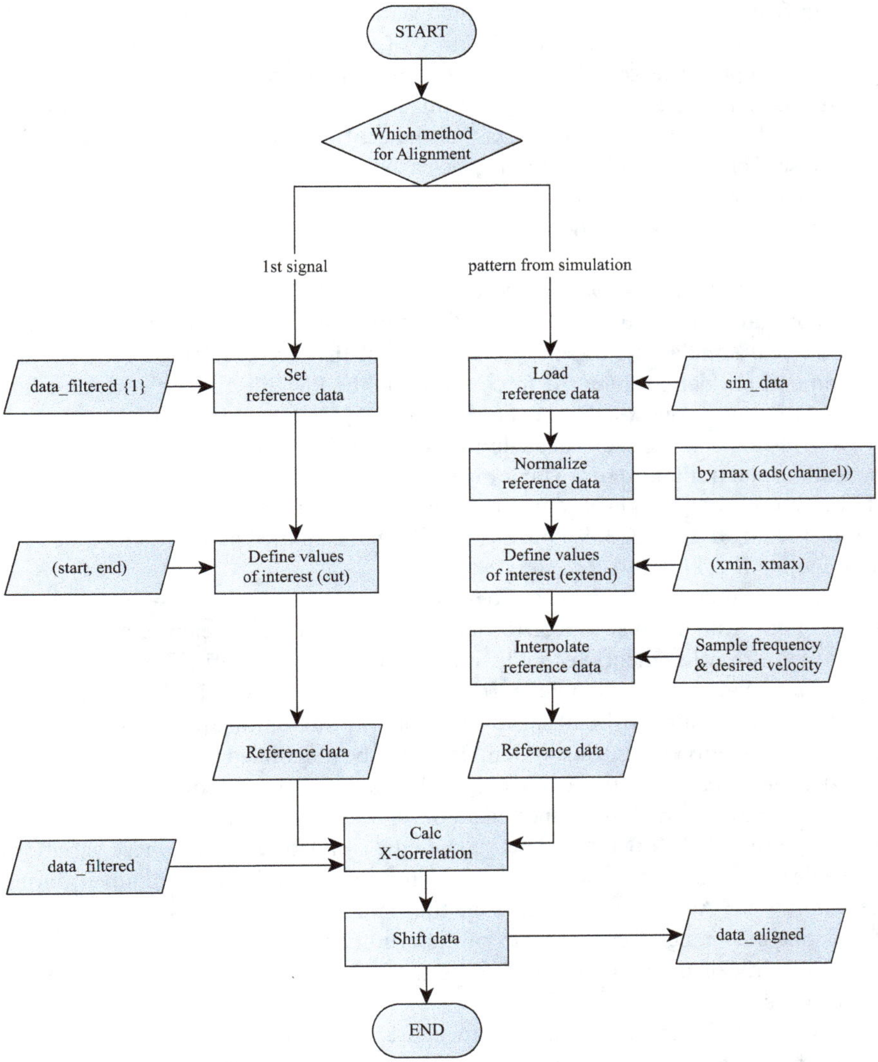

Figure 4.11 Flowchart of the aligning-data-block

4.2.3.4.1 Arbitrary ensemble member as reference signal

The first procedure uses an arbitrary member of the artificial signal ensemble as the reference signal for the estimation of individual time-lags τ_i of remaining ensemble members. The advantage of this procedure is that it can be applied without additional knowledge about the characteristics of the particular signal ensemble $\{x_k(t)\}$. Thus, no information regarding the used PM or the shape and surface of the UUT are necessary. The procedure results in a well-aligned artificial signal ensemble $\{x_k(t)\}$

if a significant pattern is included, i.e. the maximum of R_{ii} is an unbiased estimator of the time-lag τ_i. To increase the reliability of the alignment, the chosen reference signal can be preprocessed with additional digital filters to reduce impact of random disturbances. The main disadvantage of this approach is the missing possibility to properly define the range of the x-axis for the artificial signal ensemble $\{x_k(t)\}$ due to the insufficient realization for a trigger-based signal alignment (Section 4.2.1.2). As a rough approximation, the above-mentioned characteristic 50% threshold in the x-component of the measured force can be used.

4.2.3.4.2 QSA simulation as reference signal

The second alignment procedure uses for the estimation of individual time-lags results of numerical simulations as the reference signal that are performed for a similar experiment. For determining the force acting on the PM, the finite element method (FEM) [172] with a quasi stationary approach (QSA) [102] is used. The QSA approach delivers results close to the exact solution if the UUT has constant cross section and the leading and trailing edge are far away from the PM. The software package Comsol Multiphysics v.4.4 [151] is used in this framework. The FEM results used throughout this chapter are based on [309]. Further details on the applied numerical methods can be found in [102,112,309].

This alignment procedure has two major advantages. Firstly, the reference signal is free of any kind of noise and distortions, which, most likely, leads to an increased reliability of the alignment. Secondly, it provides a properly defined range of x-axis for the artificial signal ensemble $\{x_k(t)\}$. This x-axis range can be used as a global reference for the entire artificial signal ensemble and thus eliminates the need of using the external position measuring system. It should be mentioned that the evaluation of recorded incremental position encoder signals has shown nonequidistant steps, which results in position variations of individual observations, as well as the corresponding velocity of the UUT. If the x-axis range from the simulation is used, these variations are neglected. Therefore, the measured incremental positions are retained during the whole DSP. As it can be seen later, this information will expose the velocity $v(t)$ as one of the significant sources of distortion and will provide important hints on the validity of the assumption about statistical independence of the successive experiments.

The major disadvantage of this procedure is the necessity of using "suitable results" from numerical simulations, whereas no clear definition of "suitable" can be provided. Besides comparable geometries, it is not evident, what kind of deviations are acceptable for a robust alignment procedure. It is shown in Chapter 2 that the remanence B_r does not affect the electromagnetic similarity, i.e. it can be considered as a scaling factor of the second power for the generated force components. Because τ_{est} in (4.13) is invariant to the magnitude of the correlated signals, it is evident that τ_{est} is invariant to B_r. Furthermore, the product σv is expected to have only marginal influence on the characteristic shape of the Lorentz force signal for low R_m, thus τ_{est} of the x-component of the Lorentz force might also be invariant under these conditions. All other parameters, e.g. permeability of the PM or anisotropy of the UUTs conductivity, are expected to have an influence on the characteristic shape of the x-component of the Lorentz force and as they are not known precisely, no prediction on the general robustness of the procedure can be given.

The step of signal alignment results in the artificial signal ensemble $\{x_k(t)\}$, which is reduced to the ROI and is low-pass filtered to be more suitable for interpretation. The sets of signals are stored in the data base and are available for statistic evaluation of the signal ensemble. All measurement results shown in this chapter are aligned by using results of numerical simulations as the reference signal.

4.2.3.5 Statistic evaluation of the signal ensemble

Having defined the artificial signal ensemble $\{x_k(t)\}$, it is possible to calculate an estimate of the statistical properties of the nonstationary signal along the ensemble plane (cf. Figure 4.8). According to the International Organization for Standardization (ISO) *Guide to the Expression of Uncertainty in Measurements* [310], it is recommended to use classical (frequentist) statistics [311,312] for the evaluation of uncertainty from measurements (*Type A*). This interpretation of the concept of probability contradicts *Bayesian statistics*. However, as shown by [313], in practice the results from classical statistics are almost equal to a consistent Bayesian approach.

The first statistical property of the nonstationary signal is the expected value $\bar{x}(t)$ of the measurand $x(t)$. For a series of K measurements, the expected value is calculated as

$$\bar{x}(t) = \frac{1}{K}\sum_{k=1}^{K} x_k(t), \tag{4.20}$$

where $x_k(t)$ denotes the kth observation.

The *experimental standard deviation* $s[x(t)]$ (sample standard deviation) is the most common quantity for characterizing the dispersion of the measurement result. It is given by

$$s[x(t)] = \frac{1}{\sqrt{K-1}}\sqrt{\sum_{k=1}^{K} [x_k(t) - \bar{x}(t)]^2}. \tag{4.21}$$

It is worth to mention that using Bessel's correction ($K - 1$ instead of K in the denominator) s^2 is the *unbiased* estimator of the population variance while its positive square root s is a *biased* estimator of the population standard deviation. This bias can be significant for small numbers of observations K and thus has to be considered in the DOE.

In order to estimate the standard deviation of the distribution of $\bar{x}(t)$, the *experimental standard deviation of the mean* (eSDM) is defined as

$$u[\bar{x}(t)] = s[\bar{x}(t)] = \frac{s[x(t)]}{\sqrt{K}} \tag{4.22}$$

also termed *standard uncertainty* [310].

The set of signals and basic statistic properties are stored in the database, together with the underlying reference signal and a log file including important parameters of the DSP, e.g. parameters of applied filters.

The last function block of the DSP is the plotting procedure, which is a problem specific task adapted for each individual investigation.

4.2.4 Experimental study

In this section, one study is presented to give an overview of the measurement performance of the developed experimental setup for applications including artificial defects. The focus in this study is on the investigation of defect responses.

The study is performed on a UUT made of stacked aluminum sheets which allows to include artificial defects at different depths. The investigation illustrates principal characteristics of the defect responses for the measured force $\mathbf{F}(t)$ and the induced voltage $V_z(t)$ from the 1-axis DiLET sensor. The artificial defect used in the study is chosen to be especially large to assure a significant defect response signal (DRS) which can serve as a reference for the characteristics of typical defect responses.

The first UUT is a stack of 24 aluminum sheets, each with size of 250 mm × 50 mm × 2 mm. Due to size tolerance and shape deviations, the complete stack adds up to a height of about 50 mm. The average electrical conductivity measured for each sheet is equal to $\bar{\sigma} \pm 2u(\bar{\sigma}) = (30.48 \pm 0.02)$ MS/m at 20°C. It is determined at six measurement points for each sheet using an ECT device Elotest N300 [314] by *Rohmann GmbH*. Because all sheets are naturally coated with a thin aluminum oxide layer, they are treated as electrically insulated from each other. In consequence, the UUT is assumed to have an anisotropic electrical conductivity with $\sigma_{zz} = 0$ which is especially important for the discussion of the DRS. The used PM is of cylindrical shape with diameter $D = 22.5$ mm and height $H = 17.6$ mm. It is axially magnetized and the nominal magnetic remanence provided by the manufacturer is $B_{r,N} = 1.43$ T.

Figure 4.12(a) shows a technical drawing of a single sheet with an artificial defect of 5 mm× 5 mm × 2 mm. Two of these defected sheets are used in the second and third layer from top of the stack, what results in a defect depth of $d = 2$ mm. Figure 4.12(b) shows the picture of the UUT mounted on the slide of the linear drive. The UUT is oriented in the longitudinal direction with respect to the direction of travel.

4.2.4.1 Design of experiment

The experimental study consists of repeated observations at different y-positions of the UUT. This collection of observations forms a *scan* from $y_{\text{min}} = -30.00$ mm to $y_{\text{max}} = 30.00$ mm at steps of $\Delta y = 1.00$ mm. In the ROI regarding the y-position (ROIy), i.e. from $y_{\text{D,min}} = -5.00$ mm to $y_{\text{D,max}} = 5.00$ mm, the scan is refined using a step size of $\Delta y_{\text{D}} = 0.50$ mm. The measurements are performed $K = 5$ times outside the ROIy and $K_{\text{D}} = 25$ times inside the ROIy. The distinct increase of repetitions in the close vicinity of the defect is expected to further reduce the experimental standard deviation. All parameters of the experimental study of a stacked aluminum sheets are summarized in Table 4.1.

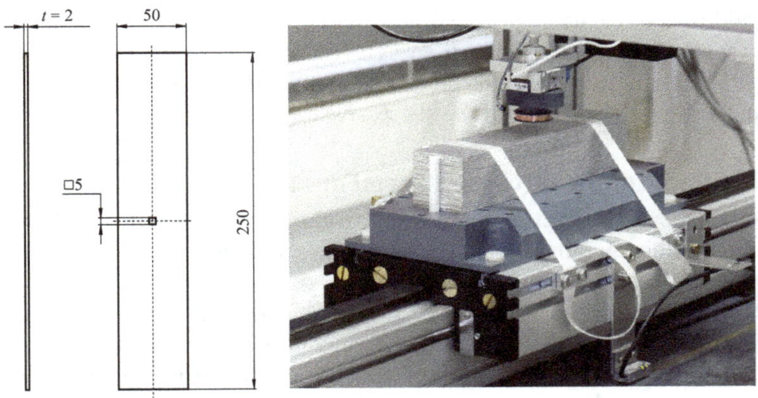

(a) (b)

Figure 4.12 Technical drawing of defected sheet (a) and fully assembled UUT (b). The UUT consists of 22 defect-free aluminum sheets and 2 defected sheets of the same material in the second and third layers from top. The resulting defect is 5 mm × 5 mm × 4 mm at depth d = 2 mm

*Table 4.1 Parameters of the experimental study of a stacked aluminum sheets**

Parameter		Value	Unit	Description
DOE	K	5		No. of observations
	K_D	25		No. of observations at ROIy
	v	0.50	m/s	Velocity of the UUT
	h	1.00	mm	Lift-off distance
	y_{max}	30.00	mm	Max. y-position of the PM
	y_{min}	−30.00	mm	Min. y-position of the PM
	$y_{D,max}$	5.00	mm	Max. y-position at ROIy
	$y_{D,min}$	−5.00	mm	Min. y-position at ROIy
	Δy	1.00	mm	Step size of the scan
	Δy_D	0.50	mm	Step size of the scan at ROIy
UUT	$\sigma_{xx,yy}$	30.48	MS/m	Electrical conductivity of each sheet
	μ_r	1		Relative permeability
	l_X	250.0	mm	Length of the UUT
	l_Y	50.0	mm	Width of the UUT
	l_Z	50.0	mm	Height of the UUT
PM	$B_{r,N}$	1.43	T	Nominal remanence
	D	22.5	mm	Diameter of the PM
	H	17.6	mm	Height of the PM

*Values rounded to significant figures with respect to the standard deviation.

In addition to the main purpose of illustrating the principal characteristics of the defect responses, this study serves as a reference for DRSs of UUTs with anisotropic electrical conductivity. Besides that, the study unintentionally contests the assumption about statistical independence of the successive experiments.

4.2.4.2 Result of the force measurement $\mathbf{F}(t)$

Figure 4.13 shows the measurement result of the force $\mathbf{F}(t) = [F_x(t), F_y(t), F_z(t)]^\mathrm{T}$ for the monolithic aluminum bar. The three surface-plots show the expected value of the respective force component $\bar{F}_i(t)$, where $i \in \{x, y, z\}$. The black curves in each plot show a complete measurement result as the estimate of the 95% confidence interval $[\bar{F}_i(t) \pm 2u(\bar{F}_i(t))]$ for selected experiments with $\Delta y = 5$ mm. These data are plotted against y, the lateral position of the PM, and the expected value $\bar{x}(t)$ of the centered position of the slide of the linear drive, based on the records delivered by the incremental encoder. The positive x-positions are sampled first in time.

The second important aspect of the comparison with the result of the monolithic aluminum bar is that the signal perturbations of all three components of the measured force at the edges of the UUT are more pronounced for the stacked aluminum sheets, although the used PM as well as the external dimensions of the UUT are the same. A possible explanation of this observation is the anisotropic electrical conductivity of the stacked aluminum sheets, which prevents the generation of ECs in the z-direction.

4.2.4.3 Result of the force DRS $\Delta\mathbf{F}(t)$

Figure 4.14 shows the estimated DRS of the measured force $\Delta\mathbf{F}(t)$ from $x_{\mathrm{D,min}} = -30.00$ mm to $x_{\mathrm{D,max}} = 30.00$ mm. The DRS of each force component $\Delta F_i(t)$ is estimated by subtracting the respective average of the force at $x_{\mathrm{D,min}}$ and $x_{\mathrm{D,max}}$, which is an estimate of the force $F_{i0}(t)$ without a defect present.

All three components $\Delta F_i(t)$ show significant responses to the artificial defect. For the x-component of the force DRS $\Delta F_x(t)$ a maximum absolute value of 77 mN is found. Considering the maximum value of the force in x-direction $|F_x(t)|_{\mathrm{max}} = 2.35$ N, a *relative* DRS is given as $|\Delta F_{x,\mathrm{rel}}(t)|_{\mathrm{max}} = 3.3\%$. The y- and z-components show maximum absolute values of $|\Delta F_y(t)|_{\mathrm{max}} = 19$ mN and $|\Delta F_z(t)|_{\mathrm{max}} = 52$ mN, respectively. In comparison to $\Delta F_x(t)$, the DRSs $\Delta F_y(t)$ and $\Delta F_z(t)$ are less affected by surrounding distortion patterns and thus better recognizable. All three components $\Delta F_i(t)$ are spatially restricted in x-direction to a length of ≈ 40 mm and ≈ 30 mm in y-direction. The spatial distribution of the defect response depends more on the size of the PM than on the defect size. This observation corresponds to the *point spread function* of simple PMs which is discussed in [102] and is a typical result in cases where the characteristic length of the defect is smaller than that of the PM.

All three force DRS components are superimposed with significant oscillations, which are similar to the measured accelerations (not shown). The applied hill shading

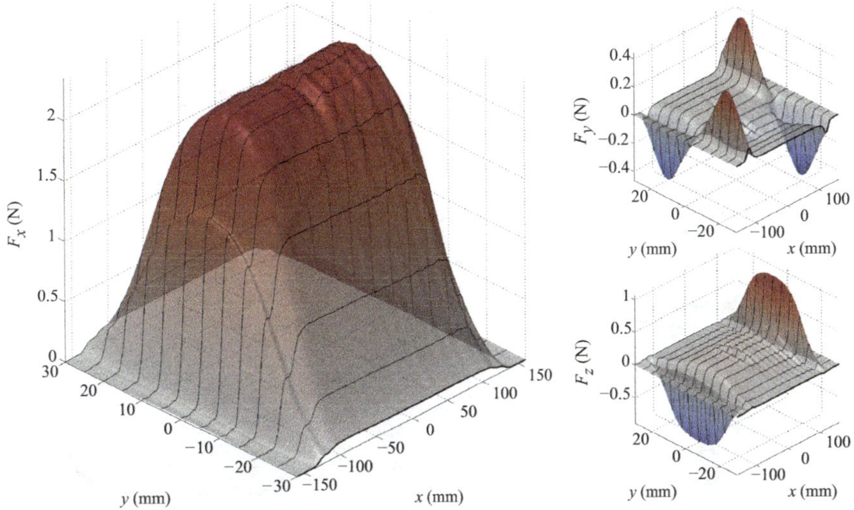

Figure 4.13 *Measurement result of the force* $\mathbf{F}(t)$ *for the stacked aluminum sheets including a 5 mm × 5 mm × 4 mm defect at depth* $d = 2$ *mm. The surface-plots show the expected value of the measurand* $\bar{F}_i(t)$. *The black curves show the complete measurement result as the estimate of the 95% confidence interval* $[\bar{F}_i(t) \pm 2u(\bar{F}_i(t))]$ *for selected experiments with* $\Delta y = 5$ *mm*

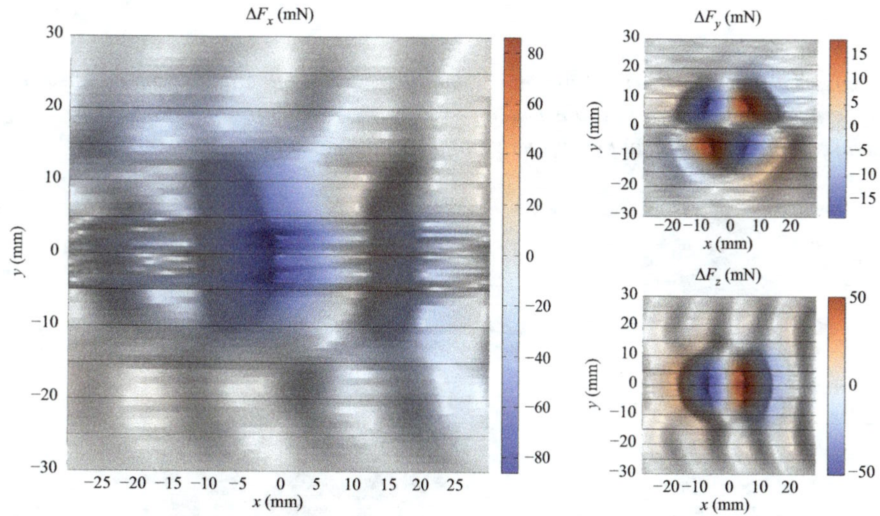

Figure 4.14 Measurement result of the force DRS $\Delta\mathbf{F}(t)$ *from* $x_{D,min} = -30.00$ *mm to* $x_{D,max} = 30.00$ *mm for the stacked aluminum sheets including a 5 mm × 5 mm × 4 mm defect at depth* $d = 2$ *mm. Significant defect responses in all three force components are observable*

technique increases the perception for this effect, although the amplitudes of these oscillations are much smaller (<8 mN) than the respective DRS. The oscillations appear to propagate from the first edge of the UUT (x =125 mm) and are sustainably perturbed by the defect response of the respective component (see $\Delta F_z(t)$).

A surprising observation is made when comparing the two regions with $K = 5$ and $K_D = 25$. While for $\Delta F_y(t)$ and $\Delta F_z(t)$ no significant difference in the distortion characteristics can be observed, so it is in $\Delta F_x(t)$. In the range of $y_{D,min} = -5.00$ mm to $y_{D,max} = 5.00$ mm, a diagonal pattern dominates the signal before and after the defect. This pattern was identified to be proportional to the number of runs in each artificial ensemble as well as to the number of previous observations. The pattern is identified to be an artifact from the calculation of the expected value of the signal ensemble where in each member the underlying distortion is slightly shifted with respect to its respective predecessor. Furthermore, the distortion is proportional to $F_x(t)$ and completely vanishes before and after the UUT passes the PM.

This effect seems to contest the assumption about statistical independence of the successive experiments, which is an assumed property for the developed signal alignment procedure. However, the distortion pattern shows no significant influence on the very low relative experimental standard deviation because the relative amplitude of the distortion is much smaller than the corresponding x-component of the force.

4.2.4.4 Result of the DiLET voltage measurement $V_z(t)$

Figure 4.15 shows the measurement result $V_z(t)$ of the 1-axis DiLET sensor (a) and the corresponding DRS $\Delta V_z(t)$ (b) for gain $g_{DF} = 10$. Since $V_{z0}(t)$ is typically zero mean in the region where the DRS is estimated, $V_z(t)$ and $\Delta V_z(t)$ are equal.

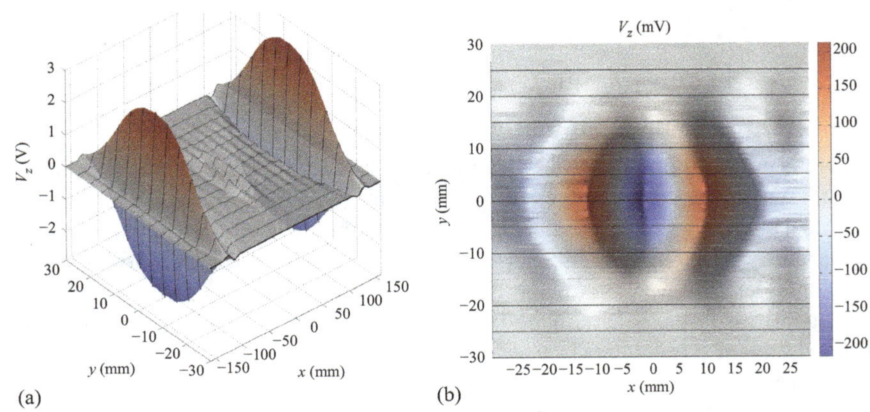

Figure 4.15 Measurement result of the 1-axis DiLET sensor $V_z(t)$ (a) and its DRS $\Delta V_z(t)$ (b) from $x_{D,min} = -30.00$ mm to $x_{D,max} = 30.00$ mm for the stacked aluminum sheets including a 5 mm × 5 mm × 4 mm defect at depth $d = 2$ mm

The measurement result of the induced voltage $V_z(t)$ shows significant responses at the edges of the UUT in the range of $V_z(t) \approx \pm 3$ V. This corresponds to an estimated effective variation of the magnetic flux density of $\partial B_{\mathrm{eff},z}(t)/\partial t \approx \pm\, 0.1049$ T/s. The observed effect is approximately three times stronger than for the monolithic aluminum bar, which corresponds to a dependency described by $V_z(t) \propto R_{\mathrm{m}}^{2.71}$. However, based only on these two studies no solid hypothesis can be formulated and further investigations of the effect are necessary.

The estimated DRS $\Delta V_z(t)$ has a maximum absolute value $|\Delta V_z(t)|_{\max}$ of $\approx$ 0.212 V and shows a similar spatial disproportion as the force DRS. It can be observed that the signal is also superimposed with oscillations of smaller amplitude than 20 mV which results in an estimated signal to distortion ratio of about

$$\mathrm{SDR} = 10 \log_{10}\left(\frac{0.212}{0.02}\right) \approx 10.3\, dB \tag{4.23}$$

4.2.4.5 Concluding remarks

The study of the stacked aluminum sheets with a large artificial subsurface defect has enabled to define the principal characteristics of the DRSs estimated by the measured force $\mathbf{F}(t)$ and the induced voltage $V_z(t)$. The study has built an expectation for the shape and relative magnitude of DRSs of the different sensor components and will serve as a reference experiment for the last experimental study presented in this chapter.

Furthermore, the study revealed a violation of one of the assumptions formulated for the alignment procedure, which will be examined in the last study of this chapter.

To complete this investigation, the key values of the experimental study are summarized in Table 4.2. This table includes also the key values of the measured acceleration to enable the comparison with previous studies, even if they were not discussed explicitly.

4.2.5 Uncertainty analysis

4.2.5.1 Introduction and motivation

Especially in the framework of NDT, the analysis of uncertainties plays an important role during the design process of new systems [315,316]. Although by means of numerical simulations, it is possible to predict the Lorentz force profile but the intrinsic variability of the input parameters was not accounted yet. Hence, one cannot rely on a single deterministic simulation if a comparison to experimental data is intended and a quantification of uncertain model data on the resulting force profiles is essential. As a consequence, it is necessary to identify prior sources of uncertainty to improve any experimental or industrial setup.

To determine the statistical information of the output quantities of a system, such as the mean or the variance, one of the most common methods is Monte Carlo (MC) sampling. The MC method is based on repetitive calculations of the forward model, while defining the random inputs according to their probability distributions. This results in an ensemble of solutions from which statistical properties can be derived.

*Table 4.2 Results of the experimental study of a stacked aluminum sheets**

Parameter		**Value**	**Unit**	**y-position (mm)**
Force	$\max(F_x(t))$	2.350	N	0.00
	$\max(F_y(t))$	0.445	N	18.00
	$\min(F_y(t))$	−0.464	N	−16.00
	$\max(F_z(t))$	1.068	N	−1.00
	$\min(F_z(t))$	−0.886	N	0.00
Force DRS	$\max(\|\Delta F_x(t)\|)$	0.077	N	−0.50
	$\max(\|\Delta F_y(t)\|)$	0.019	N	−7.00
	$\max(\|\Delta F_z(t)\|)$	0.052	N	0.50
Acceleration	$\max(\|A_x(t)\|)$	17.3	mm/s^2	−0.50
	$\max(\|A_y(t)\|)$	2.3	mm/s^2	13.00
	$\max(\|A_z(t)\|)$	23.4	mm/s^2	−0.50
DiLET	$\max(V_z(t))$	3.04	V	−0.50
	$\min(V_z(t))$	−2.85	V	0.00
DiLET DRS	$\max(\|\Delta V_z(t)\|)$	0.212	V	0.50

*Values rounded to significant figures with respect to the standard deviation.

However, a large number of simulations is needed due to the slow convergence rate of the MC method. The mean for example converges with $1/\sqrt{N_d}$, where N_d is the number of deterministic forward calculations [317]. This limits the applicability to problems with low computational cost. Nevertheless, different approaches were developed to improve MC type methods in terms of convergence for example Latin Hypercube sampling (LHS) [318] or Quasi Monte Carlo (QMC) [319,320]. On the other hand, perturbation methods, which are based on Taylor expansions, and operator-based methods, based on Neumann expansion, are restricted to small uncertainty intervals in the in- and output variables of around 10% [321] and will not be further discussed here.

The primary focus of this work rests on spectral methods [322], which are based on the determination of a functional dependence between the probabilistic in- and output of a system by means of a series of suitable selected functionals. The practical realization of spectral methods can be further subdivided into intrusive and nonintrusive approaches. Intrusive approaches are based on Galerkin methods, where the governing equations have to be modified to incorporate the probabilistic character of the model parameters. This includes the determination of the stochastic weak form of the problem according to the given uncertainties [323]. On the contrary, nonintrusive approaches are based on a reduced sampling of the probability space without any modification of the deterministic solvers. Those methods are more flexible and thus more suitable for universal application. Typical applications can be found in the fields of computational fluid dynamics [324–326], heat transfer [327,328], multibody dynamics [329,330], robust design optimization [331], or in biomedical engineering [157,332]. During the last years, spectral approaches are becoming increasingly popular in an engineering framework. However, those are not the reference tools yet and still unknown for many people. For that reason, particular emphasis is placed to describe the nonintrusive version of the generalized polynomial chaos (gPC) before it is applied and results are presented.

At first, the theoretical background of the gPC expansion is described in Section 4.2.5.2. It is followed by the definition of the stochastic problem at hand in Section 4.2.5.3, including a description how the formalism can be applied to an LET scenario. In Section 4.2.5.4, the model parameters are investigated in terms of their variabilities, which is a vital part of every uncertainty analysis. Finally, the results are presented and discussed in Section 4.2.5.5 and conclusions are drawn in Section 4.2.5.6.

4.2.5.2 The Generalized Polynomial Chaos Method

At first, the theoretical background of the gPC method is explained before it is adopted to the LET problem.

The d parameters of interest, which are assumed to underlie a distinct level of uncertainty, are modeled as a d-variate random vector denoted by $\boldsymbol{\xi} = (\xi_1, \xi_2, \ldots, \xi_d)$. It is defined in the probability space (Θ, Σ, P). The event or random space Θ contains all possible events. Σ is a σ-Algebra over Θ, containing sets of events, and P is a function assigning the probabilities of occurrence to the events. The number of random variables d determines the dimension of the uncertainty problem. It is assumed that the parameters are statistically mutually independent from each other. In order to perform a gPC expansion, the random variables must have a finite variance, which defines the problem in the L_2-Hilbert space.

The probability density function (pdf) $p_i(\xi_i)$, with $i = 1, \ldots, d$, has to be defined for each random variable ξ_i. In the present framework, it is advisable to transform the pdf from the original space into the normalized space by shifting and scaling, assuming standardized random variables. For example, uniform- or β-distributed random variables are shifted and scaled from a range of (a, b) to $(-1, 1)$. In the case of uncorrelated random variables, the joint pdf is given by:

$$p(\boldsymbol{\xi}) = \prod_{i=1}^{d} p_i(\xi_i). \tag{4.24}$$

The probability space in case of, for example, uniform- or β-distributed random variables is an N-dimensional hypercube of size $(-1, 1)^d$.

The quantity of interest (QOI), which will be analyzed in terms of the random variables $\boldsymbol{\xi}$, is $y(\mathbf{r})$. It may depend on some external parameters $\mathbf{r} = (r_0, \ldots, r_{R-1})$ like space, where $R = 3$, or any other situation dependent parameters. Those are treated as deterministic and are not considered in the uncertainty analysis. A separate gPC expansion has to be performed for every considered parameter set $\mathbf{r}$. The discrete number of QOIs is denoted as N_y.

The basic concept of the gPC is to find a functional dependence between the random variables $\boldsymbol{\xi}$ and the solutions $y(\mathbf{r}, \boldsymbol{\xi})$ by means of an orthogonal polynomial basis $\Psi(\boldsymbol{\xi})$. In its general form, it is given by:

$$y(\mathbf{r}, \boldsymbol{\xi}) = \sum_{\alpha \in \mathscr{A}(\mathbf{P})} u_\alpha(\mathbf{r}) \Psi_\alpha(\boldsymbol{\xi}). \tag{4.25}$$

Table 4.3 Connection between probability distributions and polynomial basis functions [321]

Type	Distribution	Orthogonal polynomials	Range
continuous	uniform	Legendre	(a, b)
continuous	β	Jacobi	(a, b)
continuous	Gaussian	Hermite	$(-\infty, +\infty)$
discrete	Poisson	Charlier	$(0, 1, \ldots)$

The terms are indexed by the multi-index $\boldsymbol{\alpha} = (\alpha_0, ..., \alpha_{d-1})$, which is a d-tuple of non-negative integers $\boldsymbol{\alpha} \in \mathbb{N}_0^d$. The sum is carried out over the multi-indices, contained in the set $\mathscr{A}(\mathbf{P})$. The composition of the set depends on the type of expansion and is parameterized by a parameter vector $\mathbf{P}$, which will be explained in a later part of this section.

The function $\Psi_{\boldsymbol{\alpha}}(\boldsymbol{\xi})$ is the joint polynomial basis functions of the gPC. They are composed of polynomials $\psi_{\alpha_i}(\xi_i)$.

$$\Psi_{\boldsymbol{\alpha}}(\boldsymbol{\xi}) = \prod_{i=1}^{d} \psi_{\alpha_i}(\xi_i). \tag{4.26}$$

The polynomials $\psi_{\alpha_i}(\xi_i)$ are defined for each random variable separately according to the corresponding pdf $p_i(\xi_i)$. They have to be chosen to ensure orthogonality. Table 4.3 provides examples of the relationship between the families of orthogonal polynomials and different types of pdfs. The set of polynomials for an optimal basis of continuous probability distributions is derived from the Askey scheme [333]. The index of the polynomials denotes its order (or degree). In this way, the multi-index $\boldsymbol{\alpha}$ corresponds to the order of the individual basis functions forming the joint basis function.

4.2.5.2.1 β-distribution

Pdf is defined in interval (a, b) as:

$$p(x) = B(a, b, p, q)^{-1}(x-a)^{p-1}(b-x)^{q-1}, \tag{4.27}$$

$$B(a, b, p, q) = \frac{\Gamma(p)\Gamma(q)}{\Gamma(p+q)}(b-a)^{p+q-1}. \tag{4.28}$$

Thus, pdf in interval $(-1, 1)$ is given by:

$$p(x) = B(p, q)^{-1}(x+1)^{p-1}(1-x)^{q-1}, \tag{4.29}$$

$$B(p, q) = \frac{\Gamma(p)\Gamma(q)}{\Gamma(p+q)}2^{p+q-1}, \tag{4.30}$$

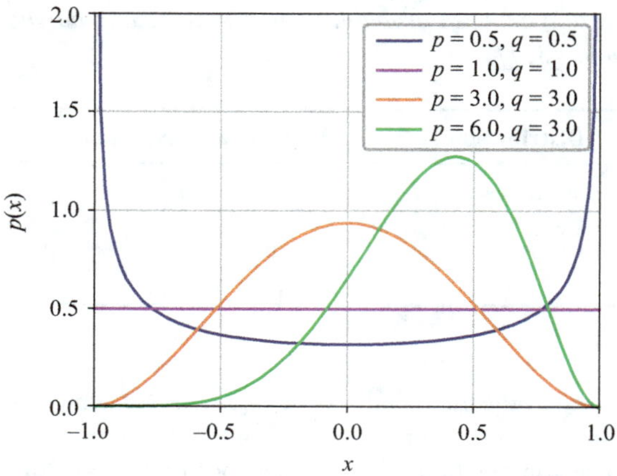

Figure 4.16 Pdfs for different β-distributions parameterized by the shape parameters p and q

with shape parameters $p, q > 0$ (Figure 4.16). The uniform distribution can be obtained for $p = q = 1$.

Jacobi polynomials $P_n^{p,q}(x)$: Orthogonal with respect to weight function $(x+1)^{p-1}(1-x)^{q-1}$ on interval $[-1, 1]$ (Figure 4.17).

$$\int_{-1}^{1} P_n^{p,q}(x)P_m^{p,q}(x)(x+1)^{p-1}(1-x)^{q-1}\,\mathrm{d}x$$

$$= \frac{2^{q+p-1}}{2n+q+p-1}\frac{\Gamma(n+q)\Gamma n+p}{\Gamma(n+q+p-1)n!}\delta_{mn}, \tag{4.31}$$

$$P_0^{(1,1)}(x) = 1, \tag{4.32}$$

$$P_1^{(1,1)}(x) = x, \tag{4.33}$$

$$P_2^{(1,1)}(x) = \frac{1}{2}(3x^2 - 1), \tag{4.34}$$

$$P_3^{(1,1)}(x) = \frac{1}{2}(5x^3 - 3x). \tag{4.35}$$

4.2.5.2.2 Orthogonality

Orthogonality of the polynomial basis for the univariate case is given by:

$$\int_{\Theta} \psi_m(\xi_i)\psi_n(\xi_i)p_i(\xi_i)\,\mathrm{d}\xi_i = \delta_{mn}\left\langle \psi_m^2 \right\rangle, \tag{4.36}$$

where δ_{mn} is the Kronecker delta function, which equals 1 for $m = n$ and 0 otherwise.

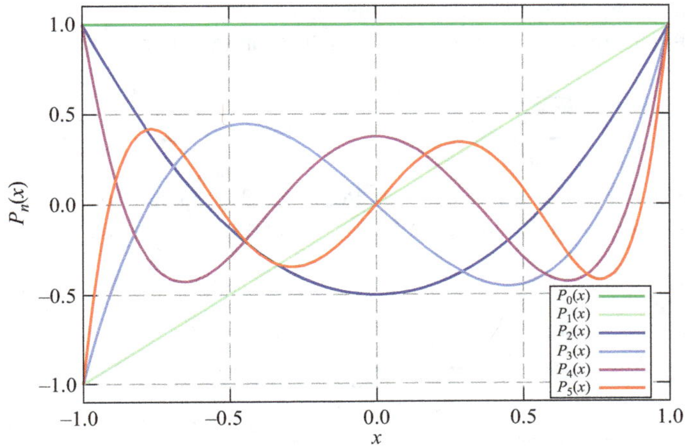

Figure 4.17 Unnormalized Jacobi polynomials for $p = 1$ and $q = 1$ (Legendre polynomials) corresponding to uniform distribution

The term $\langle \cdot, \cdot \rangle$ denotes the inner product with respect to the pdf and is defined as:

$$\langle x, y \rangle = \int_{\Theta} x(\xi) y(\xi) p(\xi) \, \mathrm{d}\xi. \tag{4.37}$$

The orthogonality also holds for the joint polynomial basis functions $\Psi_{\alpha}(\boldsymbol{\xi})$ from (4.26) even if different families of orthogonal polynomials are considered in the gPC, which means that an individual pdf can be assigned to each random variable:

$$\int_{\Theta} \Psi_{\mathbf{m}} \Psi_{\mathbf{n}} p(\boldsymbol{\xi}) \, \mathrm{d}\boldsymbol{\xi} = \prod_{i=1}^{d} \int_{\Theta} \psi_{m_i}(\xi_i) \psi_{n_i}(\xi_i) p_i(\xi_i) \, \mathrm{d}\xi_i \tag{4.38}$$

$$= \prod_{i=1}^{d} \delta_{m_i n_i} \left\langle \psi_{m_i}^2 \right\rangle \tag{4.39}$$

$$= \delta_{\mathbf{mn}} \left\langle \Psi_{\mathbf{m}}^2 \right\rangle, \tag{4.40}$$

where $\mathbf{m} = (m_1, \ldots, m_d)$ and $\mathbf{n} = (n_1, \ldots, n_d)$ are multi-indices. As soon as one pair of indices is not equal $m_i \neq n_i$, the right-hand side vanishes and orthogonality is ensured. The only nonvanishing combination is when $\mathbf{m} = \mathbf{n}$, i.e. $(m_1 = n_1, \ldots, m_d = n_d)$.

4.2.5.2.3 Normalization

It is advantageous to normalize the polynomials to form an ortho*normal* basis. The normalization factors of the polynomials are given by the square root of the $\left\langle \psi_i^2 \right\rangle$ terms in (4.39).

For the probabilists Hermite polynomials $He_n(\xi)$, which are orthogonal with respect to the normal distribution, the normalization factor can be derived by:

$$\int_{-\infty}^{\infty} [He_n(\xi)]^2 \underbrace{\frac{1}{\sqrt{2\pi}} e^{-\frac{\xi^2}{2}}}_{p(\xi)} \, d\xi = n! \tag{4.41}$$

For the Jacobi polynomials $P_n^{(q-1,p-1)}(\xi)$, which are orthogonal with respect to the β distribution, the normalization factor in terms of the shape parameters p and q is determined by:

$$\int_{-1}^{1} \left[P_n^{(q-1,p-1)}(\xi)\right]^2 \underbrace{\frac{1}{B(p,q)}(1-\xi)^{q-1}(1+\xi)^{p-1}}_{p(\xi)} \, d\xi \tag{4.42}$$
$$= \frac{1}{B(p,q)} \frac{2^{q+p-1}}{2n+q+p-1} \frac{\Gamma(n+q)\Gamma(n+p)}{\Gamma(n+q+p-1)n!}.$$

Each polynomial is normalized individually by dividing it by the corresponding normalization factor before the joint polynomial basis function from (4.26) is constructed. This considerably simplifies the postprocessing of the gPC expansion, which will be described in a later part of this section. From now on, it is assumed that all polynomials are normalized.

4.2.5.2.4 Set of basis functions

In general, the set $\mathscr{A}(\mathbf{P})$ of multi-indices can be freely chosen according to the problem under investigation. Two example sets of basis functions in case of $d = 3$ random variables and $P_g = 10$ are shown in Figure 4.18. The blue boxes correspond to polynomials included in the gPC expansion. The coordinates of the boxes correspond to the multi-indices $\boldsymbol{\alpha}$ which correspond to the polynomial degrees of the individual basis functions forming the joint basis functions. The set in Figure 4.18(a) corresponds to a full order gPC.

The number of basis functions, and hence, coefficients to determine, increases exponentially in this case ($N_c = (P+1)^d$). In practical applications, the more economical maximum total order gPC is preferably used. In this case, the set $\mathscr{A}(P_g)$ includes all polynomials whose total order does not exceed a predefined value P_g.

$$\mathscr{A}(P_g) = \left\{ \boldsymbol{\alpha} : \sum_{i=1}^{d} \alpha_i \leq P_g \right\} = \left\{ \boldsymbol{\alpha} : \|\boldsymbol{\alpha}\|_1 \leq P_g \right\}. \tag{4.43}$$

The reduced set of basis functions is shown in Figure 4.18(b). The number of multi-indices, and hence, the dimension of the space spanned by the polynomials, is [334]:

$$N_c = \binom{d+P_g}{d} = \frac{(d+P_g)!}{d!P_g!}. \tag{4.44}$$

For elliptic PDEs, the gPC expansion (4.25) converges in the mean square sense as P_g increases [335].

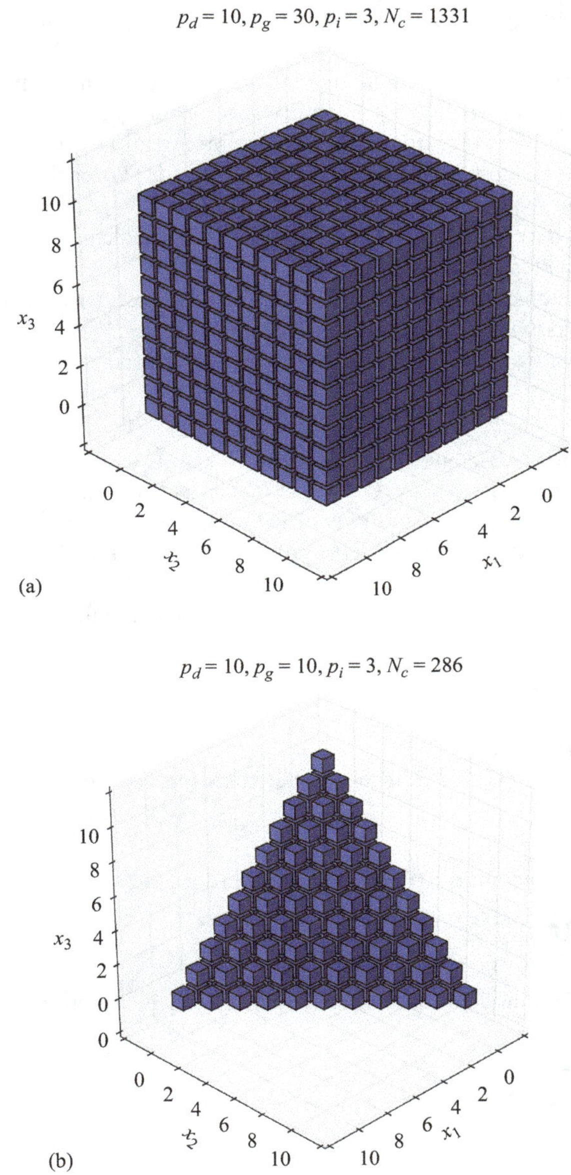

Figure 4.18 *Set of basis functions in case of $d = 3$ random variables: (a) full-order gPC including all polynomial combinations up to an order $P = 10$ and (b) truncated gPC restricted to the maximum global expansion order of $P_g = 10$*

4.2.5.2.5 Regression approach to determine the gPC coefficients

After constructing the polynomial basis, the corresponding gPC-coefficients $u_{\alpha}(\mathbf{r})$ have to be determined for each output quantity. In this regard, the output variables are projected from the d-dimensional probability space Θ into the N_c-dimensional polynomial space $\mathscr{P}_{N_c}$. In this regard, an analytical approximation of the solutions $y(\mathbf{r}, \boldsymbol{\xi})$ as a function of its random input parameters $\boldsymbol{\xi}$ is derived. This enables computational efficient investigations on its stochastics. Different methods exist to calculate the gPC-coefficients $u_{\alpha}(\mathbf{r})$, whereas in the following, the efficient and straight forward regression approach is described.

In the regression approach, the gPC coefficients are estimated using the method of least squares. Assuming that the model is evaluated $N_g > N_c$ times on some grid $\mathscr{G}$, the gPC expansion from (4.25) can be written in matrix form as:

$$[\mathbf{Y}] = [\boldsymbol{\Psi}][\mathbf{U}], \tag{4.45}$$

whereas $[\mathbf{U}]$ is the coefficient matrix of size $[N_c \times N_y]$, whose columns are the N_c gPC coefficients for each of the N_y QOIs, $[\mathbf{Y}]$ is the solution matrix of size $[N_g \times N_y]$, whose columns are the model solutions of the N_y QOIs, and $[\boldsymbol{\Psi}]$ is the gPC matrix of size $[N_g \times N_c]$, whose columns are the polynomial basis functions evaluated at the grid-points in $\mathscr{G}$.

The gPC-coefficients for each QOI (columns of $[\mathbf{U}]$) can then be found by using iterative solvers or by using the Moore-Penrose pseudo inverse of the gPC-matrix:

$$[\mathbf{U}] = [\boldsymbol{\Psi}]^{+}[\mathbf{Y}]. \tag{4.46}$$

4.2.5.2.6 Postprocessing

After calculating the gPC-coefficients, the statistical moments of the QOIs $y(\mathbf{r}, \boldsymbol{\xi})$ can be calculated.

Mean: The expectation $\mu(\mathbf{r})$ is given by the first gPC coefficients:

$$\mu(\mathbf{r}) = u_{\alpha_0}(\mathbf{r}) \tag{4.47}$$

Variance: The variance $\sigma(\mathbf{r})^2$ can be calculated by summing the remaining coefficients squared:

$$\sigma(\mathbf{r})^2 = \sum_{\alpha \in \mathscr{A} \setminus a_0} u_{\alpha}(\mathbf{r})^2. \tag{4.48}$$

Pdf: The pdfs of the QOIs can be determined by applying sampling strategies such as MC, LHS or QMC with N_{MC} realizations. Since a deterministic functional dependence is calculated by the gPC, a large number of samples such as $N_{\mathrm{MC}} \approx 10^5 - 10^6$ can be used with vanishingly small computational effort compared to direct sampling. The drawn samples can be fitted to an appropriate distribution to parametrize it for further evaluations. Additional statistical tests can be applied to proof the null-hypothesis H_0 between the QOI and the assumed distribution.

Sobol indices: An important part during the investigation of new and partially known systems is an analysis regarding its sensitivity. If it is possible to quantify the sensitivity of a system with respect to its input parameters, it is possible to identify the most important variables which influence the output quantities.

The Sobol indices are variance-based sensitivity measures, indicating the influence of the individual random input variables ξ_i and combinations thereof on the total variances $\sigma(\mathbf{r})^2$ of the QOIs. This concept is also known as analysis of variance (ANOVA). The Sobol indices are determined by a Sobol decomposition [336]. For example, the *first-order*.[2] Sobol indices provide an estimate about the influence of the *single* variables ξ_i on the output. In most cases, they are the most significant indices. The second-order Sobol indices extends the concept by considering combinations between two random variables and so on. The Sobol indices are normalized with respect to the total variance. Hence, the sum over all Sobol indices equals one.

The gPC expansion provides all information to perform the Sobol decomposition in the joint polynomial basis functions and the associated multi-indices [337]. For each Sobol-index, a separate subset of multi-indices $\mathcal{A}_i \subseteq \mathcal{A}$ is constructed, which contains multi-indices that target those polynomials that only depend on the parameter combination under consideration. For example, the set $\mathcal{A}_{(1,0,\ldots,0)}$ consists of multi-indices, which target polynomials basis functions, which depend only on the first random variable and would contain the multi-indices: $\boldsymbol{\alpha}_0 = (1, 0, \ldots, 0)$, $\boldsymbol{\alpha}_1 = (2, 0, \ldots, 0)$, and so on. The Sobol-index $S_i^{(\sigma)}(r)$ is then calculated in the same way as the variance but over the reduced set $\mathcal{A}_i$:

$$S_i^{(\sigma)}(r) = \frac{1}{\sigma^2} \sum_{\alpha \in \mathcal{A}_i} u_\alpha(r)^2. \tag{4.49}$$

Derivative-based sensitivity coefficients: The global derivative-based sensitivity coefficients $S_i^{(\partial)}(r)$ are a measure on the change of the QOIs with respect to the ith random variable. They can be determined by means of the gPC-coefficients and the corresponding partial derivatives of the basis functions. They are given by the following expression [321]:

$$S_i^{(\partial)}(r) = \mathbb{E}\left[\frac{\partial y(\mathbf{r})}{\partial \xi_i}\right] = \sum_{\alpha \in \mathcal{A}} \left(u_\alpha(\mathbf{r}) \int_\Theta \frac{\partial \Psi_\alpha}{\partial \xi_i} p(\boldsymbol{\xi})\, \mathrm{d}\boldsymbol{\xi} \right). \tag{4.50}$$

Example: The general principle of the gPC is illustrated considering the following two-dimensional model function, which represents the exemplary system to be investigated:

$$Y(\xi_1, \xi_2) = 3(1-\xi_1)^2 \mathrm{e}^{-\xi_1^2-(\xi_2+1)^2} - 10\left(\frac{\xi_1}{5} - \xi_1^3 - \xi_2^5\right) \mathrm{e}^{-\xi_1^2-\xi_2^2}$$
$$-\frac{1}{3}\mathrm{e}^{-(\xi_1+1)^2-\xi_2^2}. \tag{4.51}$$

[2]The expression *first order* in this context, means that only polynomials with *one* independent variable are addressed. Those polynomials may be of any degree (order).

In the present case, there exists only one output quantity. The model function is shown in Figure 4.19(a). The input parameters ξ_1 and ξ_2 are modeled as β-distributed random variables. The shape parameters are set to $[p_1, q_1] = [10, 5]$ and $[p_2, q_2] = [3, 12]$ in the region $[a_1, b_1] = [-1, 1]$ and $[a_2, b_2] = [-1, 0]$ for ξ_1 and ξ_2, respectively. Both distributions are shown in Figure 4.19(b). As previously mentioned, the distributions have to be normalized to the range of $(-1, 1)$ in every dimension. Since the random variables are β-distributed, the orthogonal polynomial basis has to be constructed using nth order Jacobi polynomials $P_n^{p,q}(\xi)$ (see Table 4.3).

It is noted that the shape parameters p and q significantly influence the shape of the polynomials. If those differ between the random variables, the polynomial basis differs between the dimensions as well.

In the next step, the grid of sampling points $\mathcal{G}^{\text{gPC}}$ is constructed. In the present example, a gPC expansion of the order $p = 5$ is performed. According to (4.44), this results in a total number of $N_c = 21$ gPC-coefficients. The number of grid points is defined to be the same in every dimension and is set to 6. This results in a total number of $N_g = 36$ grid points. For both random variables ξ_1 and ξ_2, the roots $x^{(1)}$ of the Jacobi polynomials $P_6^{p_1,q_1}(\xi_1)$ are determined since the β-distributions differ from each other. The computational grid $\mathcal{G}^{gPC}$ is constructed by applying the tensor product between two root vectors.

In the next step, the solutions of the model function (4.51) are computed on the grid $\mathcal{G}^{gPC}$. The gPC-coefficients of the output quantity $Y(\xi_1, \xi_2)$ are then determined by solving (4.46). In this way, a polynomial estimate of the model function is available, which is shown in Figure 4.19(c) in its denormalized form. The points indicate the location of the grid points. It can be observed that the grid obeys an adaptive character such that it is concentrated in the region, where the probability of occurrence of the input variables is high. The accuracy of the approximated model function is evaluated considering the absolute difference to the original function.

The results are shown in Figure 4.19(e). A very good agreement in the ROI can be observed, whereas the approximation is less accurate at the borders where the probability of occurrence is low. To compare the pdf of the output variable, the original and the estimated model function are both sampled 10^6 times. In the present example, the evaluation of the original model function is computational not demanding. However, in a real scenario, direct sampling of the original model function is often not practical. The resulting output pdfs of both approaches and their absolute difference are shown in Figure 4.19(d) and (f), respectively.

It can be observed that the gPC provides an accurate estimate requiring only $N_g = 36$ evaluations of the original model function compared to traditional MC approaches with 10^6 direct samples. The mean, the standard deviation, and the sensitivity measures of the output quantity are given in Table 4.4. The first-order Sobol indices $S_i^{(1)}$ indicate that the first variable ξ_1 contributes to 68.2% of the total variance. The second-order Sobol index $S_{12}^{(2)}$ quantifies the impact if both parameters are varied at the same time. As expected, its contribution is substantially lower compared to the linear Sobol indices.

In contrast, the derivative-based sensitivity coefficients $S_i^{(\partial)}$ indicate that the model is less sensitive with respect to the first variable ξ_1. This results from the shape of the model function, which obeys both an increase and a decrease over the investigated

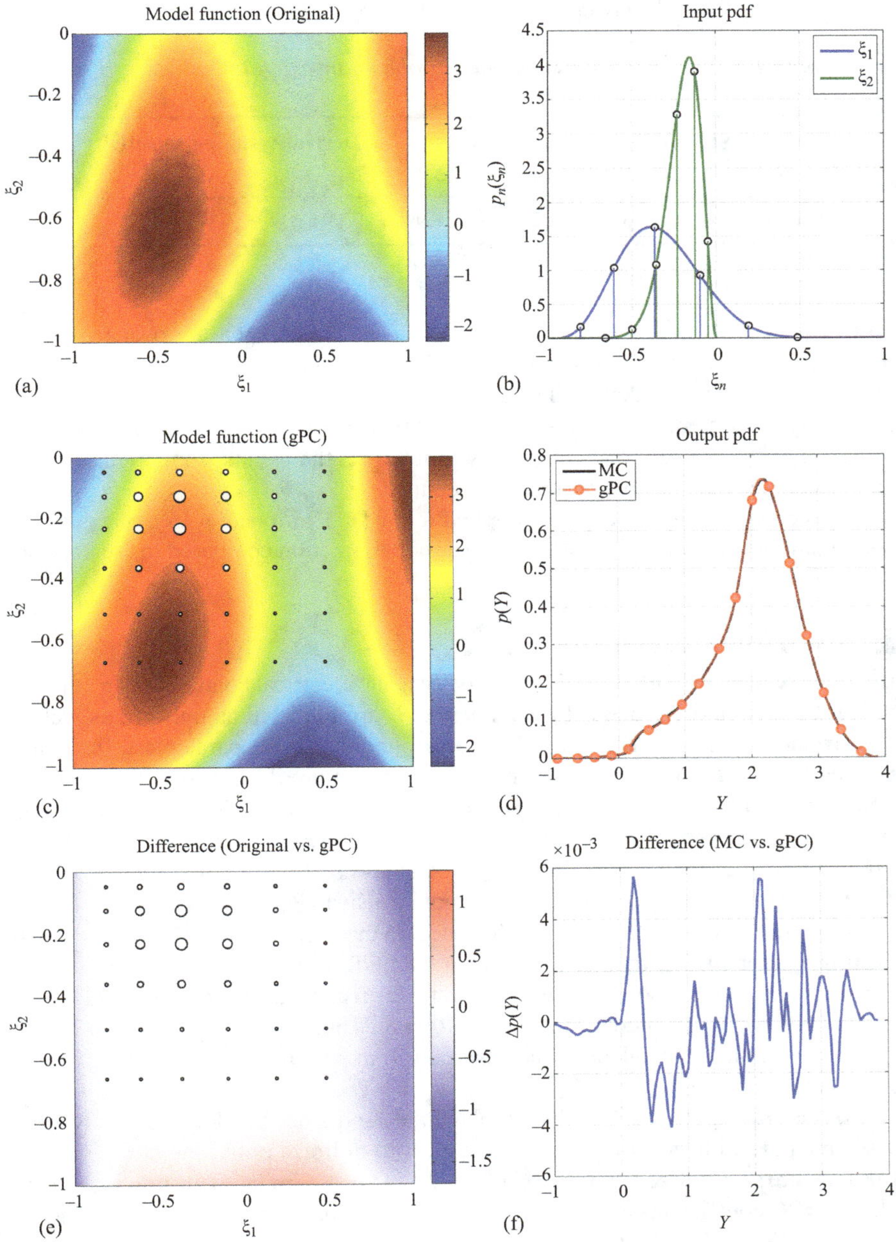

Figure 4.19 *Example containing two random variables ξ_1 and ξ_2 to illustrate the working principle of the gPC. The grid $\mathscr{G}^{gPC}$ contains $N_g = 36$ grid points and the gPC is performed with an order $p = 5$ approximation. (a) Original model function $Y(\xi_1, \xi_2)$; (b) pdfs of the β-distributed random input variables; (c) estimated model function by the gPC; (d) pdf of the output quantity $Y(\xi_1, \xi_2)$; (e) absolute difference of the model function between the original and the gPC estimate; and (f) absolute difference of the output pdf between MC and gPC. The white points in (b), (c), and (e) indicate the location of the sampling points*

Table 4.4 Results of the exemplary problem, investigated with the gPC

Mean	**Std.**	**Sobol indices**			**Derivative-based sensitivity**	
μ	σ	$S_1^{(1)}$	$S_2^{(1)}$	$S_{12}^{(2)}$	$S_1^{(\partial)}$	$S_2^{(\partial)}$
2.0621	0.6685	68.2%	29.9%	1.9%	0.191	−1.835

range of ξ_1. In consequence, the gradients do partly compensate over the whole parameter space. However, it can be observed that the model function slowly decreases in the ξ_2-direction considering the ROI, which explains the negative derivative-based sensitivity coefficient of $S_2^{(\partial)} = -1.835$. The small example demonstrates the working principle of the gPC and the complexity to interpret the results from an uncertainty and sensitivity analysis in a probabilistic context.

In the next section, the gPC is applied to the LET problem, which obeys higher computational cost to evaluate the original model function compared to the previous example.

4.2.5.3 Problem definition in LET

In the present framework of LET, the velocity v, the electrical conductivity σ, the lift-off distance h, and the magnetic remanence B_r are modeled as uniform as well as β-distributed random variables. The corresponding shape parameters $[p, q]$ and bounds $[a, b]$ are determined by experiments, which are described in the next section. In this case, the probability space has $d = 4$ dimensions. It is assumed that the specimen is analyzed along its centerline. The quantities of interest are the drag and lift component of the Lorentz force $F_{x|z}$. The analysis of the side-force F_y is omitted since it vanishes for symmetry reasons. A nonintrusive gPC is applied considering 11 characteristic positions of the PM with respect to the specimen in order to investigate the propagation of uncertainties throughout the LET system. Therefore, the magnetic convection equation is solved in its quasi-static form using the QSA, described in Section 2.4.3.3, assuming moderate magnetic Reynolds numbers ($R_m = \mu_0 \sigma v Z_s/2 \leq 1$). In this sense, time-dependent effects resulting from passing defects and the edges of the specimen are neglected.

If $p = q = 1$, the Jacobi polynomials of the β-pdfs become Legendre polynomials, which are in turn orthogonal with respect to uniform pdfs. The $d = 4$ random variables are summarized in the vector $\boldsymbol{\xi} = (\xi_1, \ldots, \xi_4)$. Similar to the previous example, the gPC coefficients are calculated by means of the regression scheme using the forces $F_{x|z}$ determined by the FEM at the points in the 4D grid $\mathcal{G}^{\mathrm{gPC}}$. As in the previous example, the grid points differ between each dimension since they depend on the corresponding input pdfs.

The results of the gPC are verified by evaluating the normalized root mean square deviation of both force components $\delta F_{x|z}$ with respect to MC simulations, which are independently determined in a random grid $\mathcal{G}^{\mathrm{MC}}$ containing N_{MC} samples. The general principle is shown in Figure 4.20.

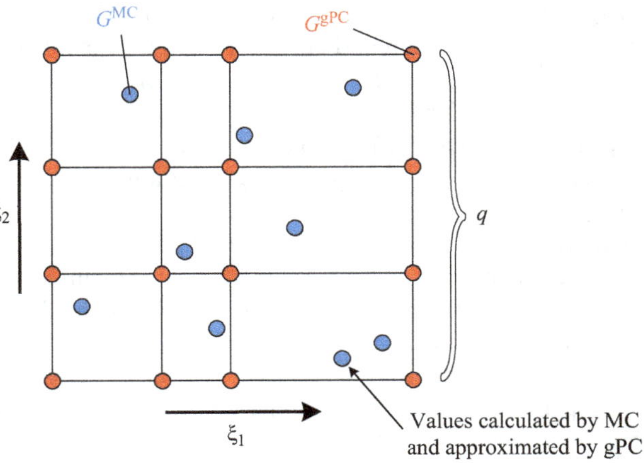

Figure 4.20 Illustration of the computational grids from the gPC and the MC approach. The blue points indicate the locations of the solutions obtained by MC sampling. After the gPC expansion is performed, the approximated solution is computed on $\mathcal{G}^{MC}$ by means of the polynomial expansion and compared to the exact values in a root mean square sense

The forces at the grid points in $\mathcal{G}^{MC}$ are approximated by the gPC expansion and termed $\tilde{F}^{gPC}_{x|z}$. In a consecutive step, the relative difference between the approximated and the exact values is determined by:

$$\delta F_{x|z} = \frac{100\%}{\max\left(F^{MC}_{x|z}\right) - \min\left(F^{MC}_{x|z}\right)} \sqrt{\frac{1}{N_{MC}} \sum_{i=1}^{N_{MC}} \left(F^{MC}_{x|z,i} - \tilde{F}^{gPC}_{x|z,i}\right)^2}, \tag{4.52}$$

$$\delta F = \tfrac{1}{2}\left(\delta F_x + \delta F_z\right), \tag{4.53}$$

where δF denotes the mean error of the Lorentz force. This error definition involves additional computations. However, it is a suitable measure to analyze the convergence properties for different approximation orders p and number of sample points $N_g = q^N$.

4.2.5.4 Uncertainty quantification of model parameters

To perform a gPC expansion, it is necessary to determine the statistical properties of the model parameters. On the one hand, the variations in the lift-off distance are approximated by the accuracy of the 2D linear stage NLS4 (Newmark Systems Inc., www.newmarksystems.com), as well as by the mounting accuracy and surface roughness of the specimen. On the other hand, the statistical properties of the remaining parameters σ, v, and B_r are determined experimentally.

The conductivity of 60 Al-sheets is measured two times at three different locations, resulting in 360 samples in total. The measurements are performed with the EC testing device Elotest N300 (Rohmann GmbH, www.rohmann.de) using the probe

KAS 7H190.03.1, operating at a frequency of 60 kHz. The measurement uncertainty is ±0.06 MS/m considering the standard deviation from the repetitions.

The velocity variations are determined by means of the incremental position encoder TONiC T1000 (Renishaw plc, www.renishaw.com). The position is sampled with a frequency of 10 kHz. After differentiating the position data with respect to the time, the velocity is obtained with a measurement uncertainty of ± 1 mm/s. The total number of samples is approximately 19.000.

The determination of the stochastic properties of the effective remanence is more difficult. To provide comparability to the cylindrical magnet applied in the LET setup, it is ensured that the magnet samples are ordered from the same distributor and are made of the same magnetic material. The statistics are determined by 100 cylindrical magnets of size $[R, H] = [5\,\text{mm}, 10\,\text{mm}]$ with a material grade of N52 delivered from the company HKCM-engineering (Z10x10Ni-N52, Art.No: 9962-2336, www.hkcm.de). The effective remanence of the magnets is determined by measuring the axial component B_z of the magnetic flux density on the cylinder axis. The experimental setup is shown in Figure 4.21. The magnetic flux is measured using the Gaussmeter FH36 (Magnet-Physik Dr. Steingroever GmbH, www.magnet-physik.de) in combination with the probe HS-MMT-6J08VH. All measurements are performed at room temperature of about 23°C. The magnets are located at a distance z_0 above the Hall-probe, which is determined during calibration. To ensure that all magnets are positioned at almost the same lateral position, a mask-sheet with a hole is used.

The distance z_0, at which the magnetic flux is measured, has a major influences on the resulting remanence due to the high sensitivity of the magnetic field and its rapid decrease with distance. The calibration of the experimental setup consists of the determination of this distance. It is done using a coil with 5000 turns supplied by a defined current of $I_0 = 3\,\text{mA}$, which is provided by the high precision current source SMU 2614 (Tektronix GmbH, Keithley Instruments, www.keithley.com). The experimental setup during calibration is shown in Figure 4.21(b)–(d). The dimensions of the coil are $D_i = 2R_i = 24.5\,\text{mm}$, $D_o = 2R_o = 30.2\,\text{mm}$, and $H_c = 5.5\,\text{mm}$. The winding body has a wall-thickness of 1 mm and the wire has a diameter of only 0.05 mm.

Due to the high number of turns, stray fluxes from the coil terminals are reduced. In this way, a well-defined magnetic flux distribution is produced, whereas its axial component as a function of the distance is given [109]:

$$B_z(z_0) = \frac{\mu_0}{2}\frac{I_0 N}{H_c(R_o - R_i)}\left[(H_c + z_0)\ln\left(\frac{R_o + \sqrt{(H_c + z_0)^2 + R_o^2}}{R_i + \sqrt{(H_c + z_0)^2 + R_i^2}}\right)\right.$$
$$\left. - z_0 \ln\left(\frac{R_o + \sqrt{z_0^2 + R_o^2}}{R_i + \sqrt{z_0^2 + R_i^2}}\right)\right]. \tag{4.54}$$

The distance z_0 is found by solving (4.54) iteratively together with the measured values of B_z. In the current setup, it is determined to be $z_0 = 1.155 \pm 0.042\,\text{mm}$. The measurement uncertainty is determined by 10^4 MC runs of (4.54) considering the tolerances of all geometrical dimensions of ±0.05 mm resulting from the caliper together with an uncertainty of ±0.25% of the magnetic flux measurements.

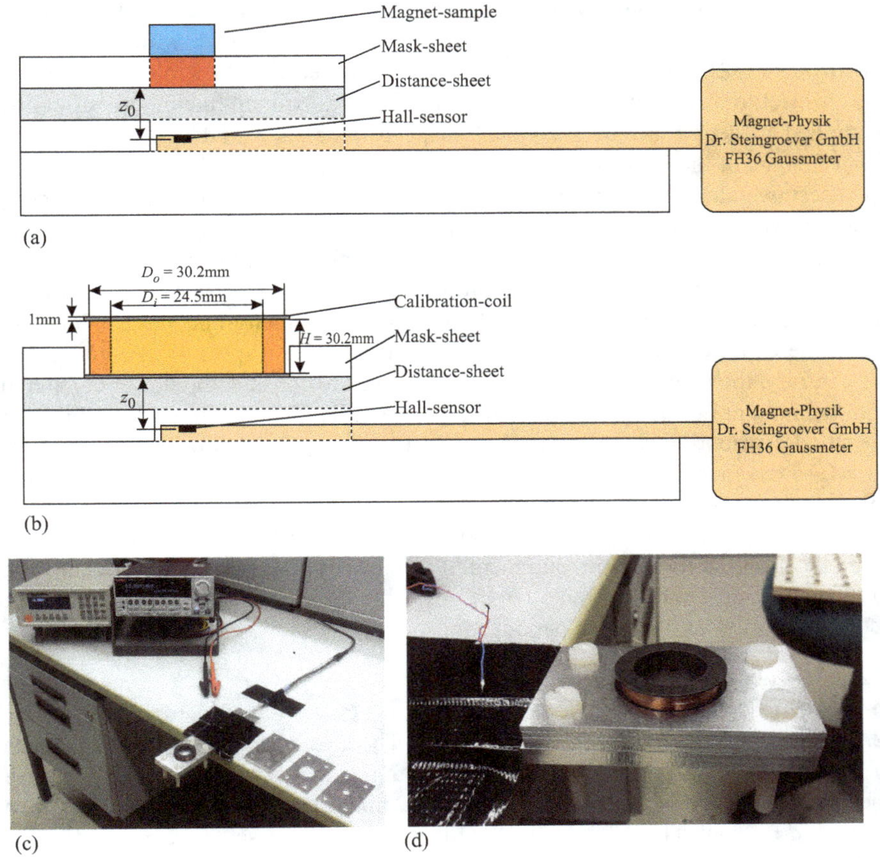

Figure 4.21 Experimental setup to determine the effective remanence. (a) Side view of experimental setup during measurements. (b) Side view of experimental setup during calibration. (c) Experimental setup with the Gaussmeter (top left) and high precision current source (to the right). (d) Detailed view on the calibration coil connected to the high precision current source

All variables are assumed to be uniform distributed. In a consecutive step, the magnetic flux density from each magnet of the 100 samples is determined. Afterward, their effective remanences are calculated by:

$$B_r = 2B_z(z_0)\left(\frac{z_0 + H}{\sqrt{(z_0 + H)^2 + R^2}} - \frac{z_0}{\sqrt{z_0^2 + R^2}}\right)^{-1}. \tag{4.55}$$

The measurement uncertainty of the B_r measurements is $\pm 0.014\,\text{T}$. It is determined by sampling (4.55) while taking into account the variability of the magnetic flux measurements and of all geometrical dimensions including z_0.

The pdfs of the model parameters are determined by fitting the β-distributions to the obtained data using the software environment "R" together with the package *fitdistrplus* [338]. The limits (a, b) are defined considering a 20% tolerance interval with respect to the extrema of the empirical data. A maximum likelihood estimation is applied to determine the individual shape parameters p and q. The upper and lower bounds of the equivalent uniform distributions are determined numerically in a consecutive step by means of the inverse cumulative distribution function (cdf), assuming that the uniform distributions cover 99% of the fitted β-cdfs.

The fitted pdfs are shown in Figure 4.22 together with their modes $\hat{v}$, $\hat{\sigma}$, and $\hat{B}_r$, which are the points of maximum probability density. The corresponding parameters are summarized in Table 4.5. The pdfs of the velocity and the conductivity are in the expected range and are fairly symmetric with respect to the modes. On the contrary, the values of the effective remanence are considerably lower compared with the nominal values provided by the manufacturer. The maximum probability of the fitted distribution is located at $\hat{B}_r = 1.32\,\mathrm{T}$, which is 7.7% lower compared with $B_r = 1.43\,\mathrm{T}$ as specified by the manufacturer. Moreover, it can be observed that the distribution is not symmetric, which motivates the choice to model the pdfs by means of β-distributions. In regard of the experimentally determined intervals from Figure 4.22, it can be observed that the measurement uncertainty is considerably lower compared to the degree of variation. Hence, it was possible to provide information about the actual variability of the model parameters.

4.2.5.4.1 Experimental LET setup

In the setup under investigation, the optimized cylindrical PM of size $[D, H] = [22.5\,\mathrm{mm}, 17.6\,\mathrm{mm}]$ is used (see Section 3.2). The layered specimen, which is made of stacked aluminum sheets, is used throughout the analysis (see Figure 1.10). One sheet contains a defect of size $[X_d, Y_d, Z_d] = [12\,\mathrm{mm}, 2\,\mathrm{mm}, 2\,\mathrm{mm}]$ and is located in a depth of $d = 2\,\mathrm{mm}$. The velocity and the lift-off distance are set to 0.5 m/s and 1 mm, respectively.

4.2.5.5 Results and discussion

The gPC is evaluated for different grids $\mathcal{G}^{gPC}$ and approximation orders p, considering 11 different positions of the magnet. The gPC is expanded as long as the number of coefficients is smaller or equal to the number of sampling points ($N_c \leq N_d$). The mean error (4.53) is computed using $N_{\mathrm{MC}} = 10^4$ direct samples. The errors are calculated for two characteristic magnet positions $P_1(x = 0)$, when the magnet is located right over the defect (highly perturbed EC profile) and $P_2(x = 65.5\,\mathrm{mm})$, when the magnet is located between the defect and the outer edge of the specimen (unperturbed EC profile). The estimated errors are summarized in Table 4.6.

Differences below 0.3% can be observed even in case of coarse grids, i.e. $q = 2$ ($N_g = 16$). The error is converging fast to the accuracy limit of the FEM indicated by the slightly higher errors if the magnet is located in P_1. The study indicates that a grid with $q = 3$ points in each dimension in combination with an approximation order of $p = 3$ is sufficient in this case to determine the uncertainties of both force components with an error of $< 0.18\%$.

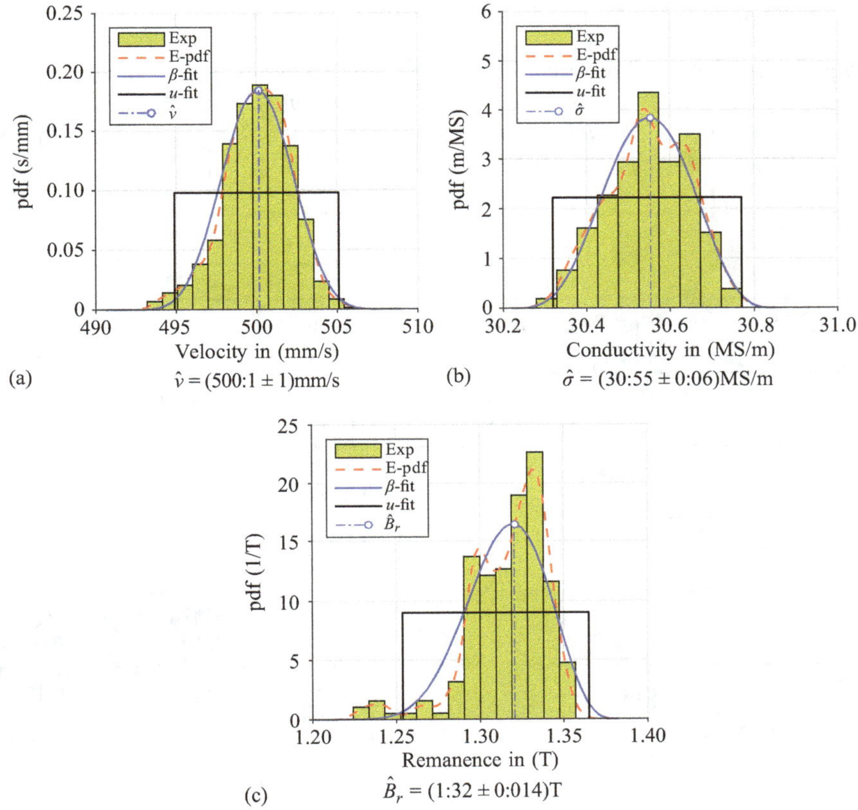

Figure 4.22 Statistical properties and modes of the input parameters: (a) velocity, (b) conductivity, and (c) effective remanence. The graphs show the histograms obtained by experiments (Exp), the associated empirical pdfs (E-pdf), and the fitted distributions used in the numerical simulations. The E-pdfs are obtained numerically using a Gaussian kernel smoother together with a bandwidth, which equals 50% of the bin width in the respective histogram. The labels u-fit and β-fit correspond to the fits to the uniform- and β-distribution, respectively [293]

Table 4.5 Limits and shape parameters of the fitted probability distributions of the different model parameters [293]

pdf	Parameter	v	σ	B_r	h
Uniform	a	0.4949	30.32	1.254	0.9
	b	0.5051	30.77	1.365	1.1
β	a	0.4906	30.22	1.201	0.86
	b	0.5090	30.85	1.383	1.14
	p	8.9109	4.6166	4.910	5
	q	9.4273	5.0283	8.336	5

Table 4.6 Averaged δF between gPC and MC for two characteristic magnet positions P_1 and P_2 in case of β-distributions (p expansion order, N_c number of coefficients, q number of grid points in one dimension, N_g total number of grid points) [293]

p	N_c	pos.	q / N_g			
			2 / 16	**3 / 81**	**4 / 256**	**5 / 625**
1	5	P_1 P_2	0.3613 0.2912	0.3366 0.2911	0.3326 0.2907	0.3311 0.2908
2	15	P_1 P_2	0.2851 0.1877	0.1705 0.0178	0.1614 0.0162	0.1583 0.0149
3	35	P_1 P_2	– –	**0.1705** **0.0172**	0.1615 0.0163	0.1599 0.0136
4	70	P_1 P_2	– –	0.2403 0.1647	0.1615 0.0164	0.1614 0.0140
5	126	P_1 P_2	– –	– –	0.1615 0.0172	0.1620 0.0141
6	210	P_1 P_2	– –	– –	0.2034 0.1263	0.1638 0.0139
7	330	P_1 P_2	– –	– –	– –	0.1655 0.0140
8	495	P_1 P_2	– –	– –	– –	0.2014 0.1680

The estimated force profiles are depicted in Figure 4.23 together with the experimental data.

The graphs show the uncertainty intervals $\mu_{x|z} \pm 2\sigma_{x|z}$ for the forces $F_{x|z}$ obtained by the gPC in case of β-distributed random variables. The intervals are shown at the 11 selected points of interest. The dashed line shows one single deterministic simulation considering the modes of the input parameters. The measurements are in the numerically predicted range when uncertainties are taken into account. The relative standard deviation with respect to the mean value of both force components is nearly independent of the observation point and amounts 3.8% and 5.2% in case of β- and uniform-distributed random variables, respectively. Moreover, the pdfs of the forces are evaluated by sampling the gPC expansion 10^6 times. They are shown in the lower part of Figure 4.23 in the event that the magnet is located right over the defect at the point P_1. The pdfs are compared to those obtained by the MC method using $N_{MC} = 10^4$ direct samples. As expected from the estimated errors, a very good agreement between both approaches can be observed.

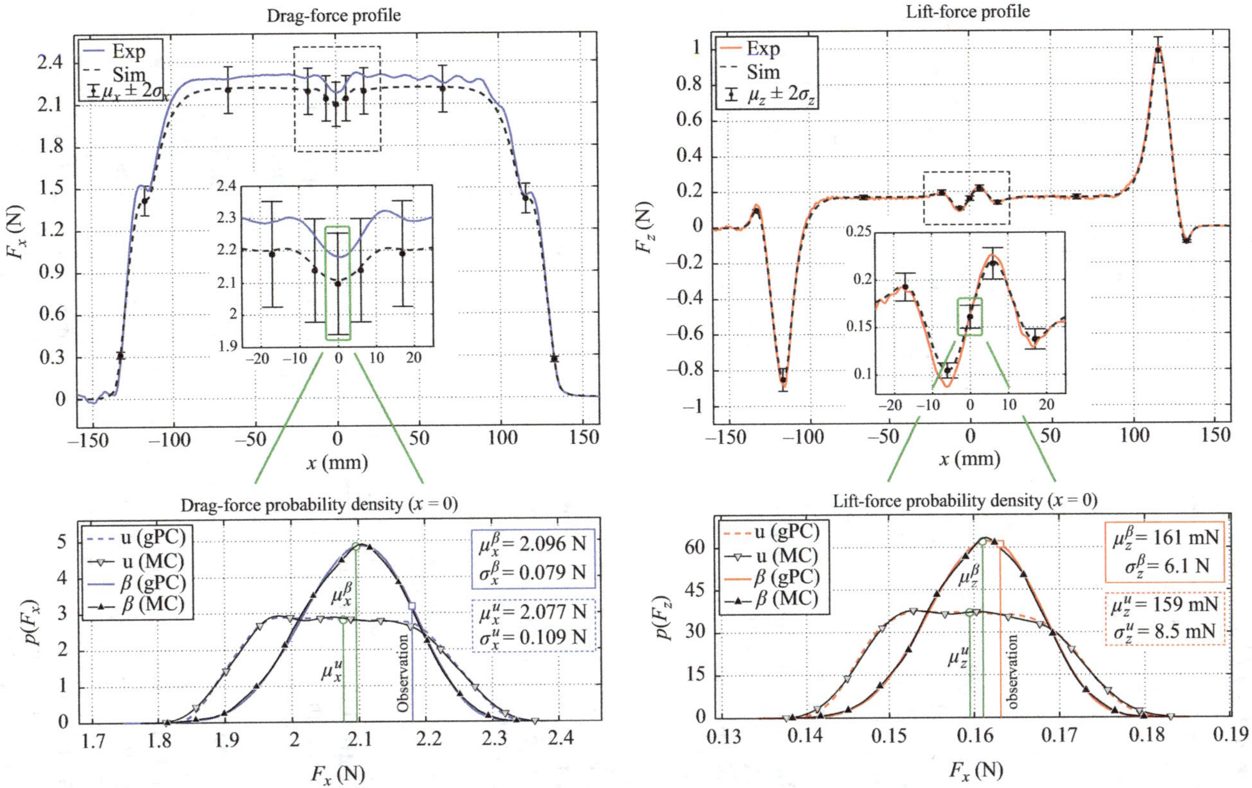

Figure 4.23 Numerical and experimental results of the Lorentz forces profiles. The uncertainty intervals are shown in case of β-distributed random variables for $\mu_{x|z} \pm 2\sigma_{x|z}$ using a grid with $q = 3$ points in each direction and an expansion order of $p = 3$. The figures at the bottom show the estimated pdfs in the point P_1 when the magnet is located over the defect at $x = 0$ for uniform distributions (u) and β-distributions (β) [293]. (a) Drag force. (b) Lift force

Table 4.7 First order Sobol indices of F_x and F_z averaged over 11 magnet positions calculated with a grid of $q = 3$ points in each direction and an expansion order of $p = 3$ [293]

pdf	$S^{(1)}_{x,B}$	$S^{(1)}_{z,B}$	$S^{(1)}_{x,h}$	$S^{(1)}_{z,h}$	$S^{(1)}_{x,v}$	$S^{(1)}_{z,v}$	$S^{(1)}_{x,\sigma}$	$S^{(1)}_{z,\sigma}$
uni	88.4	84.2	9.6	10.2	1.3	3.7	0.7	1.9
β	87.9	83.7	10.2	10.7	1.2	3.6	0.7	2.0

The first-order Sobol indices of F_x and F_z are determined at each of the 11 points and showed only minor variations with respect to the magnet position. They are averaged over all M positions and listed in Table 4.7.

The first-order Sobol indices are most significant and almost cover the total variance. Keeping in mind that the whole system is treated as a black box, assuming that the experimentalist has no detailed knowledge about the applied magnet, it is observed that the uncertainty in the magnetic remanence has with ~85% the greatest influence on the total variance of the resulting forces. In addition, the lift-off distance contributed to approximately ~10%, whereas the velocity and the conductivity showed only minor impact.

4.2.5.6 Conclusions

The present study shows that the analysis of uncertainties by means of gPC-based methods can be readily used for extended experimental validation in the framework of LET. The applied methodology is very efficient in terms of computation time compared to traditional MC approaches. The gPC already converged after $N_g = 81$ simulations using an order $p = 3$ approximation. The statistical properties of the velocity, the conductivity, and the magnetic remanence were determined experimentally. The use of β-distributions allowed to model each variable individually and in a more realistic sense compared with uniform distributions. The analysis provides the possibility to identify most influencing parameters to improve the measurement accuracy of the experimental setup. This serves as a starting point to reduce the variance of the resulting Lorentz forces and to increase the SNR, which directly enhances the reproducibility and the defect detection capabilities of the LET method.

The study shows that the magnetic remanence is the most influencing factor when considering the absolute value of the Lorentz force. If possible, it should be determined in advance using the proposed experimental setup. It could be observed that the effective remanence is generally lower compared with the values specified by the manufacturer. This should be taken into account during the optimization of magnet systems in the future. However, the use of the dimensionless scaling factor from (3.27) allows to correct the remanence variations retrospectively by increasing the velocity accordingly, while ensuring that the system does not violate the constraints and remains to operate in the linear R_m regime. Nevertheless, its value does not change over time and is thus not contributing to the noise level considering the SNR. The quantities, which may change over time, are the lift-off distance and the velocity. Since the EC density is highest at the surface of the specimen, the lift-off distance should be kept at a constant level to decrease the disturbances resulting from lift-off variations.

Despite of that, the velocity variations have an apparently small share of about 1–4% to the total variance, their time-dependent nature blurs the defect signal and reduces the SNR. Comparing the relative contribution to the experimental data of the defect depth study in Section 3.2.5, it seems inherent that the velocity variations and vibrations should be reduced as much as possible to keep the Lorentz force at a constant level. This would enable the identification of weakest force perturbations resulting from small or deep lying defects.

It is emphasized that the current study is valid for the proposed working point of the system. It is assumed that the system operates in the linear regime of low to moderate R_m, where the back-reaction of the induced ECs is moderate (QSA). If the working point considerably changes, the study has to be repeated together with an estimation of the involved uncertainties. In the same way, alternative numerical approaches have to be considered, as for example MDA or MMA described in Section 2.4.3.2.

During the present analysis, it is observed that the velocity profile obeys characteristic oscillations, which superimpose the rectilinear part of the motion. So far, the influence of sinusoidal velocity profiles on the Lorentz force has not been investigated in detail. In the next section, this effect is studied further and a full 3D analytical approach is presented to solve this kind of electromagnetic field problem.

Chapter 5
Lorentz force evaluation

Hartmut Brauer[1]

5.1 Identification of conductivity anomalies

Nondestructive material testing and evaluation is a vast interdisciplinary field as well as a challenge due to the variety of applications. Whereas the focus of NDT is to detect and localize anomalies within a specimen, the reconstruction of defect properties (dimensions, shape, structure, and composition) and their influence on the material's usability is the focus of NDE. Defect identification and assessment are very important aspects of quality assurance. Nondestructive material testing is understood as the noninvasive examination of any type of specimen without changing or altering the properties of the body under test to check whether the specimen contains anomalies. Anomalies are any type of defect or change in the material properties that can be of natural or artificial origin, influencing the usefulness or serviceability of that object [6]. NDT has turned from a rather empirical procedure dependent on the experience of the examiners into a more quantitative measurement technique that serves to determine the influence of material anomalies on the structural health of the object [339]. To classify the existing NDT techniques according to their limitations and not only according to the employed physical phenomenon, a separation in visual, surface, and volumetric methods was proposed [3]. However, often these methods have been classified as either electromagnetic or acoustic [12,36,340–344].

Biomedical applications require usually the determination of electrical conductivity of human tissues. Thus, similarities/dissimilarities between the optimized sensor setups have been investigated and compared with previous findings [345]. There exist very different strategies for the electrical conductivity measurement. The invasive procedure requires direct contact of electrodes to the body, which can lead to artifacts. An alternative approach is the use of induction coils inducing electrical currents in the tissues. This results in changes of the coil's impedances. These changes are used to obtain information about the tissue conductivity. This is in practice a very important information since the abnormal or diseased tissue has different electrical properties compared to the normal one.

A very similar approach can be used to determine the conductivity distribution of metallic specimen. If the permanent magnet system is approximated by a dipole

[1] Advanced Electromagnetics Group,Technische Universität Ilmenau, Germany

model the same strategy for the solution of the inverse problem as in magnetocardiography can be applied. This serves as the basic idea for proposing a new method for noncontact, NDE of solid conductive materials, termed LFE. In contrast to the bioelectromagnetic applications, where magnetic flux is measured at points above the volume conductor, in LFE the Lorentz force acting on a permanent magnet moving relatively to the specimen is measured.

The inverse problem of the LFE technique, which has been introduced for inverse calculations of LET signals in [98], is to characterize the geometry of the underlying defects. A straightforward approach is applied to solve this inverse problem directly and estimate the defect parameters. The geometrical parameters of the defect, such as the center of gravity, depth, and extension, can be defined as the unknown variables in $\boldsymbol{x}$. Other approaches are to determine the eddy current distribution and the conductivity distribution in the conductor. In these reconstruction approaches, the vector $\boldsymbol{x}$ composes the unknown moments of the equivalent current dipoles or the unknown voxel conductivities. Solving an inverse problem implies to minimize the error between measured $\boldsymbol{y}_m$ and forward calculated data $\boldsymbol{y}_f$ with respect to the unknown parameters composed in $\boldsymbol{x}$:

$$\min_{x}\left(\boldsymbol{y}_m - \boldsymbol{y}_f\right) = \min_{x}\left(\boldsymbol{y}_m - \mathbf{G}(\boldsymbol{x})\right), \tag{5.1}$$

where the operator $\mathbf{G}$ represents a projection of the model parameters on the measured signals. For solving such problems, a large number of inverse calculation methods is available in the literature.

One class of inverse methods are scanning methods. The simplest scanning method is the goal function scan (GFS). The value of the goal function, e.g. the error function $\boldsymbol{y}_m = \mathbf{G}(\boldsymbol{x})$, is computed for different combinations of the parameters to be determined. The inverse solution is set to the parameter combination with the smallest goal function value. The method has the advantage that the course of the goal function can be scrutinized. However, the major drawback is that the method requires a large number of forward calculations causing high computational costs. Thus, the GFS is only feasible if the number of unknown parameters is small, i.e. only one source is to be determined. The method has already been applied to reconstruct sources in the human brain from electroencephalography (EEG) and magnetoencephalography measurements [346].

When solving linear inverse problems, engineers and scientists are aware of changing of the condition number with respect to the L_2 norm with matrix dimensions. It turns out that the increment/decrement of the condition number with respect to the L_2 norm depend on the over- or under determination of the linear inverse problem. It is known that the exclusion of sensors from a sensor array decreases the condition number with respect to the L_2 norm only in the case of underdetermined linear inverse problem. Similarly, when the problem is overdetermined, exclusion of sources from a grid of dipoles decreases the condition number with respect to the L_2 norm. Influence of changing the number of sensors in overdetermined problems and number of sources in underdetermined problems depends on particular singular values of the corresponding kernel matrix. Reduction of the dependency on the smallest singular value and increment in the numerical stability of the evaluation comparing to the condition number with respect to the L_2 norm is achieved through the inverse average decay of singular values ρ, which are related to the inverse of area under the curve of singular values of a kernel matrix. The larger the area under the curve is, the smaller is the ρ, and consequently the better conditioned is the linear inverse problem. This measure of conditioning is not invariant under scaling.

When the magnetic flux density has to be measured in a number of points of a scanning plane, magnetic sensors are usually arranged uniformly and oriented in parallel. Numerical simulations have shown that random variations in the sensor directions can considerably improve the condition of the magnetostatic linear inverse problem [347]. The improvement can usually be achieved for many ill-conditioned, linear inverse problems, but this is not always the case. For a given source grid and magnetic sensor array, Petkovic has studied whether the cases when variations of sensor orientations lead to an improvement of the condition of the corresponding linear inverse problem can be defined in a strict mathematical way [345]. The possible improvement in condition leads finally to more stable inverse solutions.

5.2 Inverse solution techniques

5.2.1 Theory

Design optimization is a multidisciplinary study employing both scientific methods and technological approaches to satisfy technical, economical, and social requirements in an optimal manner. Today, because of the rapid depletion of energy resources, scarcity of economic and material resources, strong technological competition and increasing environmental awareness, engineers are under immense pressure to produce optimal designs. On the other hand, the advent of new technology and materials, as well as the imminent introduction of many mandatory international regulations on electrical products, is making it increasingly difficult to obtain an optimal design of an electromagnetic device or system using traditional analytical or synthetic approaches. In this regard, an increasing effort has been devoted to the study of inverse problems or optimizations in computational electromagnetics in the last three decades.

Generally, the design optimization in an inverse problem is to find solutions that best compromise a host of, and often conflicting, objective functions. Moreover, each objective function is usually a multimodal nonlinear one. Furthermore, highly sophisticated models such as finite element analysis are exclusively invoked in performance computations. All these complications aggregate the difficulty in developing feasible numerical methodologies for inverse problems.

The first theories of inverse problems date to the end of the 19th century and the beginning of the 20th century. One of the first inverse problems solved in the past was Newton's discovery of forces making planets move in accordance with the Kepler's laws. Determination of the body's position and shape using the values of its potential presents an inverse problem in potential theory. Research regarding the internal structure of the Earth's crust involved electromagnetic fields in the theory of the inverse problems. Nowadays, more and more applications deal with inverse problems. One of them is computerized tomography [348]. The aim is often the determination of the density distribution from profiles of its line integrals, playing an important role in biomedical engineering and NDT.

A special class of inverse problems are linear inverse problems. They can be written as $\mathbf{L}\boldsymbol{x} = \boldsymbol{b}$, where $\mathbf{L}$ is a linear operator describing the explicit relationship between the Hilbert space $\boldsymbol{x}$ and the Hilbert space $\boldsymbol{b}$. Minimizing the residual $\|\mathbf{L}\boldsymbol{x} - \boldsymbol{b}\|$, one could find the best approximate solution of the discrete linear inverse problem. In the case of less measurement data $\boldsymbol{b}$ than unknown parameters $\boldsymbol{x}$, this solution presents the minimum norm solution that minimizes $\|\boldsymbol{x}\|$ among all residual minimizers. A mathematical problem is well-posed if a solution exists, the solution is unique, and the solution depends continuously on the data.

The last requirement of continuity is related to the stability of the solution of linear inverse problems [349]. Continuity is a necessary but not sufficient condition for stability. This means that even if the problem is well-posed it may be ill-conditioned. Ill-conditioned problem means that small changes in the initial data lead to large changes in the solution. Thus, the existence of a proper figure of merit is an imperative for the estimation of ill-conditioning of linear inverse problems, as well as the availability of appropriate solution techniques. Efficient solution strategies for these inverse problems need to know whether a problem is ill-conditioned as well as its degree of ill-conditioning. A comprehensive theoretical analysis of known figures of merit has been done by Petkovic [345], considering the conditioning of linear inverse problems theoretically.

The most widely used indicator of ill-conditioning is a formulation of the condition number with respect to the L_2 norm. As a norm-wise condition measure, it gives the error bound of the worst conditioned component in the solution vector and therefore it overestimates the condition of a kernel matrix. Computation of the condition number with respect to the L_2 norm requires computation of the singular values of the kernel matrix of the corresponding linear inverse problem. This measure of conditioning strongly depends on the smallest singular value of a matrix.

Based on geometrical properties of rows/columns of a kernel matrix, new error measures have been developed. The error measure allows selecting the most linearly independent rows and performing the estimation with these rows only. This reduces computational costs and improves a stability of linear inverse solution. Numerical stability, indication of the worst conditioned component or of a mean conditioning of all components, requirement for pseudo inverting of a matrix, affection by scaling or by matrix dimensions, makes one of the figures of merit favorable for measuring of conditioning of the linear inverse problem.

5.2.2 Classification of inverse problems

Inverse problems are solved to obtain insight into the underlying system parameters based only on the observed output data. This is extremely useful in a wide range of engineering issues. For instance, in geological prospecting, the properties of the interior structure of the earth such as density and conductivity are determined based on measurements performed on the surface. In medical imaging, computer tomography is used to examine the interior structure of the human body. The attenuation of

X-rays by the human body is evaluated. Another example in the framework of medical diagnostics is the reconstruction of activated cortex regions in the human brain using EEG where measurements are recorded from the surface of the human head. In the inverse scattering method, the shape of an object is determined based on the waves scattered by it. In this framework, acoustic and electromagnetic scattering are most widely applied [350,351].

Inverse problems can be classified into nonlinear and linear problems. Their distinction can be well explained by considering the corresponding formulation of the forward problem

$$\boldsymbol{y} = \mathbf{G}(\boldsymbol{x}), \tag{5.2}$$

where the vector $\boldsymbol{y} \in R^{M\times 1}$ denotes the measured signals and M is the number of measurement points. The vector $\boldsymbol{y} \in R^{N\times 1}$ contains the N model parameters. The operator $\mathbf{G}$ represents a projection of the model parameters on the measured signals. It comprises information about the sensor and source positions and configuration. In the forward problem $\boldsymbol{y}$ is calculated using the known model parameters. In the associated inverse problem, the model parameters are unknowns to be estimated. If the problem is linear, (5.2) can be reformulated as

$$\boldsymbol{y} = \mathbf{K}\boldsymbol{x} \tag{5.3}$$

with the gain matrix $\mathbf{K} \in R^{M\times N}$ describing a linear mapping between model parameters and output data. In case of a nonlinear problem, the relation between model and data values is more complex and $\mathbf{G}$ in (5.2) is a nonlinear operator. The properties of forward and inverse problems differ in the sense that forward problems are usually classified as well-posed, whereas inverse problems are mostly ill-posed. The definition of well-posed or ill-posed problems is related to Hadamard [352,353]. A problem is well-posed if it fulfills the following three conditions:

1. For all possible data, there is a solution (Existence).
2. This solution is unique (Uniqueness).
3. This solution depends steady on the data (Stability).

If at least one of these criteria is not fulfilled, the problem is said to be ill-posed. Even if the stability condition is fulfilled, the inverse problem can be sensitive to noise in the data. A small change in the input data may have a significant impact on the solution. If this is the case, the problem is said to be ill-conditioned [354]. The degree of ill-conditioning of the inverse problem can be determined by calculating a condition number of the gain matrix.

Additionally, inverse problems can be subdivided into overdetermined and underdetermined problems. An overdetermined problem has more sensor points than unknown sources to be determined and vice versa for an underdetermined problem. Moreover, constraints can be imposed on the inverse problem. In real-world applications, often boundary and parameter constraints have to be considered. A widespread approach to solve an inverse problem is the minimum norm estimation (MNE). MNE is widely applied in biomedical engineering to localize current sources in the human brain and heart [346,355–357]. MNE is often applied to reconstruct current density distributions and especially distributed current sources. A regular grid of elementary current dipoles with fixed positions but unknown moments is defined in the region of the distributed current source. Optimal dipole moments are obtained by minimizing the norm of the vector difference between the measured data and the forward solution given in (5.3).

Further, MNE has been applied in magnetic nanoparticle imaging [358] and to detect buried ferromagnetic objects based on measured magnetic fields [359]. In the framework of NDT, this approach has been used to reconstruct pipeline defects from MFL measurement data [360]. Moreover, the authors in [361] applied the minimum L_2 norm approach to magnetic tomographic data. Further, ECT signals have been inverted with MNE to estimate flaws in metals [362]. In Figure 5.1, an overview of methods for the solution of inverse field problems is given.

Another approach to minimize the error function is to use optimization techniques. They can be divided into deterministic and stochastic methods. Widely applied deterministic methods are the simplex method [363], the Newton method [364], and the Levenberg–Marquardt algorithm [365,366]. Results of deterministic methods are reproducible provided that the search area, the starting point, and the termination criterion remain unchanged.

However, they often impose constraints such as differentiability, continuousness, and convexity of the goal function. Further, they perform a local search, i.e. they are likely to be trapped by local minima if the initial values are not chosen precisely enough. The deterministic methods are used to optimize the magnetic dipole models of the permanent magnet configurations [100]. Stochastic optimization algorithms overcome the starting point problem by defining multiple starting points on the goal function landscape. Most stochastic optimizers are zero-order algorithms, i.e. they do not depend on the derivative of the goal function. Even more, the goal function does not have to fulfill the constraints of continuity and convexity. The probability that they are trapped by local minima is smaller than for deterministic algorithms. Due to the stochastic nature, results of multiple trials scatter with a small variance around the global minimum. One drawback of stochastic optimization algorithms compared with deterministic methods is that more goal function evaluations are required and thus the computational cost is higher. Further, stochastic algorithms implicate the challenge of adjusting at least one intrinsic control parameter, e.g. weighting parameters that influence the step size of the algorithm. These parameters can have a significant impact on the result.

Stochastic optimization algorithms can be classified into the Monte Carlo (MC) method, simulated annealing (SA), evolutionary algorithms (EAs), swarm intelligence (SI) methods, and Tabu search (TS). SA algorithms adapt physical processes, e.g. the basic concept of the cooling process in metallurgy [367]. SI methods are naturally inspired and exploit the natural collective and social behavior of swarms (fishes, birds, ants, bacteria, etc.) reaching for some target, e.g. animals searching for food. Among them, the particle swarm optimization (PSO) and the ant colony optimization (ACO) are the most prominent members [368,369].

The EAs are also biologically inspired. They are based on Darwin's theory of evolution. The theory states that a population evolves by random selection, genetic

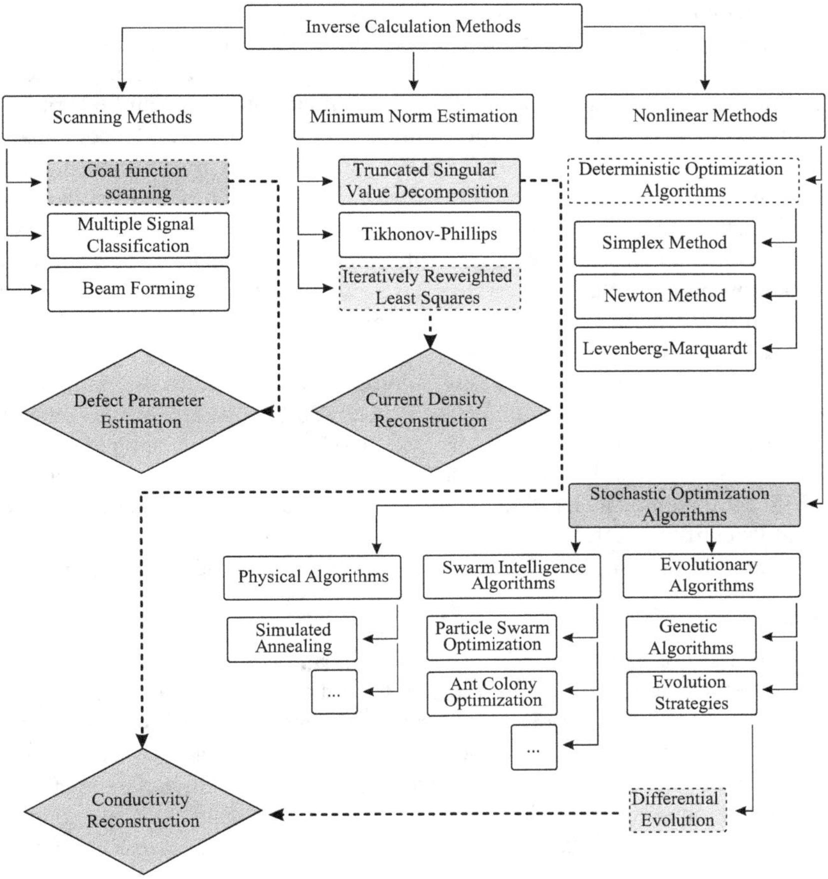

Figure 5.1 Overview of existing inverse calculation methods. The design parameters are shown in rhombuses. Inverse methods applied to these parameters are marked with light gray color and dashed lines [156]

mutation and recombination. Mutation generates innovation in the population and recombination intermingles information. Selection is determined by the survival-of-the-fittest principle. EAs can be divided into the main branches: evolution strategies (ESs) [370,371] and genetic algorithms (GA) [372,373]. In evolution strategies, the objective variables are real-valued and continuously, whereas in GA they are binary coded.

Hence, mathematical operations for evolving the population are arithmetic in ES, but logical in GA. The ES combine mutation and recombination, and in GA recombination has priority over mutation. Traditional mutation is based on probability density functions. In ESs, a survivor selection is applied, e.g. after offspring generation it is decided which population members from the combined parent and offspring population will survive to the next generation. The selection procedure in GAs is based on probability functions. Contrary, in ES selection is either deterministically or randomly (roulette wheel selection) performed, i.e. for selecting the next generation, ranking of the individuals or a tournament selection can be applied. In traditional GAs, parent selection is performed. Prior to mutation and recombination, parents that will produce off-springs are selected based on a probabilistic approach. Survivor selection as in ESs ensures that the best-so-far solution is retained, i.e. elitism can be included. In the parent selection used in traditional GA, this requirement of elitism is violated because parents that will produce off-springs are selected before the evolution process. Since researches have evaluated that elitism has significant positive influence on the performance of an algorithm, this principle has also been incorporated in modern GAs [374].

Due to their advantages, stochastic algorithms are applied in the framework of NDT and NDE for inverse calculations of material defects, where the characteristics of the goal function are inherently unknown. The PSO has been applied for defect reconstruction using ECT signals [375,376]. For the same purpose, the authors in [377,378] have applied a GA. In all applications, conductivity reconstructions were applied to identify the defect. Moreover, ESs have been of interest in electromagnetic inverse scattering problems [379,380] and for the analysis of composite materials [381].

New figures of merit have been developed, which can be applied to a large variety of linear inverse problems, e.g. to biomedical applications and NDT of materials [98]. The condition of the inverse problem depends on the sensor space (the applied grid of measurement points) and the source space (the region containing the unknown parameters). The most widely used ones include the condition number with respect to the L_2-norm. Several condition numbers have been introduced in the literature [382]. In [345], positions and orientations of magnetic sensors around a volume conductor have been optimized minimizing several figures of merit. The best results have been found for nonuniform sensors distribution on the whole conductor surface. Thus, it has been confirmed finally that such ill-posed inverse problems have no unique solutions, and quite different sensor sets can perform equally well.

The same approaches have been applied to LET. Based on this approximation method, a new inverse procedure for defect identification has been derived which is called LFE [98]. The LFE technique will be described circumstantially in Section 5.3.

In LFE, the goal function scanning (GFS) can be applied to determine the depth and extensions of defects. An enhancement of the GFS is the multiple signal classification (MUSIC) method, which originates from information theory [383]. In this method, multiple sources are determined by scanning subspaces of the goal function with single sources. Another scanning method is beamforming, which has been developed in the field of radio communication where multiple antennas are used. The working principle of beamforming algorithms is to reduce the interference between signals coming from multiple source with a spatial filter [384].

As already described previously, although an inverse problem is generally formulated as a multiobjective optimization one, it is generally converted into a single objective problem in the early stage of studies due to the gap between the strict requirement on heavy computational burdens and the state of science and technology. Since the objective function is generally a multimodal one and due to the inefficiency of traditional deterministic and stochastic optimal algorithms in finding the global optimal solution of such a problem, the attentions of the research have been given to the development of new stochastic optimal methods. The EA has become the standard for solving global optimizations in different engineering disciplines because it can find global optimal solutions that are often not attainable using traditional (deterministic) algorithms. On the other hand, there are many challenges in the next steps toward the development of more efficient EAs, which have to be addressed extensively:

- the fast computation methodology and model for computing robust performance parameters and constraint treatments,
- the robust-oriented optimizer for both robust and global optimal solutions,
- theoretical issues for existing numerical methodologies and models, and many others.

Although EA is playing an increasingly important role in the study of inverse problems, the excessive demand for computer resources often renders these methods inefficient or impractical for some practical design problems that call for repetitive usage of finite element solutions. To circumvent this problem, the response surface methodology (RSM) has been introduced to reduce the number of function evaluations that involve time consuming computer simulations without sacrificing the quality of the numerical solutions [385]. Furthermore, in practical engineering designs, imprecision and uncertainty are often inevitable and unavoidable. For example, it is usually very difficult to produce a product exactly in accordance to design specifications. The operating conditions will also vary with time. Hence, if the optimal solution is very sensitive to small variations of the designed parameters, it is possible that slight variations in the optimized variables could result in either significant performance degradation or infeasible solutions. The preferred design is thus not the global optimal solution in terms of the objective function, it should be the robust one that has good performances in terms of objective function and robust performance against slight perturbations. Consequently, the robust optimal design methodology is becoming a topical issue in computational electromagnetics.

5.2.3 Regularization

To obtain stable solutions of the inverse problems (which are usually ill-posed), regularization methods have to be applied. Regularization exploits *a priori* information of the solution and is performed by adding additional information in the form of weighting parameters or function terms to the error function in (5.1). Additional function terms are penalty terms that introduce constraints to the solution space, e.g. specific error bounds.

Regularization is performed by constraining the solution with a norm showing the desired properties. If the L_2-norm is applied, the inverse solution tends to be smeared. Common methods to calculate the solution are the truncated singular value decomposition (TSVD), the Landweber iteration, the Tikhonov-Phillips regularization, and the asymptotic regularization. In the TSVD, small singular values are omitted in order to obtain a stable solution [386]. This approach was also used to reconstruct conductivities in the first application of LFE [98]. The Landweber iteration has been applied successfully for solutions of electrical capacitance tomography [387,388]. In Tikhonov–Phillips regularization, an additional penalty term is introduced leading to a system matrix that can be inverted [389,390]. Tikhonov's method uses, in place of the problem that calculates the minimum-norm least squares, the minimization of the functional

$$\min_{x \in X} \|\mathbf{A}\boldsymbol{x} - \boldsymbol{y}\|^2 + \lambda \|\boldsymbol{x}\|^2, \tag{5.4}$$

where λ is the regularization parameter, which specifies the amount of regularization. It can often be estimated using the L-curve technique [391]. The regularization incorporates appropriate weighting of this term with respect to the error function. If another norm than the L_2 norm is applied, the inverse solution tends to be sparse. The resulting inverse problem is nonlinear and can be solved using weighted least squares algorithms.

5.3 Lorentz force evaluation

A wide variety of methods exist for NDE of electrically conducting nonferromagnetic objects. These include ultrasonics [392,393], radiography [394], thermography [395,396], tomography [361], MFL [397], and ECT [6,398–400]. LET has been introduced in 2008 [71]. This is similar to ECT and thus LET is also belong to the class of electromagnetic testing methods. The origins of LET and of electromagnetics in general goes back to the work of Faraday in 1832, Lenz in 1834, and Lorentz in 1892. Faraday's law of induction states that an electromotive force is generated in a conductor, if the conductor is moving in a magnetic field or exposed to a time-varying magnetic field [85]. This electromotive force gives rise to eddy currents flowing in the conductor. According to Lenz' law, the eddy currents are directed in such a way that the magnetic field induced by them opposes the magnetic field that produced them [401]. This justifies the law of energy conservation, which would not be fulfilled if both magnetic fields have equal direction. The electromagnetic force exerted on the conductor (i.e. the Lorentz force) has been derived first in [402]. In the 1970s, studies of forces exerted on moving coils and magnets above an electrically conducting nonferromagnetic slab have been extensively performed in the framework of the development of magnetic levitated transport systems [60,62,63,65,66,115]. The studies investigated linear moving planar objects and rotating cylinders. Later Saslow [58] provided a comprehensive analysis of the Maxwell's theory for electromagnetic suspension with special focus on motion-induced eddy currents and forces. Recent theoretical studies of the force acting on a magnetic dipole positioned above a slowly moving conductor have been performed in [116]. Further, the Lorentz force principle is also used in the LFV to estimate the velocity in an electrically conducting fluid. In this contactless method, the fluid moves across magnetic field lines and the force acting on the magnetic field lines is measured [89,90]. Even more, in Lorentz force sigmometry the electrical conductivity of metals or fluids is determined by exploiting the dependence between the conductivity and the magnitude of the measured Lorentz force (see Section 6.1) [403]. Following the idea of LFV, LET has been introduced by Brauer and Ziolkowski [71]. Material anomalies, such as changes in conductivity, defects, cracks, or inclusions, distort the eddy current distribution in the object under test and, consequently, also the Lorentz force measured at the magnetic system. Thus, defect/anomalies in the conductive material produce perturbations in the Lorentz force signals. It has been done extensive basic research to show that the direct relationship between changes in force and material anomalies can be used to detect defects.

A popular academic approach to explain the physics underlying LET is the creeping magnet problem [94,403,404], where the motion of a small spherical permanent magnet traveling in an electrically conducting pipe is investigated. Due to the movement, eddy currents are induced in the pipe. The resulting Lorentz forces exerted on the permanent magnet reduce the velocity of the falling magnet. This example is extended to ring magnets in [405], where a falling permanent magnet ring surrounds the conductor.

In conventional ECT, an excitation coil, which carries alternating current and has to be actively operated, provides the primary magnetic field. Moreover, the material under test is usually stationary. The signals measured are changes in the impedance in the pick-up coil. A limitation of ECT is the frequency-dependent skin depth which restricts the method to defects lying at the surface or close to the surface. Contrary to ECT, the primary magnetic field in LET is generated by the permanent magnet, i.e. it does not require frequency-dependent currents. The impedance measurements are replaced by measuring the Lorentz force exerting on the permanent magnet. Since eddy currents induced in the conductor are caused by the relative motion, LET is especially suitable for investigating moving objects. Hence, this NDT method is also called motion-induced eddy current testing (MIECT). If additionally the detected defect has to be identified with respect to depth, material properties, and shape reconstruction, the procedure is called Motion-induced eddy current testing and evaluation (MIECTE).

In the literature can be found several techniques, which try to avoid the frequency-dependent skin depth of the classical ECT. In [406–408], a stationary magnetic field is provided by a coil carrying a direct current and the conductor is in motion. Another modification of LET called velocity induced eddy current testing (VIECT) has been introduced [73,75,76]. This method incorporates motion-induced eddy currents in a conductor underneath a permanent magnet, but instead of force signals, changes in the secondary magnetic field generated by the induced eddy currents are measured using magnetic field sensors (Hall, GMR, or others) [13,78].

Based on the estimation of the conductivity distribution and using a dipole model like in bio-electromagnetic studies, a new method for contactless NDE of solid conductive materials, the LFE technique, has been developed [98]. As in the LET approach, the relative motion between a permanent magnet and a solid conductive specimen produces eddy currents inside the bar. When the bar has a defect, the distribution of eddy currents can be modeled as a superposition of eddy currents in the system without defect and the distribution of exactly opposite eddy currents flowing in the defect region with changed conductivity (equal to the conductivity of the bar in the case of zero conductivity of the defect). The influence of a defect with different conductivity on the Lorentz force exerted on the magnet due to eddy currents is assumed to be equal to the subtraction of the Lorentz forces caused by a defect-free system and a target with defect (Figure 5.2). In this case, the defect has to be modeled as a conductive region with conductivity of the specimen, σ_0. The resultant force signal is called defect response signal (DRS) (Figure 5.2, right). The LFE approach has been applied to the reconstruction of long and wide parallelepiped subsurface cracks (Figure 5.3) using simulated and measured data.

Because the forward solution has to be determined very often, fast calculation techniques are strongly recommended. A very simple approach is based on a defect model where the only the defect region is approximated by a set of uniformly distributed electric current dipoles. The simplest approach for generation of DRSs (Figure 5.4(a)) has been described already in Section 2.2.5 (see also Figure 2.35). The limitation of this source model is that these dipoles do not contribute to the current flow in the specimen. Thus, another forward computation of the DRs has been developed which defines the secondary current dipoles in a larger region. The extended area

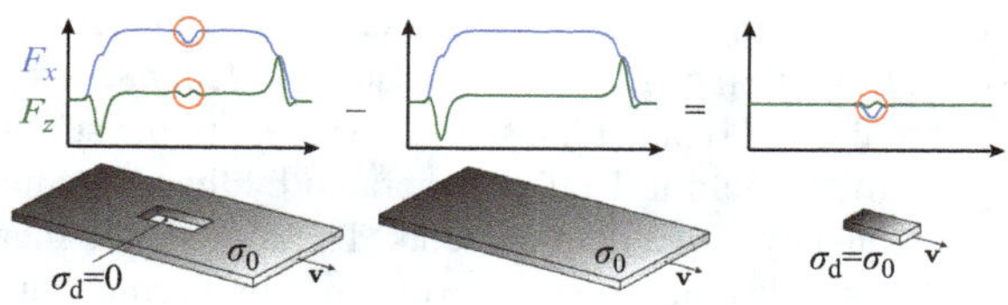

Figure 5.2 Simplest approach to model a defect (rectangular slit) in a conductive specimen (conductivity σ_0), described by the difference of a defect-free specimen and a specimen with defect

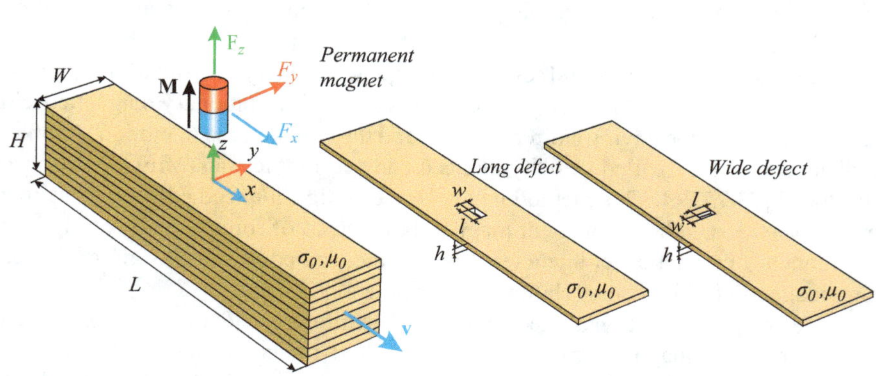

Figure 5.3 A package of thin aluminum sheets used as a test specimen moving with constant velocity below the permanent magnet [345]

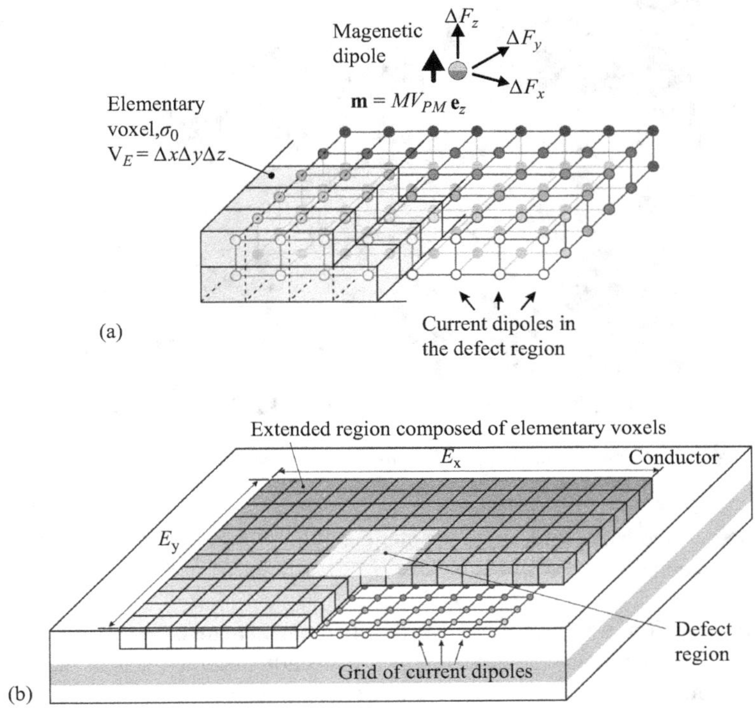

Figure 5.4 Defect modeling in an LET forward problem: (a) Simplest approach for the defect region (defect region is subdivided into voxels, each with a current dipole). (b) Extended area approach [112]

approach (EAA) is defined with respect to the region around the defect where the eddy current distribution is disturbed mostly from the defect (Figure 5.4(b)). More details about this approach can be found in Section 2.2.7.

In the LFE technique, MNE methods are often applied to the reconstruction of eddy current distributions, which are responsible for DRSs that are measured by the LET system (Figure 5.5).

Solving an inverse problem in a fully 3D source space requires the use of high computational resources, including large memory space and long computation times. To avoid the inversion of very large kernel matrices, the defect reconstruction is split into three steps:

- determination of a depth of the intermediate plane of a defect,
- determination of the length of a defect in a moving direction of a specimen, and
- reconstruction in x-y-plane, parallel to the upper surface of the bar and the scanning plane.

Determination of the depth of the long defect was precisely determined using both measurement and data obtained by finite elements method [98]. In the case of

Figure 5.5 Experimental setup for the LET measurements in the laboratory

wide defect, a depth of its intermediate plane is slightly moved toward the surface, but still being in an acceptable range of error (i.e. less than 1 mm for defects as assumed in Figure 5.6).

Determination of a subsurface defect depth requires to involve as much as possible *a priori* knowledge about the defect, e.g. it is usually assumed that in the specimen exists only one defect and the type of the defect is known.

The length of such defect with main orientation in the direction of movement of the solid bar was successfully determined in all cases, whereas the reconstruction in the x-y-plane was satisfactory as well. For the long subsurface defect, the solution using FEM data shows errors of 1 mm in y-direction and 3 mm in x-direction. For the wide defect, an error of about 1 mm in both directions has been achieved. Reconstructed conductivities in the case of long subsurface defect, obtained using measurement data are smeared out around the defect, but with an acceptable error.

The results have indicated that LET and LFE have great potential for investigating defects in nonferromagnetic conductors, especially in laminated composites. Consequently, the LFE technique has been further improved. Because the forward problem in LFE requires fast numerical calculations of the eddy currents, of the magnetic fields, and of the Lorentz forces, exhaustive investigations to improve the modeling of the permanent magnet configurations have been made [100,156]. The existing semianalytic forward calculations were further improved due to the use of the EAA [99]. Furthermore, thereby different defect simulations could be taken into account, because the computation of the eddy current distributions in the specimen has been adapted much closer to the physics of the electromagnetic field. All these measures have been contributions to the main goal, the development of better, i.e. faster inverse solution methods. Because of the complexity of the LFE method, there always has to be found a compromise of an optimal forward solver (accuracy vs. computational costs) and a fast defect identification technique. This results in the application to two examples, the inspection of multi layered structures and

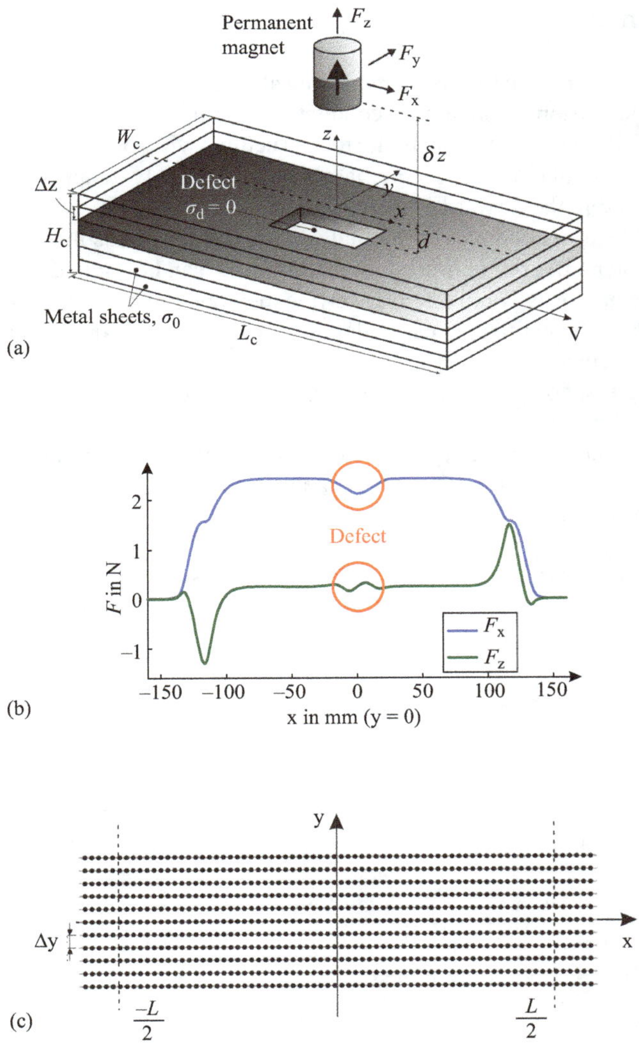

Figure 5.6 (a) Setup of a test specimen with an internal defect; (b) force profiles of an LET measurement (scan in x-direction), with drag force, F_x, and lift force, F_z; and (c) scanning area above the specimen surface together with a grid of acquisition points used in the simulations [156]

composite materials. In the first case, a special evolution strategy, the differential evolution, is applied to reconstruct internal defects in multi layered materials, e.g. Alucobond (see Section 6.2) [101]. In Section 6.3, the inspection of glass laminate aluminum reinforced epoxy (GLARE) with the goal function scan method is presented [156].

5.4 Summary

It has been shown that the LFE is an interesting alternative to the well-established ECT technique when in nonferromagnetic conductors deep internal defects/anomalies have to be identified. In such cases, complicated structures with moving components lead to very high computational costs, i.e. it is highly preferably to apply fast numerical models for solving the forward problem. Solving the inverse problem, i.e. the identification of defects or conductivity anomalies, requires efficient solution strategies. In most cases, the number of optimization methods that can be applied for this purpose is limited because usually only derivative-free methods can be used. It will be shown in the following chapter that, in particular, the use of stochastic optimization methods can lead to efficient solution strategies.

Nevertheless, application of different inversion algorithms (including improved forward modeling in the defect region), regularization techniques or usage of more complex permanent magnet configurations (to generate a maximum magnetic flux density) define the next steps to improve the reconstruction accuracy of the LET and LFE as options of the MIECTE.

Chapter 6
Applications

Robert P. Uhlig[1], Hartmut Brauer[1], Konstantin Weise[1] and Marek Ziolkowski[1,2]

6.1 Sigmometry

The following part describes how LET can be applied for nondestructive material parameter evaluation. Parts of the presented content have been published in [93] and have led to a patent of invention [409].

6.1.1 Introduction and motivation

LET as described previously in its main application is a technique for defectoscopy. Defectoscopy techniques themselves are not meant to provide absolute values describing the specimen's state but indicating the risk of the presence of a defect. Therefore, the specimen is tested by a system of sensors under certain, quasi constant conditions. The system response of the object under test is usually compared with the system response of a reference specimen containing reference defects. That procedure is allowing for a lack of knowledge about the particular testing conditions but does not allow for an evaluation of absolute defect parameters. These remain in best cases estimates.

In case that there is a need for absolute values describing the specimen's state, e.g. defect size, defect kind, characteristic material properties of the specimen must be known. The characteristic key material property of any application utilizing eddy currents is the electrical conductivity of the specimen. In fact, one further feature of the before presented LET is the ability of the nondestructive and contactless determination of the specific electrical conductivity of the specimen.

Up to now, the determination of the electrical conductivity is performed, applying only a few established measurement techniques [410–412]. While fluids are usually treated with amperometric and potentiometric measurements, the established solid-state body measurements are the impedence spectroscopy and the four-point-method. All named methods suffer from the fact that there is the immanent need for contact with the specimen, which might be difficult to provide, e.g. in hot metals or for fast

[1] Advanced Electromagnetics Group, Technische Universität Ilmenau, Germany
[2] Applied Informatics Group, West Pomeranian University of Technology, Poland

moving specimens. Available contactless methods, as e.g. the eddy current method [411], are strongly dependent on deviations in lift-off distance and cover only the subsurface region which might differ from the rest of the specimen due to oxide layers. Methods applying alternating electro-magnetic fields cannot provide conductivity measurements from within the specimen since they are limited due to the frequency dependent skin effect.

The application of LET as NDE technique overcomes the above-mentioned disadvantages. It provides contactless measurements deep inside the material, no matter whether it is a fluid or solid body. Since the Greek letter σ is used internationally to abbreviate the electrical conductivity, and the exploited physical effect is the Lorentz force, the method shall be called "Lorentz force sigmometry" (LoFoS) in the following section.

6.1.2 Basic principle

The basic problem under investigation is sketched in Figure 6.1. A conducting, non-magnetic plate of arbitrary thickness is moving in the magnetic field generated by a permanent magnet or a set of current-driven coils. The magnet is located at a constant lift-off distance δz above the plate with the magnetization direction transversal to the moving direction.

It has been shown in the previous parts that the drag component of the Lorentz force F_x, i.e. along the movement direction, depends linearly on the relative velocity v, whereas the lift force component F_z is proportional to the square of v for small v. Due to the fact that the measurement setup is neither significantly changing in size and speed nor in location, these dependencies can be rewritten utilizing the magnetic Reynolds number as

$$F_x \propto R_m, \tag{6.1}$$

$$F_z \propto R_m^2. \tag{6.2}$$

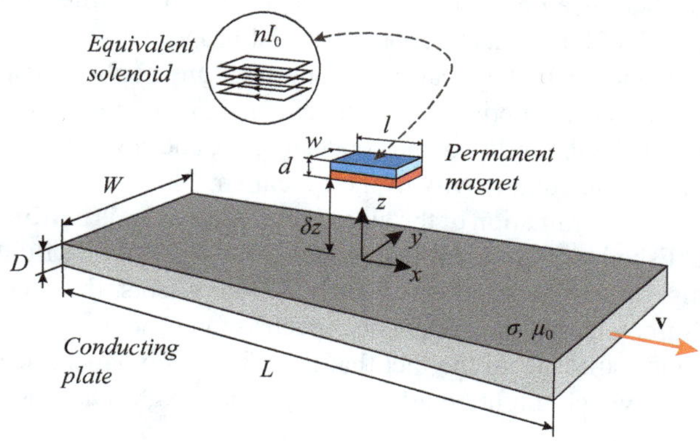

Figure 6.1 The sample problem—conducting plate moving with a constant velocity beneath a permanent magnet or a solenoid

The magnetic Reynolds number R_m is a nondimensional value that specifies the ratio between convection and the diffusion of the magnetic field [236]. It can be calculated according to

$$R_m = \mu_0 \sigma v L, \tag{6.3}$$

where μ_0 is the vacuum permeability, σ is the electrical conductivity, v is the relative velocity between the conductor and the magnetic field source, and L is the characteristic length scale. The advantage of describing the measurement conditions with the magnetic Reynolds number is the summary of physical dependencies of the measurement result within one value.

Obviously, the determination of electrical conductivity is possible by measuring the drag force component F_x. Since it depends linearly on the Reynolds number, and the Reynolds number itself depends linearly on the conductivity, there must be a calibration coefficient β that fulfills the following equation:

$$\sigma = \beta F_x. \tag{6.4}$$

The calibration coefficient is usually determined by calibration with reference specimens, but it can be provided by numerical calibration as well. It mainly depends on the geometry of the used magnet system, the relative velocity and strongly on the lift-off distance, and on the magnetic flux density **B** [91].

As stated above, the strong dependency on lift-off distance is a main uncertainty of contactless evaluation techniques that creates high uncertainties in measurement results. Manufacturing tolerances and mechanical oscillations during the measurement procedure lead to deviations in lift-off distance of several 100 μm resulting in measured force deviations of several millinewton depending on the set lift-off distance. Additionally, the magnetic field strength of the permanent magnet is difficult to determine due to the fact that magnetization direction and mounting errors have to be taken into account. Furthermore, the magnetic properties are strongly depending on temperature.

A modified measurement approach is supposed to help overcome the disadvantages stated above while measuring the drag force component only. As a consequence of (6.1) and (6.2), the lift-to-drag force ratio F_z/F_x depends linearly on R_m as well:

$$\frac{F_z}{F_x} \propto R_m. \tag{6.5}$$

For thin plates of infinite extension, it is found that, regardless the particular shape of the magnetic field source, the lift-to-drag (force) ratio can be written as

$$\frac{F_z}{F_x} \propto \frac{v}{w}, \tag{6.6}$$

where w is the characteristic velocity of the plate (given by $w = 2/(\mu_0 \sigma D)$, where D is the thickness of the thin plate) [60]. The estimation done applying (6.6) is limited to plates and sheets as long as their thicknesses do not exceed the motional skin depth δ given by

$$\delta = D\sqrt{\frac{1}{2R_m}}, \tag{6.7}$$

and while the relative velocity is kept in low R_m-range which is the case up to moderate velocities ($v \leq 5$ m/s) [61]. Considering $L = D/2$ to be the characteristic length scale of the present assembly, (6.5) simplifies to

$$\frac{F_z}{F_x} = R_m. \tag{6.8}$$

The result implies that the lift-to-drag ratio neither depends on the magnetic field $\vec{B}$ of the magnetic field source nor on the lift-off distance δz [cf. (6.3)]. Assuming that the conductivity of the plate under test is homogeneous and isotropic, (6.8) can be rewritten by analogy to (6.4) as

$$\sigma = \alpha \frac{F_z}{F_x}, \tag{6.9}$$

where the calibration coefficient α depends only on the geometry of the magnetic field source, the relative velocity, and (weakly) on the distance between the plate and the magnetic field source.

Since the calibration coefficient α is unknown *a priori*, it has to be determined by the utilization of at least two reference specimens. The conductivity of these specimens can be given by means of certificated methods, such as the Van-der-Pauw-method [410,413] and should be approved by the National Metrology Institute.

6.1.3 Semianalytical and numerical calibration

The so far described application is only valid for thin plates and sheets. To extend the validity of the findings, the calibration is performed for the typical application problem shown in Figure 6.1. In total, two methods are applied to calculate the drag force component F_x and the lift force component F_z acting on a moving parallelepipedial magnet: a semianalytical one presented in Section 2.2 that can be applied to configurations with small magnets compared to the tested specimen ($L, W \gg w, l, d, D$), and a fully numerical one presented in Section 2.4, without these restrictions.

To validate the obtained calibration coefficients experimentally later on, the geometry of the existing experimental setup according to Table 6.1 is taken into account. The magnetization of the permanent magnet is considered to be coaxial with the z-axis of the defined coordinate system. Due to model restrictions the plate is considered to be infinite, which means that the chosen dimensions are big enough in case of the fully numerical model to not show effects of edges on the results.

The lift-to-drag ratio is analyzed as a function of the specimen's conductivity. The comparison between the semianalytical method and the fully numerical method is shown in Figure 6.2. To prove the stated dependencies, the lift-to-drag ratio is calculated for different lift-off distances δz and different relative velocities v. Both models show a very good agreement in the whole range of considered conductivities. This result verifies the use of the semianalytical model for simplified model configurations, where edge effects are absent. The numerical model should be utilized if specimens with finite dimensions, or if more complex magnet systems are used.

The obtained calibration curves shown in Figure 6.2 are the basis for LoFoS implementation.

Table 6.1 *Geometrical and material properties used for comparison of the analytical with the numerical model and experiments*

Parameter	**Figure 6.2**	**Experiments**
Specimen length L	0.25 m	0.25 m
Specimen width W	0.25 m	0.05 m
Specimen depth D	0.05 m	0.05 m
Magnet length l/Diameter Ø	0.015 m	Ø 0.015 m
Magnet width w	0.015 m	–
Magnet height d	0.025 m	0.025 m
Remanent flux density B_r	1.17 T	1.17 T

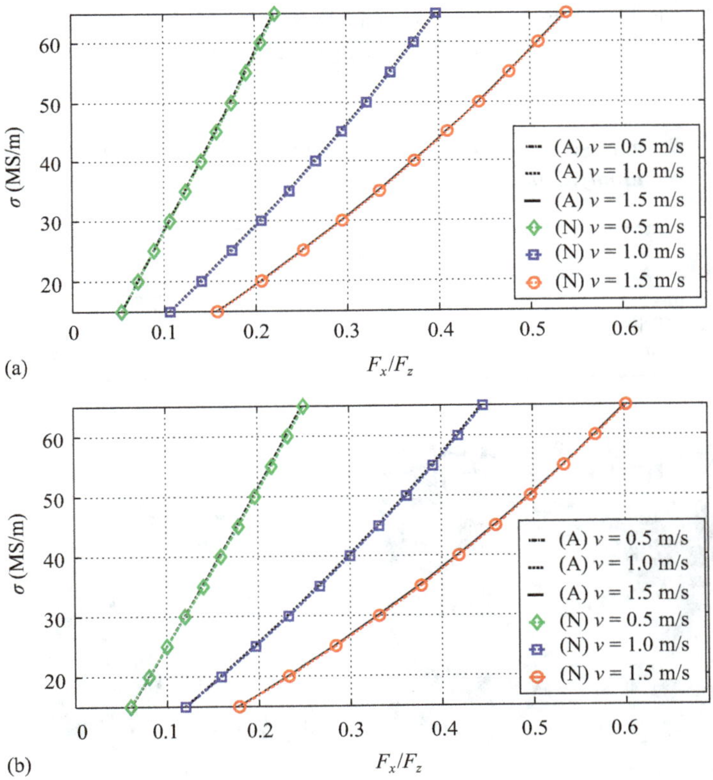

(a) (b)

Figure 6.2 *Calibration curves for plates obtained semianalytically (A) and numerically (N) for various velocities: (A) infinite plate with thickness 50 mm; (N) $W \times H \times L = 250 \times 50 \times 250$ mm. (a) Lift-off $\delta z = 3$ mm. (b) Lift-off $\delta z = 5$ mm*

The assumption of a linear dependency is verified in certain limits. As the velocity rises the linearity of the calibration curves reduces. This fact has been foreseen as the analytical model in (6.8) is valid only for low velocities and conductivities. Nevertheless, a basic linear fitting is applied to the calibration curves by fitting the following equation to the data points applying the method of least squares:

$$\sigma = \alpha \frac{F_z}{F_x} + \sigma_{\text{off}}. \qquad (6.10)$$

The resulting calibration coefficients α and the corrective offset σ_{off} are presented in Table 6.2 within the simulation range $\sigma = 20$–60 MS/m.

6.1.4 Experimental validation

The validity of the numerical predictions presented above is to be shown by means of experiment. The experimental setup is shown in Figure 6.3. Due to disturbing effects of inertia, and since this application is most common in industrial environment, the relative motion between the permanent magnet and the material under test is realized by a linear belt-driven drive carrying the specimen rather than the permanent

Table 6.2 Basic linear fitting coefficients for plate calibration applying the semianalytical model

Coefficients		**$\delta z = 3$ mm**			**$\delta z = 5$ mm**		
v	(m/s)	0.5	1.0	1.5	0.5	1.0	1.5
α	(MS/m)	299.95	171.51	129.89	266.35	153.42	117.06
σ_{off}	(MS/m)	−1.44	−4.11	−6.56	−1.50	−4.34	−6.96

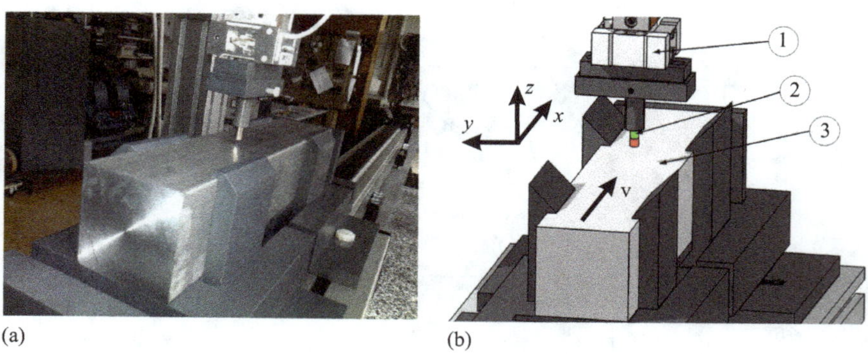

Figure 6.3 Experimental measurement setup: (a) photography of the experimental setup; (b) sketch of the experimental setup identifying the components: (1) 3D force sensor, (2) permanent magnet, and (3) specimen mounted on a linearly belt-driven sledge

magnet. The velocity range is $v = 0$–3.75 m/s which is sufficient to reach intermediate magnetic Reynolds numbers for highly conductive materials.

The acting Lorentz force is measured by a three-dimensional force sensor that is realized by a deformation body to which three resistive strain gages are applied. The absolute force in longitudinal and transversal direction of the specimens motion is limited to $|F_{x,y}| \leq 3$ N whereas in vertical direction the limitation is $|F_z| \leq 10$ N. In vertical direction, the mass of the sensor system needs to be considered as acting force offset. The force component transversal to the motion direction F_y is used as a quality measure for the alignment of the permanent magnet with respect to the object under test in the presented experiment. At the symmetry line, the transversal force component should vanish, $F_y = 0$. This statement is valid for any specimen whose shape is symmetrical to the movement direction.

6.1.4.1 Methodology characterization

The calibration of the conductivity measurement has to be performed with at least two specimens of known conductivity. For that, two solid-state bars are considered. As known from the above presented numerical models, there are two key performance parameters for the uncertainty of measurement: (i) the velocity v and (ii) the lift-off distance δz. While the velocity v should have an almost linear dependency on the lift-to-drag ratio and thus, on the calibration coefficient, the influence of the lift-off distance δz should be neglectable.

The predicted linear drag force component and the corresponding quadratic lift force component behavior by [60] can be seen in Figure 6.4. Especially for the Al-alloy bar with a known conductivity of σ_{Al} =20.4 MS/m, the linear and quadratic models are well fitting. The copper bar shows a three times higher conductivity and therefore allows for higher Reynolds number regimes. For high velocities, the predicted behavior is not valid anymore; the drag force component is rising nonlinearly whereas the lift force component does rather linearly than quadratic.

Surprisingly, the lift-to-drag ratio of both the Al-alloy specimen and the copper bar show a quite linear dependency to velocity for a much bigger range of velocity than the single components (cf. Figure 6.5). The experiments with the copper bar validate the linear behavior even on the bounds of the existing conductivity range.

As described earlier, controlling the strong dependency of the Lorentz force on the lift-off distance is crucial for the industrial application of since the measured conductivity is influenced directly due to that deviation. Concerning the lift-off distance, there exists a conflict of objectives: on the one hand a smaller lift-off distance δz increases the total measured Lorentz force whereas on the other hand the sensitivity to changes in lift-off distance rises significantly, e.g. surface roughness.

Some authors argue that a permanent magnet can be modeled by an equivalent magnetic dipole [90]. For a dipole a reduction in Lorentz force closely to δz^{-3} is expected. Results shown in Figure 6.6 imply that the dipole model is valid only if the permanent magnet is sufficiently far away from the specimen. In the vicinity of the permanent magnet of arbitrary shape, the Lorentz force is scaled with approximately $\delta z^{-4.7}$.

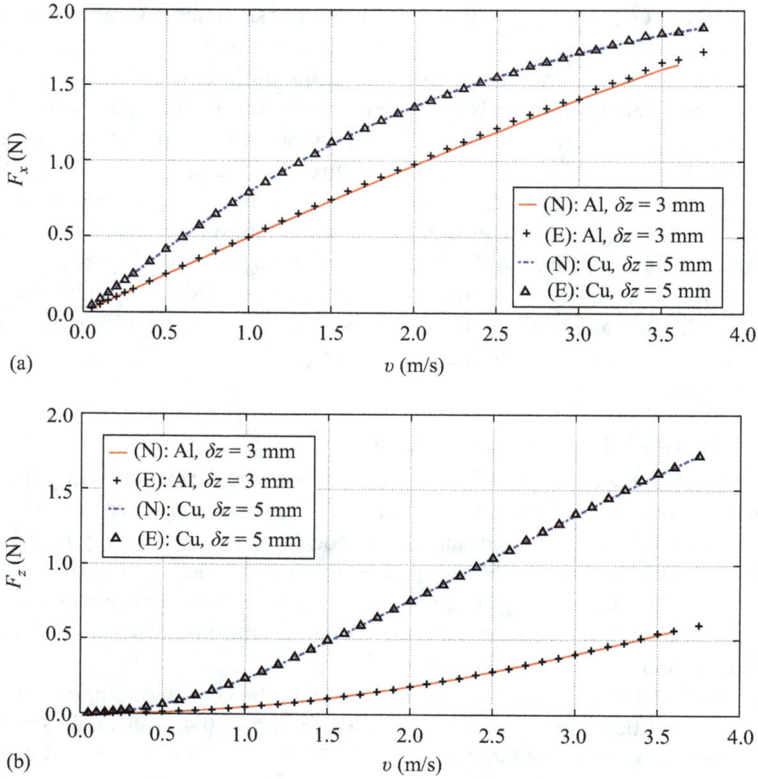

Figure 6.4 Acting Lorentz force depending on relative velocity between specimen and permanent magnet at a lift-off distance of $\delta z = 3$ mm (Al-alloy) and at $\delta z = 5$ mm (Cu); (N) numerical results, (E) experiment. (a) Drag force. (b) Lift force

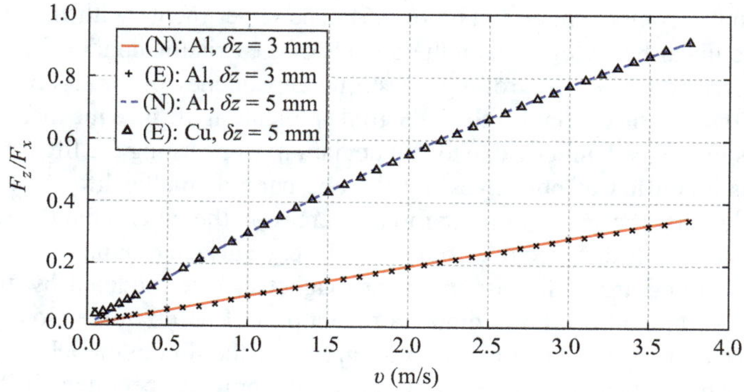

Figure 6.5 Lift-to-drag force ratio depending on velocity and material at a lift-off distance of $\delta z = 3$ mm (Al-alloy), and $\delta z = 5$ mm (Cu); (N) numerical results, (E) experiment

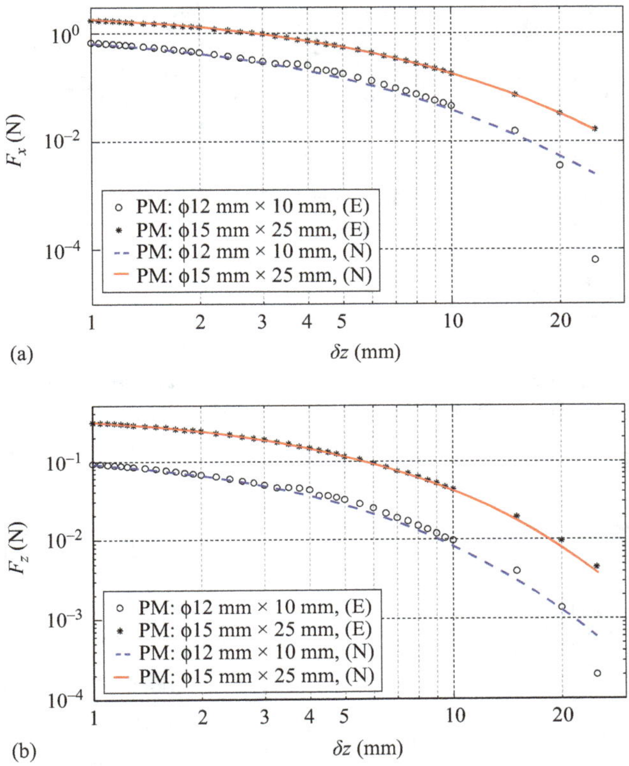

Figure 6.6 Lorentz force depending on lift-off distance for Al-alloy specimen (velocity $v = 2$ m/s). (a) Drag force. (b) Lift force

The strong sensitivity to varying lift-off distances is significantly reduced by utilizing the lift-to-drag ratio as shown in Figure 6.7. Edge effects of the specimen, and the arbitrary-shaped magnetic field source with orientation misalignment are most likely responsible for the remaining slight increase in lift-to-drag ratio.

Both effects the dependency on velocity and lift-off distance lead to the conclusion that utilizing the lift-to-drag ratio rather than the single Lorentz force components has huge advantages concerning the sensitivity to measurement setup imperfections. The overall measurement error can be estimated by the total derivative of the lift-to-drag ratio

$$d\left(\frac{F_z}{F_x}\right) = \frac{F_x\, dF_z - F_z\, dF_x}{F_x^2}, \tag{6.11}$$

where the values denoted by a d symbolize the measurement uncertainties. The overall measurement error of the presented experiment can be read from Figure 6.8 for fixed

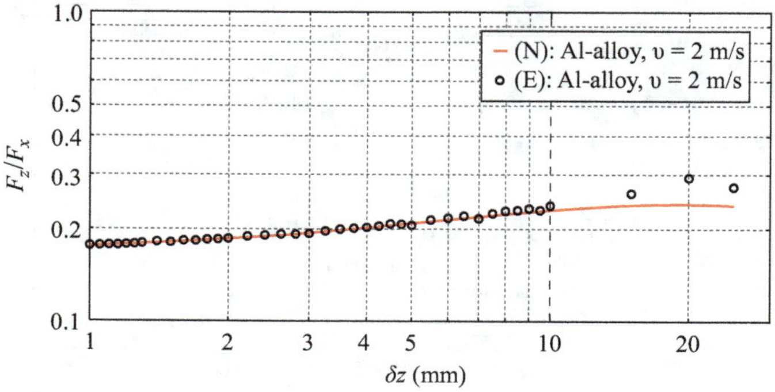

Figure 6.7 Lift-to-drag force ratio depending on lift-off distance for Al-alloy, where the permanent magnet is PM ø15 × 25 mm

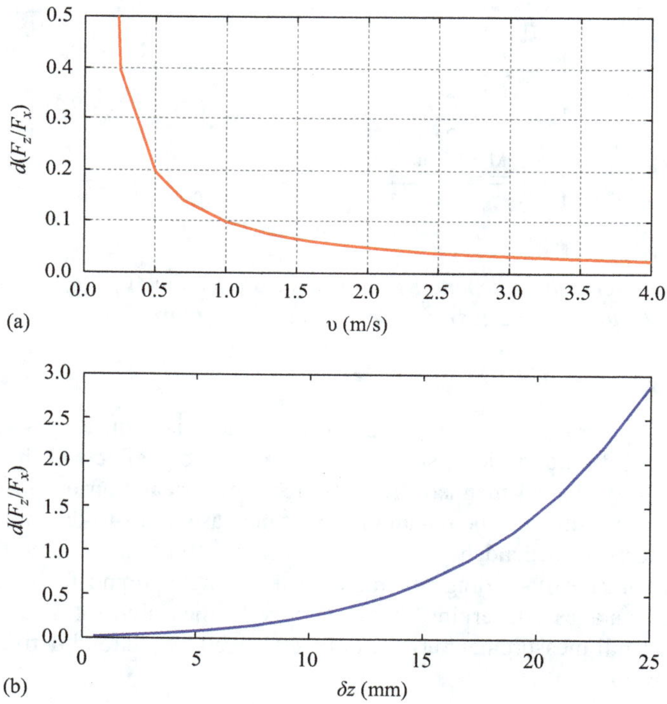

Figure 6.8 Total measurement error of the acting lift-to-drag ratio depending on (a) velocity at a fixed lift-off distance of $\delta z = 3$ mm and (b) lift-off distance at a fixed velocity of $v = 2$ m/s (PM ø15× 25 mm)

velocity and lift-off distance. Since velocity and lift-off distance influence the error in opposite way, the use of medium velocities and small lift-off distances is proposed to minimize measurement uncertainties without exceeding the medium R_{m}-range.

6.1.4.2 Conductivity measurement

Lorentz force measurement depends on the operational conditions, i.e. relative velocity, lift-off distance, and magnetic field. Therefore, a calibration with standard conductor targets is required. Since the force sensor itself can be calibrated by national mass standards, the result of the Lorentz force measurement can be traced back to SI units.

For simplicity, the calibration is performed with two materials, e.g. copper and aluminum. It is definitely required to calibrate with more standard materials and impose a higher order dependency to obtain more accurate results. Calibrating the measurement setup applying different velocities is recommended as well. The presented results use calibrations at velocities $v_1 = 0.5$ m/s, $v_2 = 1.0$ m/s, and $v_3 = 0.5$ m/s that suit best for the particular measurement setup ensuring a medium magnetic Reynolds number R_{m}.

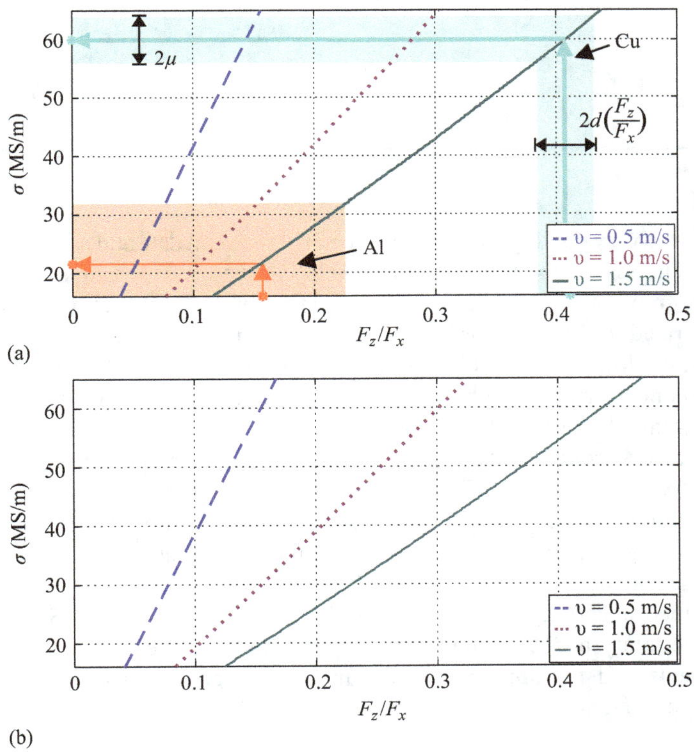

Figure 6.9 Calibration curves for the used bar geometry ($W \times H \times L = 50$ mm $\times$ 50 mm $\times$ 250 mm) for different velocities obtained numerically: (a) lift-off distance $\delta z = 3$ mm, the particular example from Table 6.4 is visualized colorfully, (b) lift-off distance $\delta z = 5$ mm

Table 6.3 Basic linear fitting coefficients for bar calibration applying the numerical model

Coefficients		$\delta z = 3$ mm			$\delta z = 5$ mm		
v	(m/s)	0.5	1.0	1.5	0.5	1.0	1.5
α	(MS/m)	417.91	216.26	151.43	389.46	201.54	140.89
σ_{off}	(MS/m)	−0.26	−1.04	−2.18	−0.25	−1.01	−2.10

Table 6.4 Obtained conductivity values using LoFoS

Parameter	Unit	Aluminum		Copper
v	m/s	1.5		1.5
δz	mm	3		3
F_x	N	0.735		1.92
dF_x	N		0.015	
F_z	N	0.115		0.789
dF_z	N		0.050	
$\frac{F_z}{F_x} \pm d\left(\frac{F_z}{F_x}\right)$	–	0.157 ± 0.065		0.411 ± 0.022
$\sigma \pm u$	MS/m	21.59 ± 9.82		60.08 ± 3.46

Due to the very good agreement between experimental data and numerical results, a purely numerical calibration is valid as well. For the used measurement setup, the calibration graphs are shown in Figure 6.9 and the obtained calibration coefficients α are summarized in Table 6.3 for linear fitting. In the present example, reference bars of aluminum alloy ($\sigma_{\text{Al}} = 20.4$ MS/m) and copper ($\sigma_{\text{Cu}} = 57.92$ MS/m) are utilized. The given conductivities have been approved with the eddy current device *Sigmatest 2.0069* (Institut Dr. Foerster GmbH & Co. KG).

For the final conductivity determination, the purely numerically obtained calibration curves are used (cf. Figure 6.9). The measurement uncertainties and the intermediate results are given in Table 6.4. Applying the linear calibration coefficients from Table 6.3 and taking into account the uncertainty calculation according to (6.11) the conductivities $\sigma_{\text{Al}} = 21.59$ MS/m and $\sigma_{\text{Cu}} = 60.08$ MS/m are measured. The combined uncertainties of $u_{\text{Al}} = 9.82$ MS/m and $u_{\text{Cu}} = 3.46$ MS/m are based on the uncertainty of the applied force sensor and could be reduced by repeating measurements and the application of a more appropriate sensor concerning the measurement range.

6.1.5 Findings

The presented application called *LoFoS* has a few advantages over present available and established conductivity measurement techniques. The main advantage is that LoFoS is contactless. Therefore, it can be applied continuously during production process. LoFoS can be applied to a variety of specimens including these whose physical condition is fluid. The only limitation from measurement point of view is given by the minimal measurable Lorentz force components of the applied force sensor.

In difference to other noncontact methods, the considered conductivity is not only limited to the surface of the specimen. Since direct magnetic fields are used, the penetration depth is not limited by the source frequency but by the relative velocity between specimen and magnetic field source.

Due to the exploitation of the lift-to-drag ratio, the measurement procedure is quite robust to changes in lift-off distance, velocity, and magnetic field strength. For a wide range of conductivities, the ratio is linearly dependent on the conductivity which offers accurate results even for two-point-calibration. Finally, the measured conductivity is directly traceable to SI units if calibration has been provided by appropriate references.

6.2 Defectocscopy of multilayered structures

6.2.1 LET measurements of alucobond specimen

In the framework of NDT, the detection and reconstruction of defects in laminated composites is of major interest. Composites consists of alternating metal and isolating layers and have a high strength and damage tolerance. Therefore, they are especially useful for weight-critical, safety-related components such as aircraft fuselages [414]. Due to periodic and extensive in-service loads, fatigue fractures in the metal layers can occur. Alucobond is a multi layered composite (facade material) usually consisting of two thin sheets of aluminum (0.5 mm each), separated by an isolating kernel of polyethylene (3–5 mm, $\sigma = 0$) (Figure 6.10).

It is very stable as well as flexible at the same time. The bottom aluminum cover sheet contains an artificial cylindrical hole (diameter = 5 mm) at the center of the specimen.

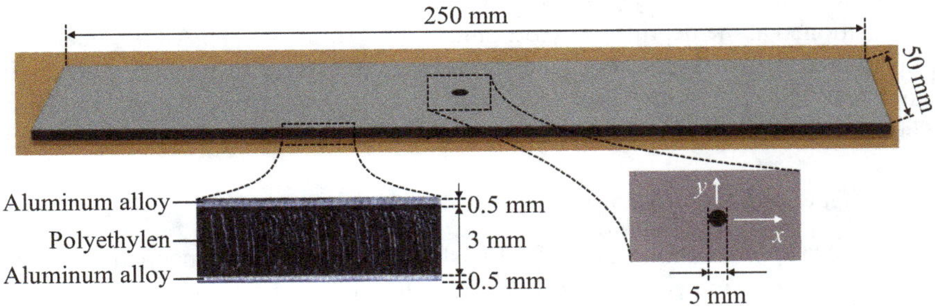

Figure 6.10 Three-layered alucobond composite specimen used in the experimental setup and view of the artificial defect in the lower cover sheet

Table 6.5 Parameters of the LFE setup

Conductor (alucobond)	
Dimension ($L \times W \times H$)	250 mm × 50 mm × 4 mm
Thickness of metal sheets	0.5 mm
Conductivity of metal sheets	30.1 MS/m
Velocity v	0.5 m/s
Defect	
Shape	Cylinder
Radius R_d	2.5 mm
Height H_d	0.5 mm
Depth d	3.5 mm
Permanent magnet	
Shape	Cylinder
Radius R_p	7.5 mm
Height H_p	25 mm
Remanence	1.17 T
Lift-off δz	1 mm

As in the usual experimental LET setup described previously, the test specimen moves under the fixed cylindrical permanent magnet, with a small liftoff distance between magnet and specimen's surface. The detailed parameter values for the conductor, defect, and permanent magnet are summarized in Table 6.5.

To identify the defect, the Lorentz force evaluation (LFE) has been applied [345]. LFE evaluates perturbations in Lorentz force signals obtained from LET. The LFE reconstruction scheme was applied to two sets of simulated Lorentz force data. Two different laminated specimens, consisting firstly of stacked aluminum sheets having each a thickness of 2 mm whereas one sheet contains a cylindrical defect with a radius of 2.5 mm (at different depths, $d = 2$ mm or 4 mm). The depth corresponds to the upper surface coordinate of the layer. For depth reconstruction, it is applied a set of voxel grids positioned at the depths $d = \{0, 2\text{ mm}, 4\text{ mm}, 6\text{ mm}, 8\text{ mm}\}$. All optimizations are performed with 50 repetitions. The second specimen was the alucobond shown in Figures 6.10 and 6.11. Simulated data are obtained using a finite element model of the laminated specimen moving under the cylindrical permanent magnet.

6.2.2 Forward simulations

In the forward approximations, only LET systems with permanent magnets of simple geometries are considered. A typical configuration of the LET system is shown in Figure 6.11.

Using the simplest possible approximation, the permanent magnet is represented by a single magnetic dipole positioned at $\mathbf{r}_p = [x_p, y_p, z_p]^\mathrm{T}$ (Figure 5.5). The magnetic

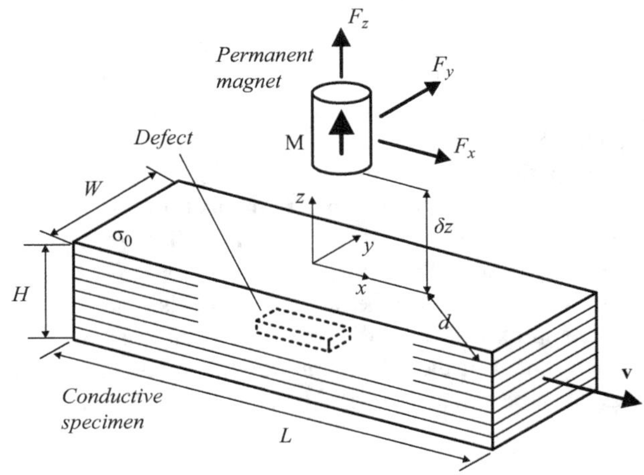

Figure 6.11 LET simulation setup for an anisotropic specimen

dipole has the magnetic moment $\mathbf{m} = m\mathbf{e}_z = V_p\mathbf{e}_z$, with V_p denoting the volume of the permanent magnet. The magnetic dipole produces at any point $\mathbf{r} = [x, y, z]^{\mathrm{T}}$ the magnetic flux density

$$\mathbf{B} = \frac{\mu_0}{4\pi}\left[3\frac{\mathbf{m}\cdot(\mathbf{r}-\mathbf{r}_p)}{|\mathbf{r}-\mathbf{r}_p|^5}(\mathbf{r}-\mathbf{r}_p) - \frac{\mathbf{m}}{|\mathbf{r}-\mathbf{r}_p|}\right]. \tag{6.12}$$

Previous investigations have shown that the condition outlined above is fulfilled, if the specimen has a characteristic length of $L = H_c \leq 50$ mm, is composed of aluminum alloy, and the measuring velocity is not larger than 0.5 m/s [241]. Then, the influence of the secondary magnetic field resulting from the induced eddy currents on the total magnetic field can be neglected [weak reaction approach (WRA)]. Consequently, the magnetic field in the conductor can be reduced to the primary magnetic field of the permanent magnet. In these LFE studies only low velocities are considered, since otherwise the violation of the limitations assumed for the LET measurements is likely to have noticeable effects. A validation of the WRA for low velocities can be found in [156]. Moreover, defects are assumed to be ideal, i.e. the conductivity of the defect equals zero and no eddy currents flow in the defect region. Furthermore, the forward solutions do not take into account the boundary conditions at the edges of the conductor, i.e. the conductor is considered to have infinitively large x- and y-extensions. This simplification holds if the defect is much smaller than the conductor, $\{d_x, d_y\} \ll \{L_c, W_c\}$. Taking into account the WRA and the fact that with respect to the global coordinate system the conductor is in motion (Figure 5.6(a)), eddy currents can be calculated using Ohm's law for moving conductors

$$\Delta\mathbf{J} = \sigma\left(-\nabla\varphi + \mathbf{v}\times\mathbf{B}\right), \tag{6.13}$$

where φ is the electric scalar potential. For this simple approximation, the real defect region of conductivity $\sigma_d = 0$ is substituted with the electric conductivity σ_0 and forms a fictitious conducting region. The eddy currents are only considered in the defect region and flow in exact opposite direction as for a conductor without a defect, i.e. $\Delta\mathbf{j} = -\mathbf{j}_0$. To calculate the simple approximation of the DRS, $\Delta F^{(1)}$, a finite volume discretization of the fictitious conducting defect region is applied. The defect region is substituted with a regular grid of volume elements (voxels). The voxels have the elementary volume $V_E = \Delta x \Delta y \Delta z$ and conductivity σ_0 (Figure 5.4(a)). In each voxel, a current dipole $\mathbf{p}_k$ is placed at the center of gravity described by $\mathbf{r}_k = [x_k, y_k, z_k]^T$ with $k = 1, \ldots, K$, where K denotes the number of voxels. The entirety of all current dipoles is a discrete approximation of the continuous eddy current distribution in the defect region. The moment of the kth current dipole equals $\mathbf{p}_k = \Delta\mathbf{j}_k V_E$. The eddy current density $\Delta\mathbf{j}_k$ can be calculated with the help of (6.13) as

$$\Delta\mathbf{j}_k = \sigma \left(\nabla\varphi_k - \mathbf{v} \times \mathbf{B}_k\right), \tag{6.14}$$

where $\mathbf{B}_k$ is the magnetic flux density at the position of the kth current dipole calculated using (6.13). To find an analytic expression for φ_k calculated in (6.12), it is exploited that the interfaces between the metal layers are electrically isolating and $\Delta z \ll \{L_c, W_c\}$. The conductor is assumed to be anisotropic, i.e., the diagonal conductivity tensor $[\sigma] = \mathrm{diag}(\sigma_{xx}, \sigma_{yy}, \sigma_{zz})$ has the entries $\sigma_{xx} = \sigma_{yy} = \sigma_0$ and $\sigma_{zz} = 0$. Thus, the eddy currents flow only in the x-y–planes and the z-component vanishes, i.e. $j_z = 0$. Under the condition that the permanent magnet is modeled with one magnetic dipole, the potential φ_k at the point $\mathbf{r}_k = [x_k, y_k, z_k]^{\mathrm{T}}$ can be determined from $\partial\varphi/\partial z = v$ by analytically. Then, the DRS profile can be approximated with the help of the electric current dipoles as

$$\begin{aligned}\Delta\mathbf{F}^{(1)} &= \int_{V_d} \Delta\mathbf{j} \times \mathbf{B}\,\mathrm{d}V \approx V_E \sum_{k=1}^{K} \Delta\mathbf{j}_k \times \mathbf{B}_k \\ &= \sigma_0 V_E \sum_{k=1}^{K} \left(\nabla\varphi_k - \mathbf{v} \times \mathbf{B}_k\right) \times \mathbf{B}_k.\end{aligned} \tag{6.15}$$

To improve the forward solution, the extended area approach (EAA) has been applied [112]. For the extended approximation of the DRS, the eddy currents in the area outside the defect produced by current dipoles in the defect region are taken into account as well (see Figure 5.4(b)). Current dipoles in the K voxels covering the defect region are calculated as described previously. Then, the distortion current density $\Delta\mathbf{j}_e$ of the eth current dipole in the extended region outside the defect can be approximated by superimposing the current densities from the defect region and the surrounding specimen. Finally, the DRS can be approximated with respect to (6.14) as

$$\Delta\mathbf{F}^{(2)} = \int_V \Delta\mathbf{j} \times \mathbf{B}\,\mathrm{d}V \approx \underbrace{\sum_{k=1}^{K} \Delta\mathbf{j}_k \times \mathbf{B}_k\,\mathrm{d}V}_{\text{defect region}} + \underbrace{\sum_{e=1}^{E} \Delta\mathbf{j}_e \times \mathbf{B}_e\,\mathrm{d}V}_{\text{region outside defect}}. \tag{6.16}$$

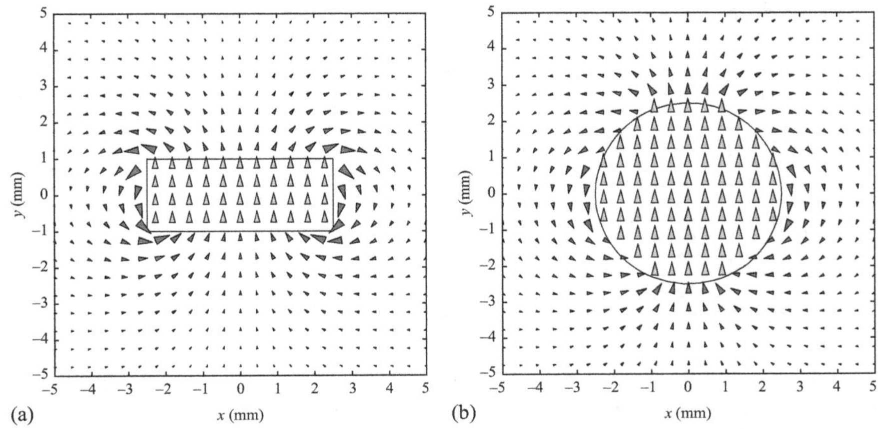

Figure 6.12 Defect response eddy current distribution in the defect region (light red cones) and in the surrounding extended region (red cones) [156]. (a) Cuboidal defect. (b) Cylindrical defect

This approach results in a current flow as shown in Figure 6.12 for a cuboidal and cylindrical defect, respectively. Here is always assumed that the defect is located in one layer only. The solution is finally verified with solutions using very fine finite element models. The distributions are calculated in the x-y–plane for the center of the defect ($d = -3$ mm) in that moment where the magnetic dipole is positioned just above the center of the defect ($x_p = y_p = 0$). For better visualization, the axes are limited to the range $-5\,\text{mm} \leq (x, y) \leq 5\,\text{mm}$. It can be observed that the continuity equation $\nabla \cdot \mathbf{j} = 0$ is not fulfilled, if only eddy currents in the region covering the defect are taken into account for forward simulations (light red cones). Thus, the EAA leads to a much better approximation of the physics of the field.

6.2.3 Defect identification

In general, the problem of defect identification can be formulated as an optimization problem, which searches for an optimal defect profile by minimizing the normalized root mean square error (NRMSE) (2.34) of measured and simulated force signals. For this purpose, optimization strategies based on deterministic algorithms have been extensively used. However, they often fail to converge to the global optimum if the objective function has multiple local minima. To overcome this drawback, an optimization scheme based on a global stochastic optimization algorithm, the differential evolution (DE) is proposed. It belongs to the class of Evolutionary Algorithms that mimic Darwin's theory to describe the biologically evolution process [415]. DE is a population-based, real-valued, and continuous optimization algorithm. An initial population evolves by generating off-springs in terms of random genetic mutation, crossover, and selection of population members (Figure 6.13).

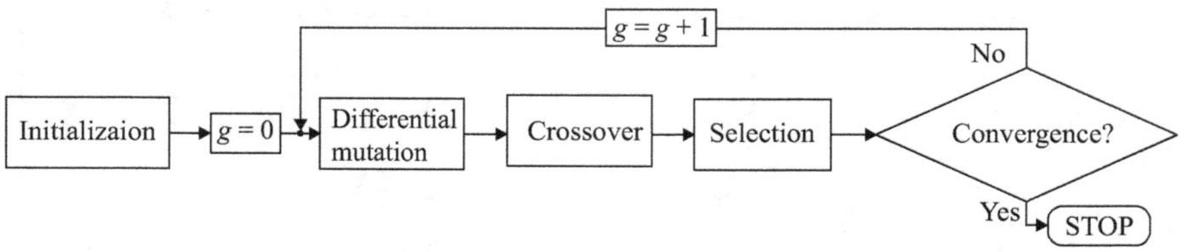

Figure 6.13 Flow chart of DE. The current generation is denoted by index g

The population is initialized by calculating N_p random population members. The number of elements in each member is equal to the number of voxels. Mutation describes the process of change or perturbation in random elements of the individual. For each population member, a mutant vector $\mathbf{x}^m$ is calculated by the sum of one randomly sampled member $\mathbf{x}^{(1)}$ and the difference between two other random vectors, $\mathbf{x}^{(2)}$ and $\mathbf{x}^{(3)}$, weighted with the mutation scale factor F as:

$$\mathbf{x}^m = \mathbf{x}^{(1)} + F(\mathbf{x}^{(2)} - \mathbf{x}^{(3)}). \tag{6.17}$$

Then, in binomial crossover, the parent and mutant exchange elements. The kth element of the resulting trial vector $\mathbf{x}^t$ is equal to

$$\mathbf{x}_k^t = \begin{cases} x_k^m & \text{if } r_k \leq C_r \\ x_k & \text{otherwise.} \end{cases} \tag{6.18}$$

where C_r denotes the crossover probability. Furthermore, the next generation is selected based on the principle of the survival of the fittest. Selection is performed as a one-to-one competition between parent and corresponding trial vector. The member having the smaller objective value is passed on to the next generation. Since the conductivity in the conductor can only be equal to zero (no defect) or one (defect), the objective values are calculated based on parameter vectors having a binary conductivity distribution. The binary objective variables are obtained by thresholding the continuous variables with a predefined threshold. The optimization is terminated if the NRMSE between measured and forward calculated data did not change for $g = 500$ generations.

6.2.3.1 Inverse solution strategy

The aim of the inverse calculation scheme is to reconstruct the conductivity distribution in the laminated specimen. The first approximation of the defect response signal as outlined in Section 6.2.2 is applied for forward calculations. Since the defect is represented with a fictitious conducting region, regions with high conductivity in the inverse solution are interpreted as defect regions. The applied defect reconstruction strategy consists of four steps. First, the depth of the defect is estimated, i.e. the layer at which the defect is expected is determined. Each layer of the laminated composite is modeled as an individual, two-dimensional regular grid of voxels, where the unknown conductivities are the variables of the objective function. Then, the DE is applied to minimize the NRMSE of measured and forward calculated data, separately for each layer. After convergence, the corresponding NRMSE is obtained for each depth. Finally, the defect depth corresponds to that layer with the minimum error among all layers.

The procedure starts with the definition of source space as a regular grid of voxels (one layer) with the unknown conductivities. This has to be done for each of the N_L metal layers of the composite, where the defect is assumed to be located in. Voxels are defined with sizes of $\Delta x \times \Delta y \times \Delta z$ and they are distributed equally in quadratic source spaces, which are positioned in the x-y–planes of the metal layers and have the dimensions $-l_s \leq (x,y) \leq l_s$. The depth of the source space and the height of the voxels are equal to the depth and thickness of the metal sheets, respectively. The number of voxels N_S in the source space is defined as $N_S = [2l_s/\Delta x][2l_s/\Delta y]$. The unknown conductivity distribution for each metal sheet is summarized in a vector $\sigma = [\sigma_1, \sigma_2, ..., \sigma_{N_S}]$ with respect to the number of voxels in the defect region.

Using the DRS profile $\Delta \mathbf{F} = [\Delta F_x^1, \Delta F_y^1, \Delta F_z^1, ..., \Delta F_x^M, \Delta F_y^M, \Delta F_z^M]^{\mathrm{T}}$ measured at M points, i.e. $\Delta F \in R^{3M \times 1}$, Eq. (6.15) can be reformulated into the following linear system of equations $\Delta F = \mathbf{K}\sigma$, where $\mathbf{K} \in R^{3M \times N_s}$ is the gain matrix between the N_S unknown conductivities of voxels from the source space and the DRS at M measurement points. The goal function $f(\sigma)$ that needs to be minimized in the LFE is assigned to the NRMSE between the forward calculated Lorentz force profiles $\Delta F^{(A)}$ and the observed $\Delta F^{(\mathrm{obs})}$ formulated as

$$f(\sigma) = \mathrm{NRMSE}(\sigma) = \frac{\sqrt{\frac{1}{3M}\sum_{i=1}^{3}\sum_{m=1}^{M}\left[\Delta F_{i,m}^{(\mathrm{A})}(\sigma) - F_{i,m}^{(\mathrm{obs})}(\sigma)\right]}}{\max\limits_{m=1\ldots 3M}\left[\Delta \mathbf{F}_m^{(\mathrm{obs})}(\sigma)\right] - \min\limits_{m=1\ldots 3M}\left[\Delta \mathbf{F}_m^{(\mathrm{obs})}(\sigma)\right]}. \tag{6.19}$$

In this study, $\Delta F^{(\mathrm{A})}$ profiles are calculated by the first approximation of the analytic forward procedure described in the previous chapter, whereas $\Delta F^{(\mathrm{obs})}$ profiles are either simulated numerically or measured.

6.2.4 Results and discussion

Since the choice of the intrinsic DE control parameters N_p, F, and C_r can have significant influence on the global optimum searching capability, the described reconstruction strategy has been applied for all combinations of $N_p = \{20, 40, 60\}$, $F = \{0, \ldots, 1\}$(step 0.1), and $C_r = \{0, \ldots, 1\}$(step 0.1). The results for the variation of the intrinsic control parameters are shown for a population size of $N_p = 40$ in Figure 6.14. The values are averaged over 50 optimizations. The figure shows the average NRMSE for the variation of F and C_r for each C_r and optimal F corresponding to a minimum error value. For all C_r, the minimum lies in the interval $F = [0.5, 0.7]$. If F is smaller than the optimal value, the NRMSE increases with increasing C_r. A strong dependence of the NRMSE and estimated radius can be observed. Parameter combinations with increased/decreased population ($N_p = \{20, 60\}$) did not yield any performance improvement.

Based on the outlined results, the parameter combination $N_p = 40$, $F = 0.5$, and $C_r = 0.1$ has been selected [101,156]. The results are in good agreement with

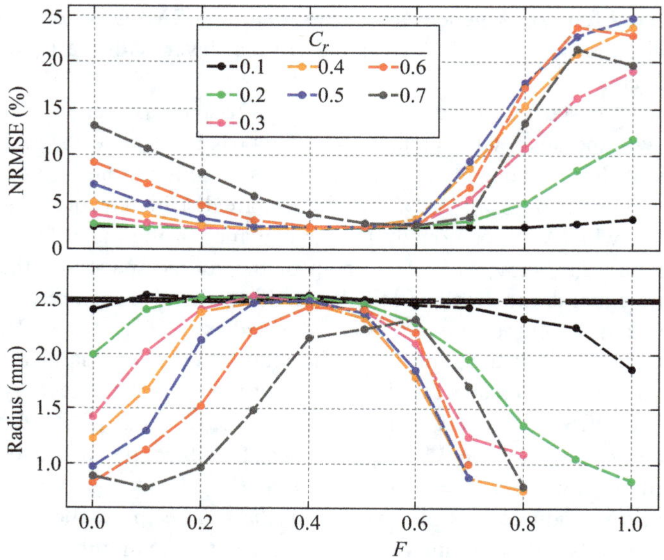

Figure 6.14 NRMSE and estimated defect radius of the parameter evaluation for the DE. The variation over F and C_r for $N_p = 40$ (population size) is shown. The dashed straight black line indicates the real radius (=2.5 mm)

recommendations found in literature, where have been proposed to set F between 0.4 and 0.9. This ensures that the trial and mutant vector are significantly different which yields an increased population diversity. Moreover, a low C_r value results into a search that changes each direction separately, which is an effective strategy to minimize decomposable functions as in LFE [416].

Figure 6.15 shows the defect response profiles $\Delta F = [\Delta F_x, \Delta F_y, \Delta F_z]^T$ caused by the defect in the vicinity of the center of the specimen ($x = 0$). Residual disturbances resulting from parameter deviations and inaccuracies that occurred during measurements can be seen outside the defect region.

In this chapter, an inverse calculation strategy based on a stochastic optimization algorithm called DE was proposed. The findings show that the reconstruction scheme yielded correct depth estimates. The geometry estimates had errors less than 4% relative to the size of the defect. Other reconstruction approaches for LFE have obtained comparable errors [345].

In the inverse strategy, the binarization of continuous-valued population members and area opening to obtain the DE solution is employed. These processing steps accelerate the convergence of the DE. Without these steps, it is likely that the DE would tend to converge to a similar final solution if the termination criterion is changed, i.e. more function evaluations would be required without changing the overall best solution.

Differential mutation and one-to-one selection are reported to be the strengths of the DE compared with other evolutionary strategies [416]. Differential mutation

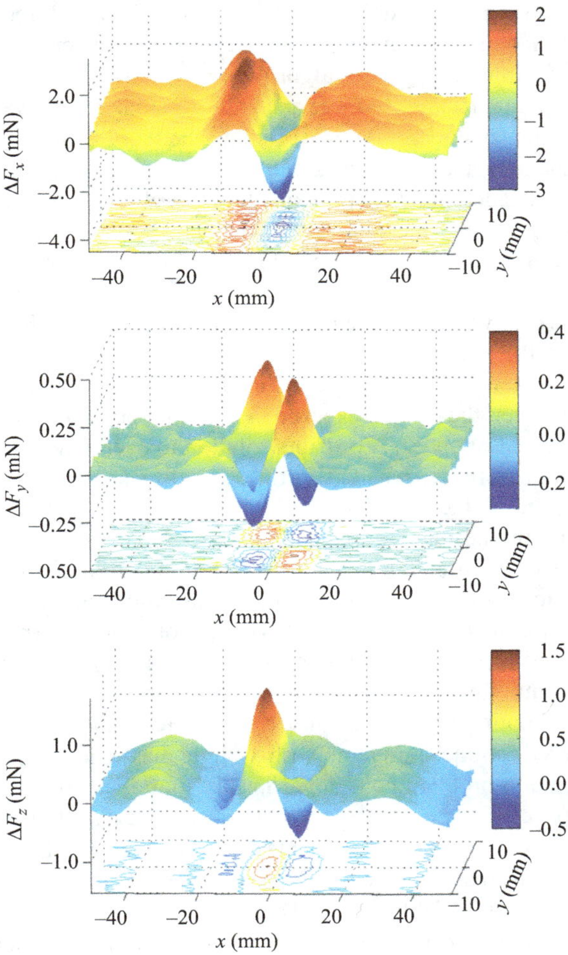

Figure 6.15 Profiles of the DRSs $\Delta F = [\Delta F_x, \Delta F_y, \Delta F_z]^T$ (top to bottom) obtained from preprocessed measured data for the alucobond composite with a defect at the bottom cover sheet

ensures a high-population diversity and improves the convergence. Individuals adapt to the search space because the scaled difference vectors gradually conform to the space. It is not necessary to use a predefined probability density function to define the step size, which is also the case for other stochastic optimization algorithms. One-to-one selection ensures that the overall best solution is retained. These aspects have contributed to selecting the DE algorithm [101,156].

Furthermore, the continuous-valued DE individuals enable to drop in future studies the assumption that the defect has zero conductivity. By omitting the binarization for selection operation, other defects can be analyzed in addition to metal fractures, such as corrosion. This extension would be more difficult to achieve with a binary-valued algorithm, e.g. genetic algorithm.

The reconstruction results are more accurate when the defect was located closer to the surface of the conductor. The reason might be that the influence of the metal layers between the permanent magnet and the layer containing the defect on the force signals were not considered in the approximate forward solution. Further, the results show that the reconstructed defect has a slightly smaller x-extension but slightly larger y-extension than the real defect. This phenomenon is likely to be explained by two aspects. First, the resolution of LET is poorer in the y- than in the x-direction. Because the eddy currents below the permanent magnet flow in the y-direction, the magnitude decay is slower in the y-direction than in the x-direction. Second, the y-extension of the sensor space is too small to provide sufficient information about the y-extension of the defect. However, the current experimental setup does not allow to increase the y-extension of the sensor space. A possible solution to overcome this drawback is to measure the specimen twice, whereas in the second measurement the specimen in rotated by 90 degrees in the x-y–plane.

In this study, the permanent magnet was modeled with one magnetic dipole, having an optimized position, although a multiple-dipole-model (MDM) with more than one magnetic dipole would provide a more accurate solution of the magnetic field. The reason was that the use of the MDM does not allow to formulate the inverse problem into a linear system of equations. Hence, this would result in a significantly increased computing time. However, the computing time was in the range of a few hours for a single optimization run. This is a significant increase compared to the about 15 s observed for one simulation with a single magnetic dipole.

The optimization problem considered in this study has 900 unknowns (voxels in the source space) if $\Delta x = \Delta y = 1$ mm. Therefore, it can be treated as a high-dimensional nonlinear problem. The complexity increases with the dimension of the problem and the performance of the DE is often expected to be degraded for problems with many optimization variables [416]. This explains the less accurate results if the voxel size is below 1 mm. In the present study, it was shown that the basic DE provides robust and satisfying inverse solutions within a reasonable computational costs when it is applied to the LFE problem with appropriate voxel size [101].

Two aspects can be addressed as limitations of the proposed method. Only the very simple approximation of forward calculated signals was considered, because the determination of the eddy currents for multiple defect regions with arbitrary shape, that are likely to occur during the optimization, is challenging. For simulation data, these errors are in the range of a few percent. However, the comparison between the measured and forward calculated DRSs shows that the presence of interfering signals yields significantly larger deviations for the measured Lorentz forces (Figure 6.16). This explains the larger NRMSEs obtained after DE convergence and impedes the depth reconstruction. On the other hand, the resolution of the system might be not good enough to obtain more accurate depth reconstruction for measured data. The resolution can be improved by applying sophisticated magnet systems.

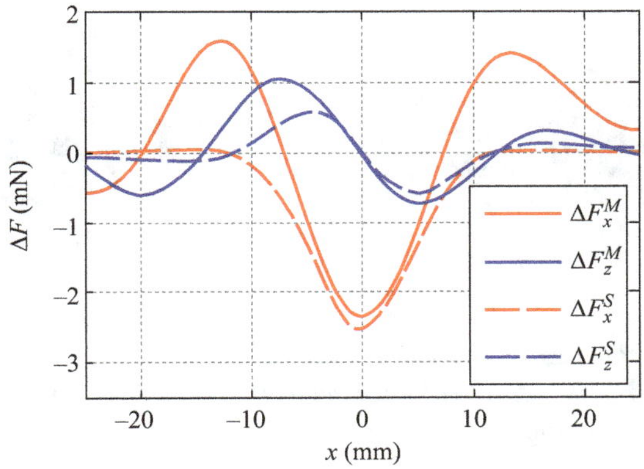

Figure 6.16 Comparison of measured (M) and simulated (S) DRS components ΔF_x and ΔF_z. Data are compared along the symmetry line of the LFE setup

6.3 Inspection of composites

6.3.1 Composite material

In industry, the market share of composites is even increasing much faster, and composites are now widely used in aerospace industry, automotive industry, marine industry, etc. Composites or composite materials are combinations of two or more materials, in such a way that the separate material phases can still be distinguished after manufacturing. Composites are typically designed with a particular use in mind, such as added strength, efficiency, or durability (Figure 6.17).

In most cases, the term *composite* or *composite material* refers to fiber-reinforced plastics, because this is by far the most important category of composite materials. In these fiber-reinforced plastics, the plastic is reinforced with fibers to make a light and strong material. The material in which the fibers are embedded is called the matrix, while the fibers are called the reinforcement.

Composites, also known as fiber-reinforced polymer (FRP) composites, are made from a polymer matrix that is reinforced with an engineered, man-made or natural fiber (such as glass, carbon, or aramid), or other reinforcing material. The matrix protects the fibers from environmental and external damage and transfers the load between the fibers. The fibers, in turn, provide strength and stiffness to reinforce the matrix and help it resist cracks and fractures. In many industrial products, polyester resin is the matrix and glass fiber is the reinforcement. But many combinations of resins and reinforcements are used in composites, and each material contributes to the unique properties of the finished product: fiber provides strength and stiffness, while more flexible resin provides shape and protects the fiber.

Many terms are used to define FRP composites. Modifiers have been used to identify a specific fiber such as *Glass Fiber Reinforced Polymer* (GFRP), *Carbon Fiber*

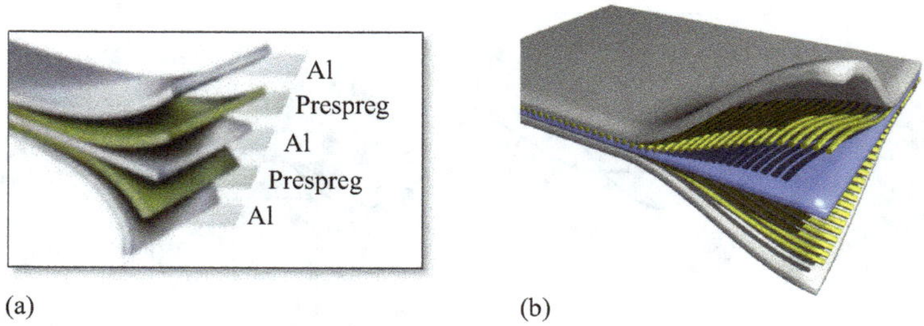

Figure 6.17 Composites are formed by combining materials together to form an overall structure with properties that differ from individual properties of the components

Reinforced Polymer (CFRP), and *Aramid Fiber Reinforced Polymer* (AFRP). Another familiar term used is *Fiber-Reinforced Plastics*. In addition, other acronyms have been developed over the years and their use depended on geographical location or market use. For example, *Fiber-Reinforced Composites* (FRC), *Glass-Reinforced Plastics* (GRP), and *Polymer Matrix Composites* (PMC) can be found in many references.

Beside the fiber-reinforced plastics, there are a wide variety of composite materials: metal matrices with metal fibers, ceramic matrices with metal or ceramic fibers, etc. Even steel reinforced concrete could be classified as a composite, although it is rarely considered to be part of the composites family. The matrix can basically be any type of plastic: epoxy, polyester, vinyl ester, polypropylene, etc. The fibers are typically glass, carbon (graphite), or aramid. The fiber reinforcement can take any form: a mat of short chopped fibers, a woven fabric, a unidirectional arrangement of fibers, and many others. One of the big advantages of fiber-reinforced composites is that the stiffness and strength can be tailored to the specific loading conditions. If the loading of the composite component is dominantly oriented in one direction, you can apply the fiber reinforcement mainly in that direction. This makes the composite anisotropic. It means that its stiffness and strength are not the same in every direction, but can differ for every direction in the material.

A special type of composite laminate is the sandwich composite, where some of the middle layers of the laminate are replaced by a very lightweight core material, typically being foam, balsa wood, or metallic honeycombs. Sandwich composites are very attractive because they are very lightweight structures with a very high bending stiffness and usually good impact and fatigue resistance. They can be found in ship hulls, floor panels, architectural claddings, etc. The composites industry continues to evolve. The use of FRP composites has already transformed the marine, automotive, and aerospace markets. There is huge potential for a similar technology shift in the architectural and building & construction segments as the industry takes advantage of the design flexibility, durability, low weight, corrosion resistance, and other properties that composites offer. Composite materials have fueled the growth of new applications in markets such as transportation, construction, corrosion-resistance, marine, infrastructure, consumer products, electrics, aerospace, appliances, and business equipment. Composites are used in industrial applications where corrosion resistance and performance are critical. Generally, premium resins (e.g. isophthalic resin or vinyl ester) are required to meet corrosion-resistance specifications. Other specialty resins may be used depending on the chemical resistance properties required. Fiberglass is almost always used as the reinforcing fiber. Industrial composite products include underground storage tanks, scrubbers, piping, fume hoods, water treatment components, and pressure vessels. The sector of the advanced composites industry is characterized by the use of high-performance resin systems and high-strength, ultra-stiff fiber reinforcement. The aerospace industry, including military and commercial aircraft, is the major customer for advanced composites.

Due to the complex structures of the composites with very different and often not precisely known material properties, it is difficult to provide appropriate techniques for testing and evaluation. If the MIECT technique should be applied, only electrically conducting materials can be evaluated, i.e. with NDT methods like LET the conductive components of the composite devices can be evaluated because there the eddy current flow is induced. Furthermore, the evaluation of composite materials and in particular the identification of defects or conductivity anomalies need the numerical modeling of the eddy current flow in these conducting components. Thus, in the following chapter laminated composite materials will be considered.

6.3.2 Glass laminate aluminum reinforced epoxy (GLARE)

New materials are developed and used to support the trend of a more ecological and economical society. Especially composite materials can withstand stronger mechanical stress but at the same time they are a big challenge for NDT and NDE. In this chapter, measurement results using LET for *Glass Laminate Aluminum Reinforced Epoxy* (GLARE) samples are presented and compared with simulations [417,418].

GLARE is a fiber metal laminate (FML) composed of several very thin layers of metal (usually aluminum) interspersed with layers of glass-fiber "pre-preg," bonded together with a matrix such as epoxy. The unidirectional pre-preg layers may be aligned in different directions to suit the predicted stress conditions (Figure 6.18). GLARE is made of thin-stacked aluminum sheets with glass fiber laminated in between the sheets. It is mainly used in aerospace industry because of its mechanical properties and low specific weight [419,420]. Typically GLARE is tested nondestructively with ultrasonic [421] or thermography [396].

Although GLARE is a composite material, its material properties and fabrication are very similar to bulk aluminum metal sheets. It has far less in common with composite structures when it comes to design, manufacture, inspection, or maintenance. GLARE parts are constructed and repaired using mostly conventional metal material techniques. Its major advantages over conventional aluminum are:

- Better *damage tolerance* behavior, especially impact and metal fatigue
- Better corrosion resistance
- Better fire resistance
- Lower specific weight

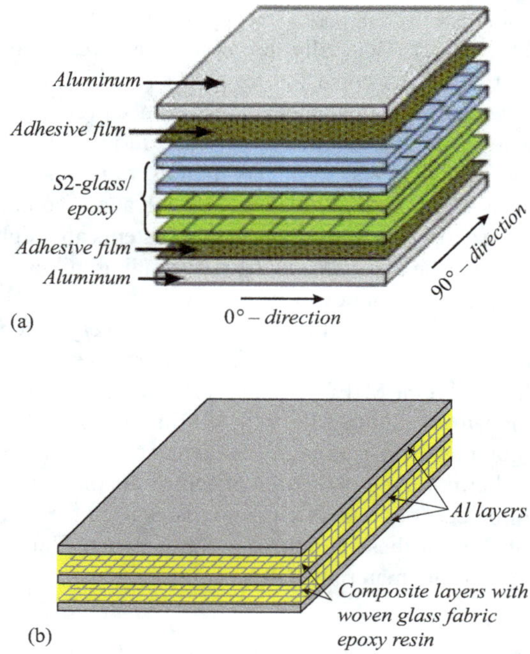

Figure 6.18 Composite materials made of Al-layers separated by glass/epoxy sheets

Furthermore, it is possible to *tailor* the material during design and manufacture such that the number, type, and alignment of layers can suit the local stresses and shapes throughout the aircraft. This allows the production of double-curved sections, complex integrated panels or very large sheets, for example.

GLARE is a successful FMLs, patented by Akzo Nobel in 1987, which has entered commercial application in the Airbus A380 (Figure 6.19) [420]. In the 1970s and 1980s, especially researchers at the Faculty of Aerospace Engineering, Delft University of Technology, have done much of the R&D on FML. Besides the applications on the Airbus A380 fuselage, GLARE has multiple "secondary" applications. GLARE is also the material used in the ECOS3 blast-resistant Unit Load Device. This is a freight container shown to completely contain the explosion and fire resulting from a bomb such as that used over Lockerbie. Other applications include among others the application in the Learjet 45 and in the past also in cargo floors of the Boeing 737.

GLARE is currently produced by Cytec Engineered Materials in Wrexham, UK who supplies it to the Airbus A380 component manufacturing facilities at Stork Fokker in the Netherlands as well as at Airbus in Nordenham, Germany. Test specimens made of GLARE (from airbus A380 fuselage) at the Premium AEROTEC GmbH in Nordenham, Germany, a supplier of Airbus, have been used for LET measurements in the university laboratory at the Technische Universität Ilmenau, Germany (Figure 6.20). While a simple manufactured sheet of GLARE will be more expensive

Figure 6.19 Airbus A380

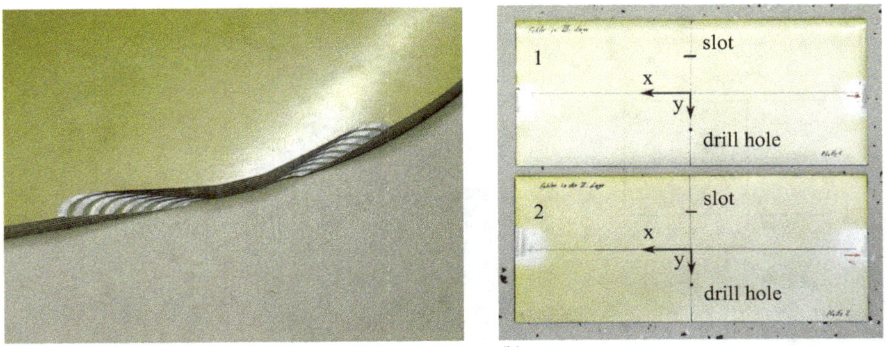

(a) (b)

Figure 6.20 Test specimen (GLARE): (a) test piece of A380 fuselage (cabin window); (b) two specimens (each 350 mm × 150 mm) with machined defects included in the second aluminum layer (2) and third aluminum layer (1); same defects in both Al layers: a drill hole (diameter = 2 mm) and a through slot (1 mm × 10 mm)

than an equivalent sheet of aluminum, considerable production savings can be made using the aforementioned optimization. A structure properly designed for GLARE will be significantly lighter and less complex than an equivalent metal structure, and will require less inspection and maintenance and enjoy a much longer lifetime-till failure, making it a cheaper, lighter, and safer option overall.

GLARE is a composite material made of laminating thin aluminum sheets with glass fiber layers in between. Figure 6.20(a) shows the composition of eight aluminum (light gray) and seven glass-fiber laminate layers (dark gray). In the present study, two GLARE samples (aluminum alloy No 3.1354, σ =17 MS/m) shown in Figure 6.20(b) are used. Both specimens are 350 mm × 150 mm and made of five aluminum sheets, each 0.4 mm thick, with 0.25 mm glass fiber laminate between the sheets, resulting in a total thickness of the specimen of 3 mm. Both specimens have as machined defects a slot (10 mm × 1 mm) and a drilled hole (diameter = 2 mm) at the marked spots in one of their aluminum layers. Specimen1 has the defects in the third aluminum layer (defect depth = 1.3 mm) and specimen2 in second layer (defect depth = 0.65 mm), counted from the top layer.

6.3.2.1 Magnetic field measurement

To measure the magnetic field formed by the applied magnet system, a Hall probe is fixed to the z-slide. Thereby, the probe can be moved in the y-z–plane. The magnet system is fixed to the x-slide; hence, the Hall probe can be positioned in all three spacial dimension in relation to the magnet system. The probe is moved along a grid and takes a reading at every gridpoint. The AS-N3DM probe by Projekt Elektronik Mess- und Regelungstechnik GmbH [422] is used to measure the magnetic field of the magnet system. It is a triaxial Hall probe with a range of ±2 T. Two permanent magnet configurations have been applied: A longitudinally magnetized cylindrical magnet (diameter = 22.5 mm, height = 17.6 mm) and a radially magnetized Halbach

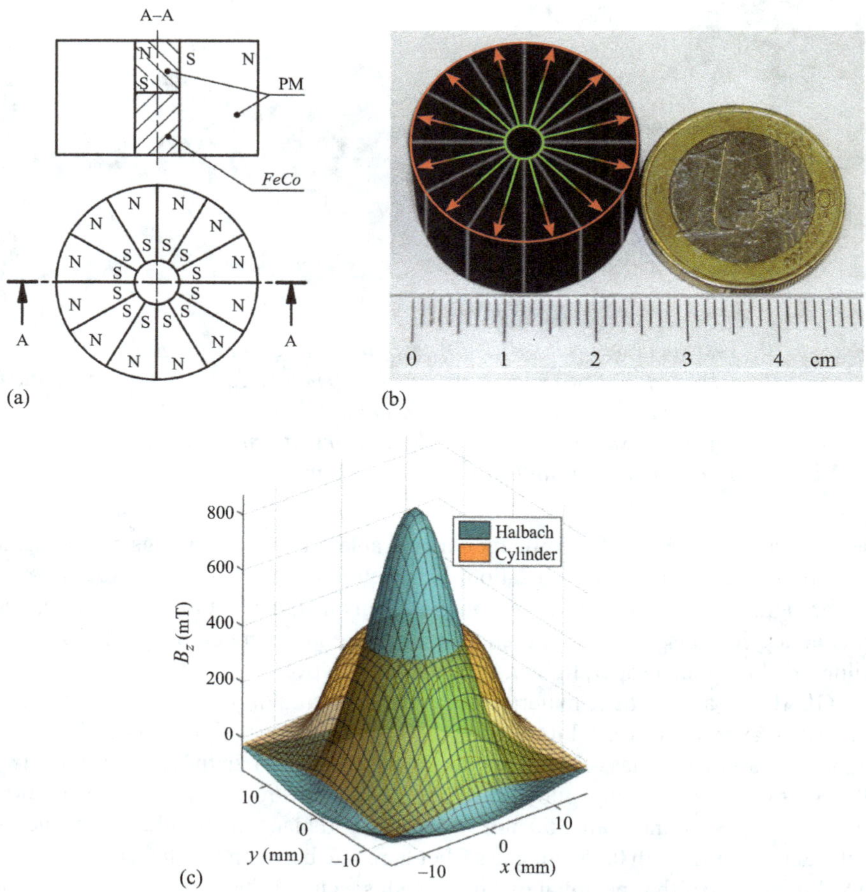

Figure 6.21 Halbach configuration for detection of subsurface defects (a and b). The bottom figure (c) illustrates the focusing effect of this design for the z-component of the magnetic induction, compared with that of a cylindrical magnet

configuration (diameter = 24.8 mm, height = 14.5 mm) with a high-permeable disc made of iron–cobalt alloy VACOFLUX 50 [256]. Both magnet systems are the results of the optimization procedure presented in [245].

The structure of the Halbach system is shown in Figure 6.21. The idea behind the Halbach configuration is to increase the magnetic flux density near the surface of the magnet system. Because of this focusing effect due to the highly saturated FeCo-part, the Halbach magnet system is expected to perform better for small defects and defects close to the surface of the specimen. The design of this magnet system has been found by means of a optimal design process which has been described previously in Section 3.2.

Figure 6.22 shows the magnetic flux density components of the used magnet systems. The top pictures show the measured data for the Halbach structure, whereas

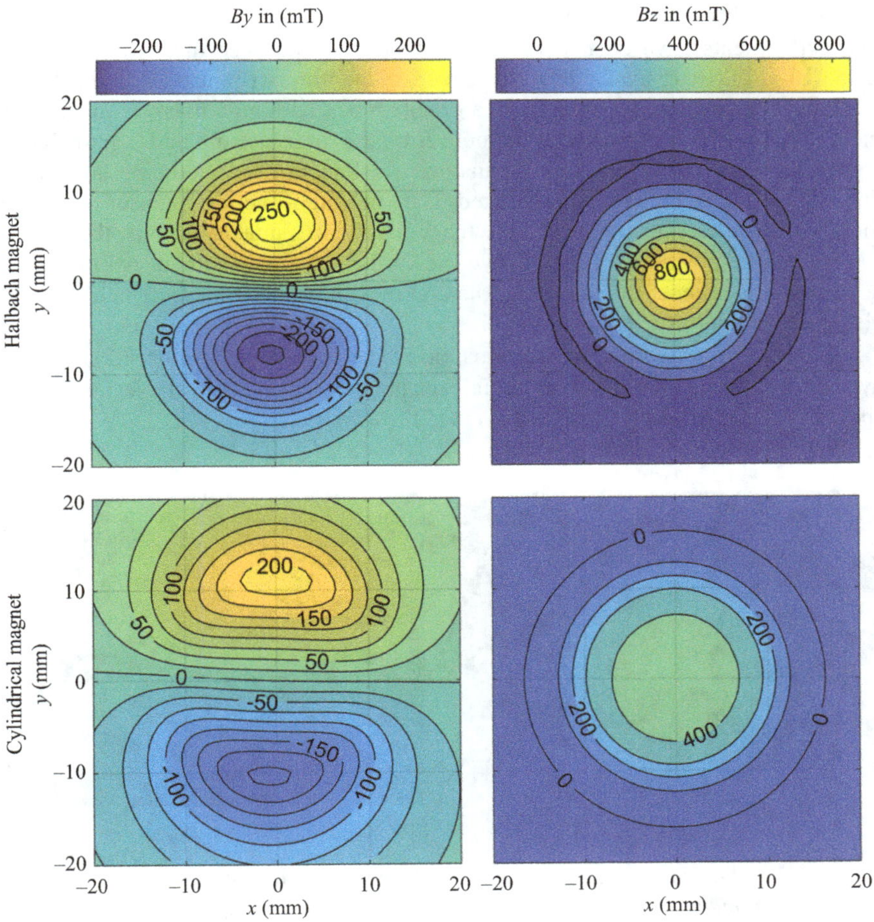

Figure 6.22 Comparison of magnetic flux density components of the Halbach magnet and the cylindrical magnet

the bottom pictures show data for the cylindrical magnet. The x-component is not shown because it is very similar to the y-component, only rotated around the z-axis by 90°. The magnetic flux density is measured at a distance of 1 mm from the housing of the probe to the surface of the magnet system. The B_z-component produced by the cylindrical magnet forms a circular plateau, whereas for the Halbach magnet it is cone-shaped. The plateau is only half ($\approx$ 400 mT) as strong as the cone tip of the Halbach magnet ($\approx$ 800 mT). It can be observed that the contour lines for the Halbach magnet are much denser than for the cylindrical magnet. A similar behavior can be observed in the B_y graphs. Thus, it can be noted that the magnetic flux density B_0 below the Halbach magnet is much more focused than for the cylindrical magnet. Thus, a specimen moved under the Halbach magnet should result in a higher measured Lorentz force.

6.3.2.2 Force measurements

The experiments have been performed on the multi-purpose measuring platform BASALT-C MMP-15 available at the university laboratory (Figure 6.23) [423].

The mechanical system is set on a granite block with high mass to increase the inertia, thereby reducing external influences on the force sensing unit. On the block, a granite portal is installed which spans over the linear drive. On the portal, spindle drives in z- and y-direction are mounted. The one end of the force sensor is fixed to the z-slide while on the other end the magnet system is attached. Thus, the magnet system can be positioned freely above the surface of the specimen. All drives are controlled via a panel PC located in the control cabinet. The measurement data are also gathered by this computer.

In this study, two magnet systems have been used (see Section 3.2), either a cylindrical permanent magnet or the Halbach magnet is positioned above the GLARE specimen at the lift-off δ (Figure 6.24) [417].

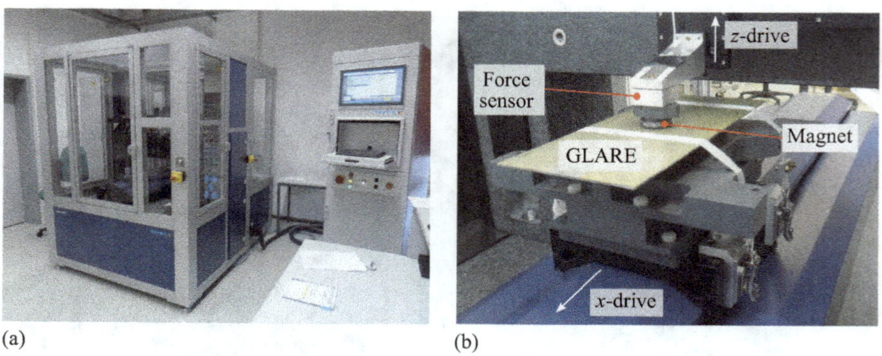

Figure 6.23 Measurement setup for GLARE specimen: (a) multipurpose measurement platform MMP-15; (b) LET measurement of GLARE specimen

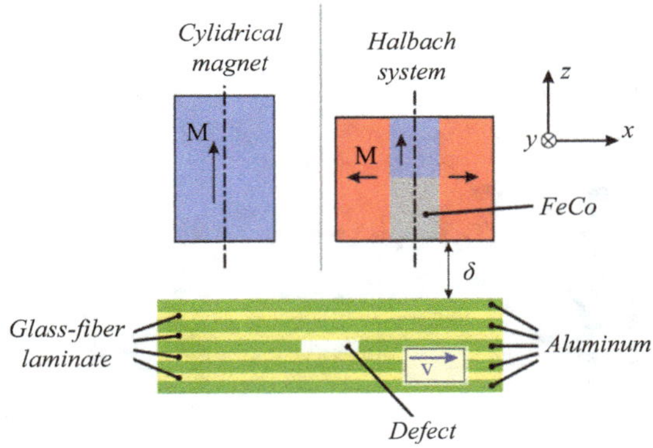

Figure 6.24 Sketch of the LET configurations for GLARE samples using different permanent magnet systems

The sample specimen is mounted onto the x-slide. The slide moves the specimen with a constant velocity crossing the magnetic field B_0 produced by the magnet system. Due to the relative motion between B_0 and conductive parts, eddy currents are induced in the specimen resulting in the Lorentz force acting on the specimen. The corresponding counterforce $\mathbf{F} = [F_x, F_y, F_z]^T$ exerts on the magnet and is measured with a three-axial strain gauge force sensor. The sensor with the magnet system can be positioned at a desired y-coordinate to perform a scan of the sample for various x-positions.

To perform LET experiments, a specimen is mounted on the x-drive and the magnet system is positioned at a desired y-coordinate above the specimen. The z-coordinate is set by moving the magnet system down, until a specified gap to the specimen (lift-off distance δ) is left. The specimen starts to accelerate from a start position outside the magnetic field of the magnet system. After reaching the desired velocity, the PC starts to collect the measured force and positional data with a sampling frequency f_S. After leaving the magnetic field, the specimen is stopped and moved back to the starting position. Now, the scan can be repeated at the same or a different y-coordinate.

The force sensor used during measurements is the K3D40 [104]. It is a triaxial sensor with $\pm$ 2 N nominal force in all three directions. As stated above, it is mounted to the z-drive. On the other end, a plate with a notch is mounted. The counterpart of this notch is part of the magnet system mounting bracket, which enables to mount the magnet system at a reproducible position.

The LET experiments have been performed with a specimen velocity of $v =$ 200 mm/s and a lift-off $\delta = 1$ mm. The y-coordinate was incremented by 0.5 mm over the 150 mm width of the specimen. For every y-coordinate 25 measurements were performed, and averaged to reduce random errors in the data. The surface scan leads

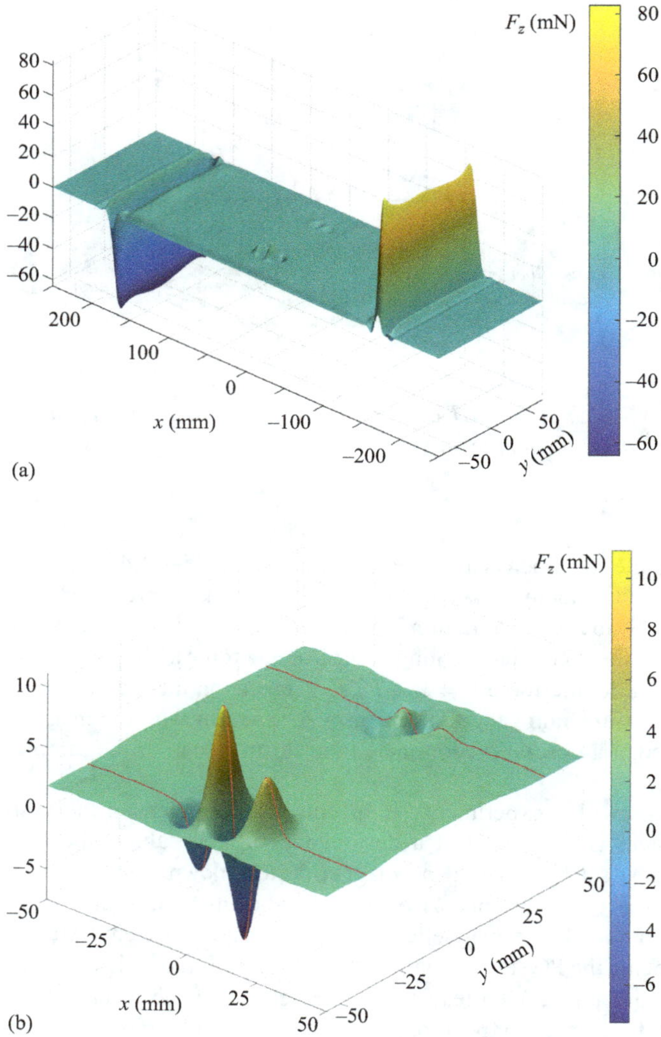

Figure 6.25 LET scan of a GLARE specimen: (a) total scan of GLARE sample with the two defects in the center (see Figure 6.20(b)); (b) zoom into the region around the defects

to a plot of the lift force profile shown in Figure 6.25 obtained for the GLARE sample presented in Figure 6.20(b).

Both the cylindrical magnet and the Halbach magnet have been used to scan the GLARE specimen. The coordinate system is set in the center of the specimen.

Figure 6.26 shows the x- and z-component of the measured forces (drag and lift force) along the centerline of the specimen, where no defect is present. The leading

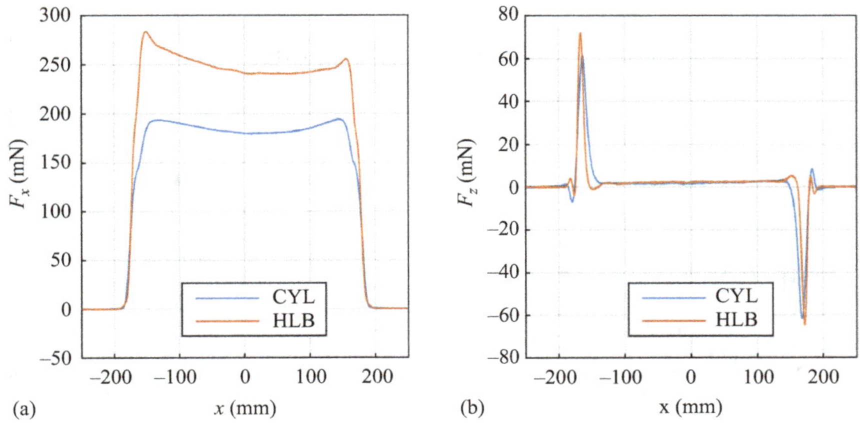

Figure 6.26 Measured forces exerting on the cylindrical and Halbach magnet. (a) Drag force. (b) Lift force

edges of the specimen entered the magnet field at $x \approx -175$ mm. The trailing edge left the magnet field at $x \approx -175$ mm. The lift force F_z has a positive and negative peak at these positions, whereas the drag force F_x rises to a plateau and drops back to zero. According to the theory, the plateau should be constant since there are no parameter changes over the specimen. Figure 6.26(a) shows that the plateau has a dent, which is caused by the uneven surface resulting in a change of the lift-off distance. Both the cylinder magnet and the Halbach magnet show this behavior.

For the Halbach magnet, the maximum forces are clearly higher and the general force profile is sharper comparing to the cylindrical magnet. This is the result of the focused magnetic flux density produced by the Halbach magnet. Because the structure of the Halbach magnet leads to the field focusing effect just below the magnet bottom, much higher magnetic flux density values could be achieved. This results in significantly higher Lorentz forces (especially for the drag force) which are about 25% – 30% higher than those for the cylindrical permanent magnet.

Figure 6.27 presents a comparison of the Lorentz force acting on the magnet depending on the defect depths. Here only the results for the Halbach magnet are shown. The plots are cropped to the range $-50\,\text{mm} \leq x \leq 50\,\text{mm}$. The solid lines correspond to the slot defect while the dotted lines to the drill hole defect. The results for GLARE specimens (1) and (2) are depicted with red and green color, respectively.

It can be noted that the deformed uneven surface affects F_x a lot and makes it difficult to find out, where the deeper defect is located [Figure 6.27(a)]. In Figure 6.27(b), it can be observed that the deflections caused by the slot defect are a little bit larger for the specimen (2) than for the specimen (1). Since the specimen (2) has the defects in the second aluminum layer, they are 0.65 mm closer to the magnet system then in specimen (1) and this indeed results in a stronger deflection in the force. The drill hole is hardly distinguishable in the z-component. From the drill hole

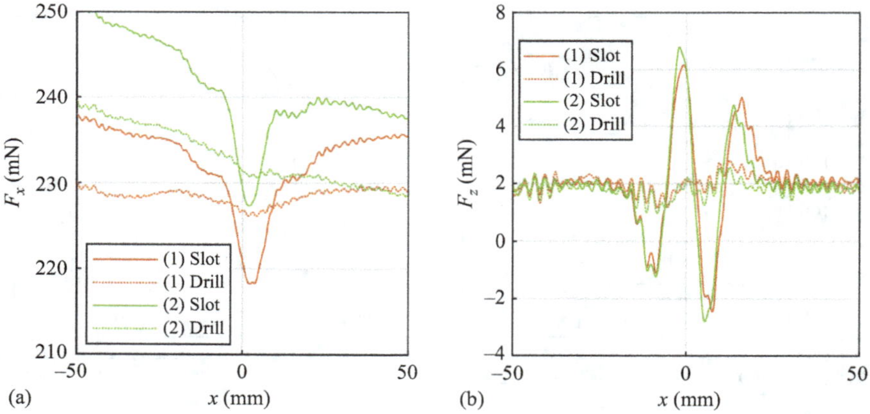

Figure 6.27 Measurements of the Lorentz force exerting on the Halbach magnet by scanning both GLARE samples (see Figure 6.20(b)) which is moving in x-direction. (a) Drag force. (b) Lift force

signals, it cannot be concluded which the deeper defect is, since the noise oscillations in the signal are too high [417,418].

6.3.2.3 Defect localization

Similar to the procedure described in Section 6.2 for alucobond samples, appropriate inverse solution strategies have to be developed to localize the defect by means of the LET measurements and modeling of the electromagnetic field distributions in the moving conductor. It is the objective of this chapter to demonstrate the defect detectability for the aluminum alloy layers in the GLARE material (see Figure 6.20(b)), containing two artificial defects in different aluminum layers, i.e. at different depths. A further aim is to evaluate the performance of Lorentz force sensor system containing the already mentioned Halbach magnet configuration (see Figure 6.21). Based on numerical simulations, it has been shown that the cylindrical Halbach structure has, compared to a cylindrical permanent magnet, a superior performance in detecting small subsurface defects [245]. Similar to the approach in Section 6.2.3 where the DE has been applied, here the goal function scan (GFS) is performed in order to reconstruct the defect [101,156,415]. The NRMSE between the numerically calculated and measured Lorentz force signals is used to define the goal function. A finite element model is used to calculate the magnetic flux density, including the EAA for the force calculation described in Section 5.3. This is necessary, because no analytic solution exists and the cylindrical Halbach structure contains a FeCo-component with a nonlinear $B(H)$ curve. This cannot be modeled with the magnetic dipoles model as it was possible in the case of Section 6.2. To verify the localization results, radiographic and through-transmission ultrasonic testing as alternative NDT techniques have been applied to the GLARE test object. The general setup of the study is shown in Figure 6.28.

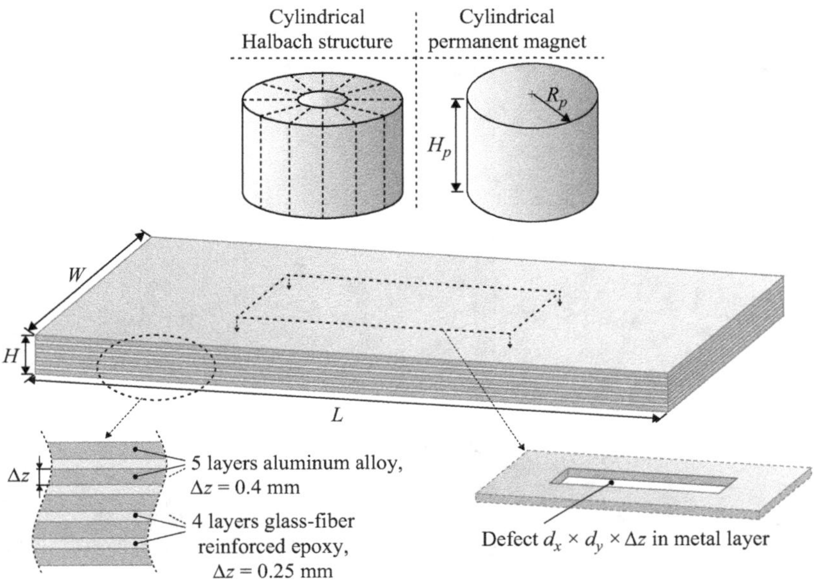

Figure 6.28 Setup for subsurface defects localization in GLARE samples [156]

The GFS method has been applied to solve the inverse problem. In this method, the landscape of the goal function is investigated for a set of grid points uniformly distributed in the search space. Similar to the previous Section 6.2, the goal function is defined as the NRMSE between forward calculated and measured data (6.19). Since the GFS method is limited to one defect, it is evident to use the geometry parameters of the defect as design variables. Then, the location of the defect, i.e., the x- and y-coordinates of its center of gravity, can be determined straightforwardly. The defect location is assigned to the position at which the ΔF_x-component of the DRS has the largest absolute amplitude. Thus, only the depth and the x- and y-extension of the defect remain to be determined. Therefore, the goal function is evaluated for all combinations of the x-extension $d_x = \{5 : 0.5 : 15\}$ mm, the y-extension $d_y = \{0.5 : 0.5 : 5\}$ mm, and all physically possible defect depths $d = \{0, 0.65, 1.3, 1.95, 2.6\}$ mm.

Figure 6.29 shows the DRS measured from the GLARE specimen containing a defect in the second aluminum layer. The DRS obtained with the cylindrical Halbach structure is compared with the DRS obtained with the cylindrical permanent magnet. The DRS of the Halbach structure has a significantly larger absolute amplitude than that of the cylindrical permanent magnet. In particular, defect perturbations in the ΔF_y-component are hardly to be distinguished from noise and interfering signals, if the cylindrical permanent magnet is used. Contrary, they are distinctly visible if the cylindrical Halbach structure is applied. Further, the slopes in the DRS are steeper if the Halbach structure is applied. Thus, the area spanned by the DRS is in either coordinate direction smaller than for the cylindrical permanent magnet.

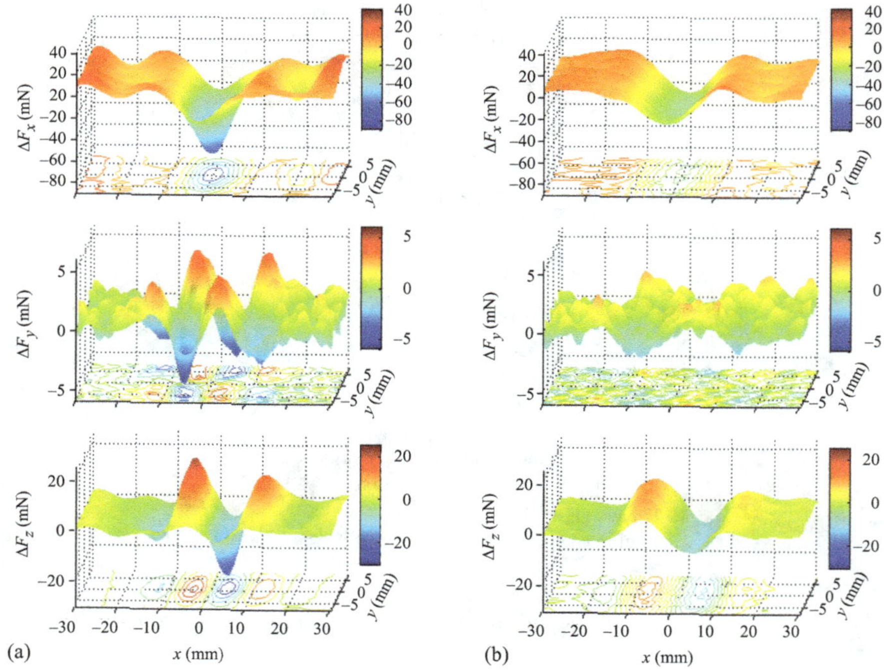

Figure 6.29 Measured and preprocessed DRS obtained from measurements of the GLARE specimen. LET experiments are performed using two magnet systems: the cylindrical Halbach structure and a cylindrical permanent magnet. The defect is positioned in the second aluminum ($d = 0.65$ mm). (a) Halbach magnet. (b) Cylindrical magnet

Figure 6.30 compares the ΔF_x- and ΔF_z-component of the forward calculated DRS to the measured data at the symmetry line $y = 0$. The forward calculated DRS has a smaller absolute amplitude than the measured data. The corresponding NRMSE of the ΔF_x- and ΔF_z-component are 8.94% and 8.18%, respectively.

The GFS is applied to each aluminum layer separately. The results show that there exists one local minimum in each single aluminum layer. Thus, if the optimization function is considered separately for each layer, it is convex. The minimum NRMSE in the five layers correspond to the correct defective second layer. The corresponding defect extensions are $\{d_x, d_y\} = \{9.5, 2.5\}$ mm. It can be observed that with increasing depth of the metal layer the defect extensions corresponding to the local minima in the single layers increase. Furthermore, because the goal function is relatively flat, the minimums cannot be distinguished precisely. All minima, global and local, differ from each other by less than 1.5% only [156].

For comparison, the NDT techniques radiographic and through transmission ultrasonic testing are applied to detect the defect in the GLARE specimen (Figure 6.31).

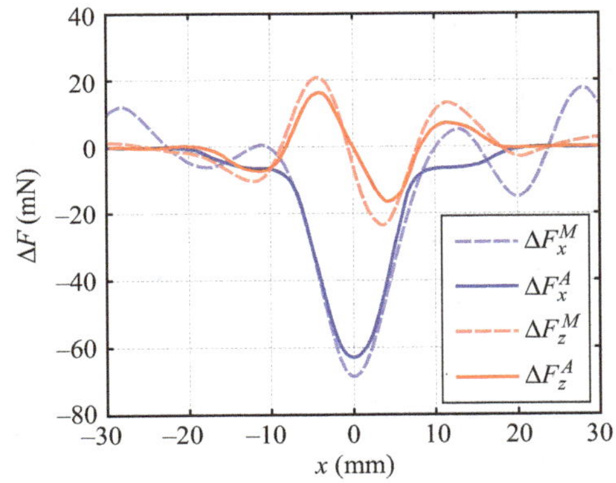

Figure 6.30 *Comparison of the DRS obtained from measured data (ΔF^M) and forward calculated Lorentz force signals using the EAA (ΔF^A) and the cylindrical Halbach structure, with the defect in the second aluminum layer*

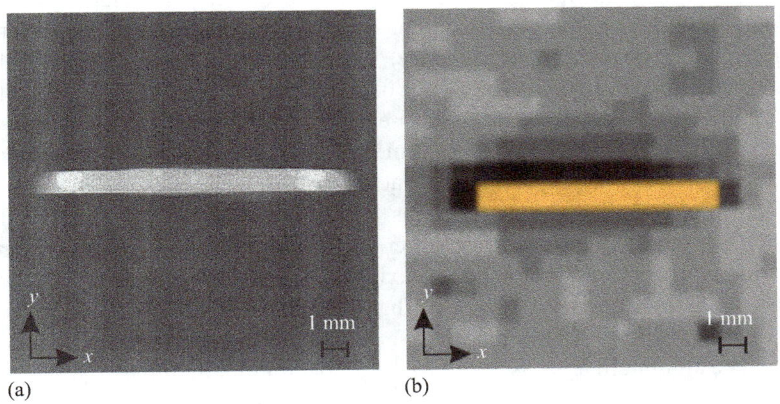

Figure 6.31 *(a) Results of radiographic testing and (b) through-transmission ultrasonic testing of the GLARE specimen (provided by Premium AEROTEC GmbH, Nordenham). Both projections are performed on the x-y–plane of the specimen*

In the radiographic testing, the used X-ray generator is operating with 15 kV and a current flow of 35 μA. The ultrasonic testing is performed by the company Premium AEROTEC GmbH Nordenham, which has manufactured the GLARE test objects. An automatic scanning system using water-coupling of the ultrasonic waves with the squirter technique is applied [424]. The probes use a frequency of 5 MHz. The resolution of the method equals 1 mm in each direction. Both methods scan the GLARE specimen in z-direction, i.e. perform projections on the xy-plane. A scanning in x- and y-direction is due to the setup and specifications of the NDT testing systems

Table 6.6 Results compared for the radiographic testing, the ultrasonic testing, and the GFS in LFE. The NDT methods are applied to estimate the parameters of the defect in the GLARE material. The nominal defect parameters equal $d_x = 10$ mm (x-extension), $d_y = 1$ mm (y-extension), and $d = 0.65$ mm (depth)

Parameters	Radiography	Ultrasonics	LET & LFE
d_x (mm)	10.6	8.3	10.5
d_x (mm)	0.9	1	5
d (mm)	–	–	1.3

not possible. The x- and y-extension of the GLARE specimen are too large to position the specimen accordingly. Further, neither the used x-rays nor the ultrasonic waves are able to penetrate the specimen in x- and y-direction. Thus, they deliver no depth information of the defect.

Table 6.6 depicts the defect parameters estimated from the images in Figure 6.31, and compares them to the results of LFE, i.e. the defect parameters assigned to the global minimum in the goal function. The x-extension found with radiography is slightly larger than the nominal value. Contrary, with ultrasonic testing the x-extension is estimated smaller. The results of the GFS in LFE are comparable to those of the radiographic testing. In case of the y-extension, the imaging techniques provide accurate results and outperform the LFE method. However, a depth estimation is not possible with either imaging technique, but solely with LFE.

The evaluation of the goal functions shows that the reconstruction of the x-extension of the defect has a very small error. However, the reconstruction error in y-direction is large. In Section 6.2, the y-extension of the defect has also been reconstructed too large. The reason might be that the resolution of LET in y-direction is worse than in x-direction, and that the y-extension of the sensor space is not large enough. The aluminum layer assigned to the global minimum of the goal function is positioned directly below the layer in which the defect is included. Thus, the error in the depth reconstruction equals 0.65 mm. However, this error can be considered as very small, because the aluminum layers in the GLARE sample are rather thin than the layers of the composites evaluated in Section 6.2. This aspect also explains why the local minima of the goal function (one local minimum in each layer) lie in the range of about 1.5% only. There are still many options to improve the MIECTE solution strategies. In future studies, the experimental setup could be enhanced with the main aim to reduce the interfering signals and the noise in the measured data. On the other hand, there is enough scope for development of more appropriate inverse solution strategies, depending on the problem under consideration, and with higher numerical efficiency.

6.3.3 Carbon fiber reinforced polymer (CFRP)

6.3.3.1 Introduction to FRP

A composite is a material that is composed of two or more constituent materials. One subgroup of composite materials are fiber-reinforced composites that consist of fibers embedded in a polymer or metal matrix. A second class are laminates that consist of a set of thin sheets of different materials bounded together. These can be metal alloys, polymers, or composites themselves. A special type, the FMLs, is hybrid composite structures made of alternating layers of a metal, mostly an aluminum alloy, and an adhesive fiber-reinforced epoxy resin. All materials in a composite remain separate and distinct, and retain their physical and mechanical characteristics. However, their combination has superior properties including higher bearing and tensile strength, better damage and fatigue tolerance, larger corrosion resistance, and less weight [414]. Due to their outstanding features, composite materials are increasingly used in weight-critical components, where high stiffness is required. Fields of application comprise naval architecture, aerospace, railway supply, automotive industry, paneling of buildings, and sporting goods. Nowadays, the airframes of commercial airplanes comprise by more than half of advanced composites. Widely applied are carbon fiber-reinforced plastics. CFRP becomes very popular when there are high demands on the mechanical properties of a part. Most CFRP parts are made by laminating different types of carbon fiber fabrics on another. This type of composite material has one of the highest strength-to-weight ratios available. Therefore, CFRPs are mostly applied in aerospace, automotive, and sports. Because the fibers are electric conductors, non-destructive eddy current–based methods have been already successfully applied on CRFPs [398]. The Airbus A350 XWB is built of 52% CFRP [425] including wing spars and fuselage components, overtaking the Boeing 787 Dreamliner, for the aircraft with the highest weight ratio for CFRP, which was held at 50% [426]. This was one of the first commercial aircraft to have the wing spars made from composites. The Airbus A380 was one of the first commercial airliners to have a central wing-box made of CFRP; it is the first to have a smoothly contoured wing cross-section instead of the wings being partitioned span-wise into sections. This optimizes aerodynamic efficiency [427]. Moreover, the trailing edge, along with the rear bulkhead, empennage and unpressurised fuselage are made of CFRP [428]. Many aircraft that use CFRP have experienced delays with delivery dates due to the relatively new processes used to make CFRP components, whereas metallic structures have been studied and used on airframes for years, and the processes are relatively well understood. But CFRP has become a notable material in structural engineering applications and in particular in lightweight design. Studied in an academic context as to its potential benefits in construction, it has also proved itself cost-effective in a number of field applications like strengthening concrete, masonry, steel, cast iron, and timber structures. Its use in industry can be either for retrofitting to strengthen an existing structure or as an alternative reinforcing (or pre-stressing) material instead of steel from the outset of a project. A recurrent problem is the monitoring of structural ageing, for which new methods are constantly investigated, due to the unusual multimaterial and anisotropic nature of CFRP [429]. During the manufacturing process, material abnormalities such as porosities and debonding of sheets can arise. Further, defects can occur during life cycle, because materials are exposed to high stress. External impacts such as lightning strokes, bird strikes, and stone chipping can give rise to surface damages. Moreover, with increasing operating time composites wear out and the strength and durability are reduced. Fatigue cracks are the consequence. Early fatigue damages usually occur in the subsurface microstructure of the material. Most likely cracks occur in the area close to welding seams and riveted joints. Modes of failures include fiber breaking and matrix cracking in the fiber-reinforced matrix, and cracks in metal layers. Ongoing in-service load yields propagation of the defects. In transport systems, material failure during operation can have significant safety implications. Therefore, quality standards are very high and regular in-service inspections and maintenance are required to comply them. Large impact damages on the surface can be visually identified. In order to detect small and subsurface defects, high-resolution NDT methods are required. Ultrasonic have been proven to be an effective tool to check composites after manufacturing, whereas the common ECT is widely applied for in-service controls of airplanes [400].

6.3.3.2 CFRP test specimens

In general, LET and LFE are suitable to detect and reconstruct defects occurring in electrically conducting parts of composites, because the working principle of LET is based on eddy currents induced in electrically conducting materials. Thus, basically LET should have great potential to inspect abnormalities in the metal layers of laminates, because the single metal layers are made of monolithic material and have isotropic electrical conductivities. On the other hand, it is not clear whether LET can also be applied to fiber-reinforced composites with a metal matrix and those with electrically conducting fibers in a polymer matrix such as carbon fiber-reinforced plastics. Due to the fibers and their directed orientations, the material is highly anisotropic. This impedes significantly the modeling of the material, which is inevitable for fundamental research.

These aspects have been the motivation for an investigation whether the MIECT technique can be applied to CFRP material. Based on the encouraging results of inspection of GLARE samples, four different CFRP specimens for testing with LET were fabricated in the university workshop (Figure 6.32). They have all the same size of 275mm × 275mm × 20mm and contain each 45 vol.% carbon fibers. The main fiber directions are in the bi-directional case alternating 0° and 90°, whereas in the quasi-isotropic case it is following the series: 0°, 45°, −45°, 90°, 45°, −45°, 90°, etc.

Whereas the same unidirectional noncrimped fabric was used for specimen-1, -2, and -3, in specimen-4 a plain weave structure was included. According to [430], the eddy currents increase when there are more contact points between the fibers in the laminate. Since specimen-4 is made from a weave, where the fibers cross each other inside the single layers and therefore a lot of contact points should exist, such specimens have been selected to be used in the study. More details about the specimens are given in Table 6.7.

6.3.3.3 Conductivity measurement

A very crucial point is the measurement of the electrical conductivity of the material under test. It is very hard to find any measurement data for the conductivity of CFRP in the literature [21,431]. Therefore, a few CFRP samples with different fiber orientations (uni-directional, bi-directional, quasi-isotropic, and plain weave structure, each containing 45 vol.% of fibers) have been manufactured in the university workshop [417]. Before they can be investigated using the LET technique, their

Figure 6.32 Four different CFRP samples used for the LET application (top-left: plain weave; top-right: bi-directional; bottom-left: uni-directional; bottom-right: quasi-isotropic)

Table 6.7 Details of the fabrication and layer structure of the CFRP specimens

Specimen No.	**1**	**2**	**3**	**4**
Layer structure	Uni-directional 	Bi-directional 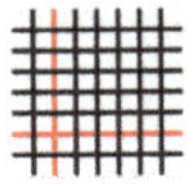	Quasi-isotropic 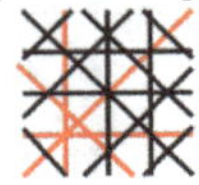	Fiber fabric 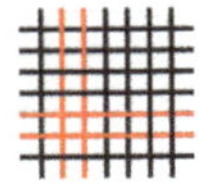
Specific fabric weight	620 g/m^2	620 g/m^2	620 g/m^2	400 g/m^2
Layer directions	0°	0°, 90°	−45°, 0°, 45°, 90°	plain weave
Number of layers	29	29	32	40
Total thickness	20 mm	20 mm	22 mm	20 mm
Estimated fiber vol.%	45	45	45	45

electrical conductivity tensor has to be measured. Depending on the manufacturing process, CFRP materials show a very anisotropic behavior. The eddy currents can only flow in the carbon fibers, i.e. there should be enough contact points between the carbon fibers enabling the flow of eddy currents and leading to the Lorentz forces which are finally measured. These contacts between the fibers are often due to the fabrication disturbed by nonconducting polymer, plastics, or epoxy resin resulting in conductivity values which are two or three orders of magnitude less than those of metallic material (e.g. $\sigma_{Al} = 37$ MS/m).

Thus, the conductivities of the CFRP samples defined in Table 6.7 were measured in all three spatial directions. Therefore, from each CFRP sample has been cropped small stripes (Figure 6.33).

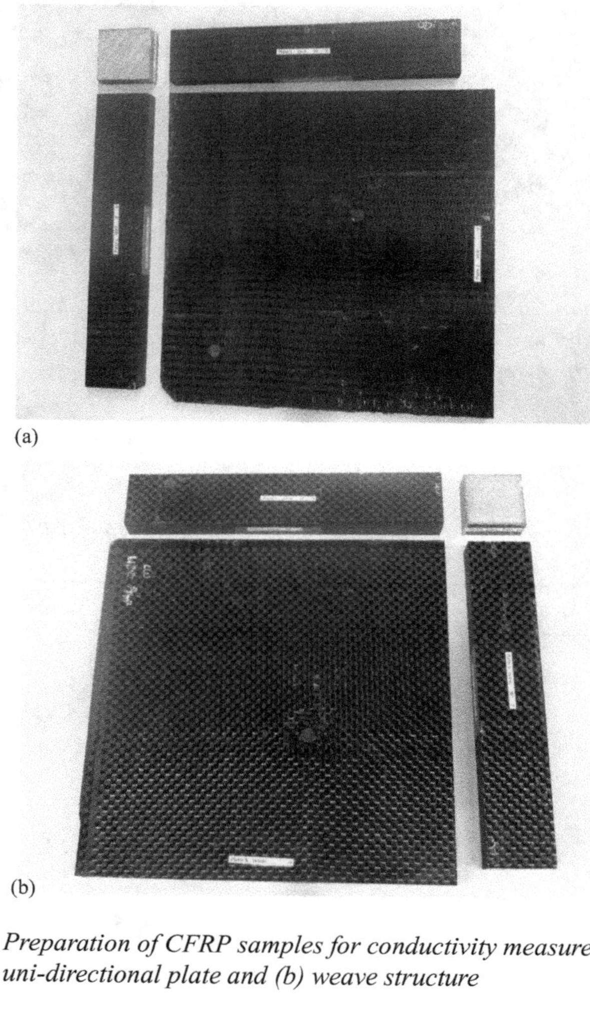

(a)

(b)

Figure 6.33 Preparation of CFRP samples for conductivity measurements: (a) uni-directional plate and (b) weave structure

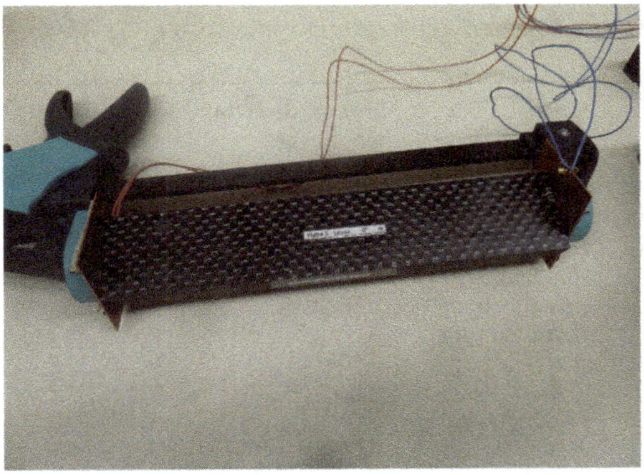

(a)

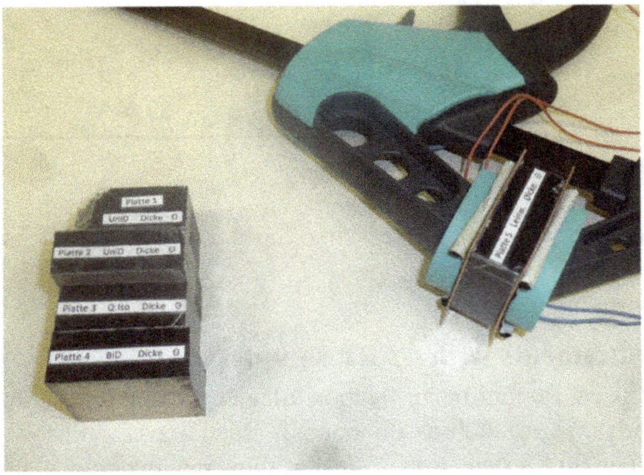

(b)

Figure 6.34 CFRP stripes clamped between two copper electrodes to measure the conductivity: (a) in length direction; (b) in thickness direction

Each stripe has been clamped between two copper electrodes (Figure 6.34) enabling the conductivity measurement in each axis. The results are given in Table 6.8. They agree very good with the values available in the literature [21,431]. It can be seen that conductivity values have been determined in fiber direction between 6000 and 14000 S/m (depending on the degree of anisotropy of the sample), and perpendicularly to the fiber direction values of about 1–2 S/m, whereas in thickness direction only values of 0.1–0.8 S/m have been found. Especially the values perpendicularly to the fiber direction are highly questionable, because this is the situation used to consider during LET measurements. This results in strong problems for the data recording of force measurements in the case of such low-conducting materials, because the signals

Table 6.8 Electrical conductivity measurement results of CFRP samples

Meas. No	**Sample No**	**Fiber orientation**	**Meas. direction vs fiber orientation**	**Conductivity (S/m) ($I = 1$ A)**	**Std. dev. (S/m)**
1	1	Uni-directional 1	0°	136, 44.371	1.121
2	1	Uni-directional 1	90°	1.232	0.008
3	2	Uni-directional 2	0°	8, 822.616	0.407
4	2	Uni-directional 2	90°	1.853	0.010
5	3	Quasi-isotropic	0°	5, 673.893	0.466
6	3	Quasi-isotropic	90°	3, 679.416	0.257
7	4	Bi-directional	0°	7, 514.567	0.725
8	4	Bi-directional	90°	5, 806.940	0.278
9	5	Weave	0°	9, 724.785	0.603
10	5	Weave	90°	8, 064.015	0.444
11	1	Uni-directional 1	thickness	0.169	$3.055E-04$
12	2	Uni-directional 2	thickness	0.234	$3.798E-04$
13	3	Quasi-isotropic	thickness	0.787	$3.028E-04$
14	4	Bi-directional	thickness	0.270	$2.283E-04$
15	5	Weave	thickness	0.099	$2.727E-04$

are up to eight orders of magnitude less than those found during measurements of an aluminum specimen.

6.3.3.4 Measurements with CFRP samples

ECT is widely used for online inspection of electrically conducting components for its unique advantages such as noncontact, speediness, and low cost, and it is notably appropriate for the detection of surface thin-opening cracks. There are already considerable published experimental results on the ECT of CFRP composites. The capabilities of ECT for CFRP include figuring out the internal structure (fiber orientation and ply sequence) and detecting cracks, delamination, inclusions, or others [432]. Eddy currents are usually created through electromagnetic induction by an excitation coil. Eddy currents are induced electrical currents that flow in a circular path through the (conductive) specimen which caused a new magnetic field reversely to the excitation field. This excited field can be measured with a second, receiving coil. Carbon fiber based materials show a low electrical conductivity and in most cases an anisotropic behavior. It depends on the specimen's characteristics whether the conductivity is sufficient to measure deviations in the material by using eddy current methods. Thus, eddy current measuring is a well-established method for characterization of surfaces (crack detection) or material differentiations, caused by conductivity or permeability variations. CFRP is characterized as layered and heterogeneous anisotropic material, which makes it very difficult to interpret ECT signals and to quantitatively analyze the defects. But discontinuities can be localized in CFRP and in other laminated materials, if the material properties are appropriate [433]. The dimensions of the smallest detectable defects are in the range of a few millimeters. Special single sensors and eddy current sensor arrays have been successfully developed [398,399]. If the sensor is not optimized for the required frequencies which are suitable to detect deeper defects, the chaotic conductivity arrangements makes it very complicate to inspect deeper areas. In such cases, larger sensors with lower resolution but higher penetration depth can be applied even at higher frequencies. It can be stated for ECT, where an electromagnetic field is usually applied orthogonal to the surface of CFRP laminate, that there is expected that the induced eddy currents should circulate in the composite. Unfortunately, this can only lead to a closed current loop, if the eddy currents can pass through the fibers in an area via the fiber–fiber contacts, i.e. a conducting material is needed for an eddy current flow. There is no other way for low frequency applications. If high-frequency currents are applied, the loops can only be closed via the capacitive coupling. In this case, there is a combination of electric current flow and dielectric displacement currents induced in the composite and thus, no longer an eddy current loop exists. In the low-frequency range (less about 1 MHz), where the ECT is usually applied, displacement currents are negligible compared to electric currents [434]. Thus, the eddy current path in CFRP composite is directly and strongly related to the material characteristics and the material modeling. If the currents circulate according to the capacitive bridges, the carbon fiber and resin matrix should be considered individually. The very few references reporting own conductivity measurements on CFRP samples support this conclusion [417,431]. The resistance in vertical direction of the fibers is so high, that there is no electrical current flow. This is simply due to the well-known fact that current always takes the path of least resistance.

6.3.3.5 Numerical modeling of CFRP

It is challenging to model the electromagnetic field in CFRP because of its specific properties. Numerical simulations are needed for better understanding of ECT applications on CFRP materials. In addition, it is also essential for the technology optimization and defect reconstruction. Menana and Feliachi [435,436] proposed a model based on an integro-differential formulation in terms of the electric vector potential. The finite element method (FEM) has also been widely used to the modeling of the ECT problems. A finite element analysis code based on the $\mathbf{A} - \varphi$ method was used by Cacciola *et al.* [437] to compute the magnetic flux density and eddy currents in the CFRP structure due to the probe velocity and presence of defects. Cheng *et al.* [431] proposed a numerical solver for simulation of ECT signals of laminated CFRP composite based on the reduced vector potential FEM approach with edge elements. For the modeling of the CFRP laminate, the anisotropic impedances in the three directions of the material are measured where the sample has a thickness of only one millimeter. It was proven that in this case the numerical solver can predict the ECT signals of CFRP with good accuracy. Barbato *et al.* [438,439] assumed for full-wave numerical modeling of CFRP composites at low frequencies to be modeled at the macroscopic scale by a complex dielectric permittivity tensor. He *et al.* [440] applied the scanning pulsed eddy current method and used the magnetic field intensity changes as well as the conductivity variations to characterize different types of defects in CFRP laminates and honeycomb sandwich panels. They observed a better defect detectability especially for deep defect. Li *et al.* [430,441] used inductive probes to characterize the CFRP samples including the identification of the fiber directions. It has been shown by experiments that the resistance of CFRP increases with the appearance of internal damage, such as fiber breakage and delamination. This indicates that conductivity is closely related to other composite properties. Li *et al.* [441] assumed *a set of contact points* between the single fibers in a ply enabling the eddy current flow, but a model of eddy current paths is not given there. In addition, they conducted 3D FEM simulations, although the modeling of commercially available CFRP samples has not been clearly demonstrated.

In the year 2013, Bui *et al.* from the University of Nantes presented a thermo-inductive NDT modeling applied to CFRP composites [442]. A 3D magneto-thermal finite-element-model has been used to calculate eddy currents and thermal distributions in pieces with or without flaw. Therefore, the electrical conductivity distribution in the CFRP composite needed for the FEM simulations was described by a randomized approximation. The carbon fibers were placed randomly in the cross section. Because these fibers are usually not straight but undulated, this fiber undulation can generate electrical contacts between fibers. These contacts are assumed to be randomly distributed in the composite and depending on the rate of filling per unit volume. Micrographic cut observations reveal that for a filling rate of the order of 60%, the mean distance between two contacts is about 500μm. The electric conductivity in fiber direction is almost independent of the number of contacts. It can be estimated with good precision, if the volume filling rate of the fibers per unit volume is multiplied by the electrical conductivity of carbon (66,000 S/m). To obtain the transversal conductivities while taking the electrical contacts into account, a volt-amperometric simulation in an equivalent impedance network has been performed. In the 3D FEM simulation, degenerated hexahedral Whitney elements were used to model the CFRP composites. These elements were used to model anisotropic multilayer composite containing flaws, thereby the meshing of the thin regions of the composite is avoided and the anisotropic multilayer characteristic of the materials as well as flaws inside their volume can be considered. The eddy current problem was solved using the $\mathbf{A} - \varphi$ formulation [443]. This 3D FEM model represents a remarkable step forward because it enables an almost realistic approximation of the electrical conductivity distribution in the CFRP composites, at least for a large part of the materials which are used in practice.

Although a lot of works have been done on the electromagnetic modeling, the electrical impedance characteristics and the origin of eddy current path in CFRP materials are still not systematically explored and analyzed. Therefore, the feasibility of modeling CFRP laminate with homogeneous anisotropic layers has not been clearly demonstrated. In addition to that, there is no numerical solver so far that is reported to be capable of dealing with the probe motion problem in CFRP testing, and the accuracy of calculated detecting signals is not verified by comparing with the real test data.

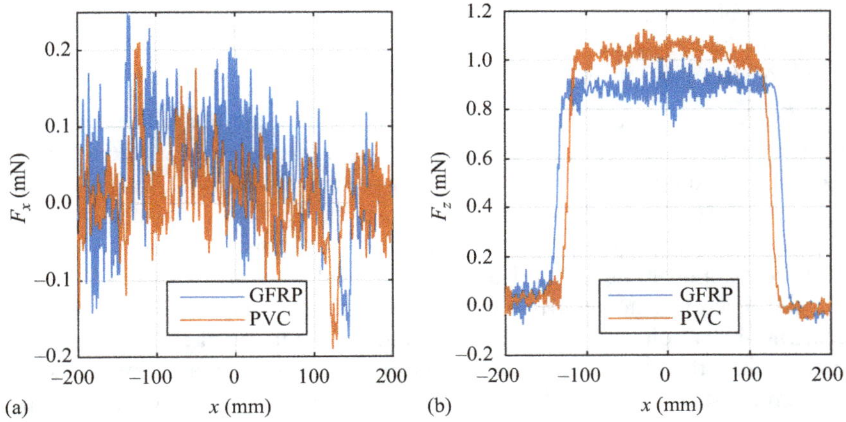

(a) (b)

Figure 6.35 Filtered force signals for CFRP and PVC. (a) Drag force. (b) Lift force

6.3.3.6 LET experiments with CFRP

In this context, it was also the motivation for our experiments, to study the usability of the LET system for the inspection of CFRP samples. The CFRP specimen 4 (plain weave, Table 6.7) in combination with the Halbach magnet configuration (see Section 3.2.4) has been used [417]. The LET experiments have been performed with a specimen velocity of v =200 mm/s and a lift-off δ =1 mm. CFRP specimen measurements at five different lateral positions (y-coordinates) over the middle of the specimen were carried out. In this case, the goal was not to find defects but rather to detect the specimen structure with LET. The results have been compared with measurements with a nonconductive PVC bar. The recorded signals are filtered using a running average filter (window width = 17 ms). The resulting force signals are presented in Figure 6.35.

It can be observed that the noise in relation to the signal is very large, therefore the F_x-component cannot be interpreted. The F_z-signal has a plateau with the length corresponding to the specimen length, which is 250 mm for PVC and 275 mm for CFRP. Because the nonconductive PVC bar creates almost the same plateau in the measured force profile, the observed force cannot be the result of eddy currents induced in the specimen. Probably aerodynamic effects exerting on the magnet system and result in the measured force.

6.3.3.7 Summary

If the MIECT technique is applied to the NDT of CFRP material, the following main issues have to be taken into account:

- Preparation of appropriate samples;
- Availability of the material characteristics (in particular the electrical conductivity);

- Numerical modeling and computation of the motion-induced eddy current field distributions;
- Appropriate (fast) inverse solution strategies for defect identification.

It has been shown that the MIECT technique can be applied successfully to the nondestructive inspection of FMLs using the experimental setup with the MMP-15 available at our university laboratory. The magnetic field distributions generated by two magnet systems, the Halbach structure and the cylindrical permanent magnet, have been measured and evaluated. The effect of various magnet systems, optimized for different applications, on LET are demonstrated. The advantages of the numerically optimized Halbach configuration for detection of small defects in rather thin specimens are exploited. Due to the fact that the computation of transient field problems including the conductor movement is still a complicated and time-consuming task, the reconstruction of defects in laminated conductive materials based on Lorentz force measurements remains a challenge.

Furthermore, carbon-fiber-reinforced polymer specimens have been investigated as well. Unfortunately, this investigation was not successful. It turned out that the inspection of CFRP using the existing LET measurement system is not possible. Due to the very low electrical conductivity of the CFRP specimen, not enough eddy currents have been induced in the sample and the resulting Lorentz force could not be measured. Furthermore, there is still a lack of knowledge how the physical properties of composite materials can be modeled in such a way that the inverse problem, i.e. the identification of defects, can be solved in a reasonable time period. It can be expected, even if it is not proved yet, that this conclusion holds not only for CFRP composites but also for other fiber-reinforced materials with low electrical conductivities. Although the results could perhaps become slightly better if a higher fiber filling rate (or fiber vol.% rate) is used, this further remains a big challenge, but not only for MIECT.

6.4 Defectoscopy of friction stir welding

6.4.1 Friction stir welding (FSW)

Friction stir welding (FSW) is a solid-state joining process that uses a nonconsumable tool to join two facing work pieces without melting the work piece material. It was invented and experimentally proven at The Welding Institute (TWI), Cambridge/UK in December 1991 [444,445]. A nonconsumable rotating tool with a specially designed pin and shoulder is inserted into the abutting edges of sheets or plates to be joined and traversed along the line of joint (Figure 6.36). The tool serves two primary functions: (a) heating of workpiece and (b) movement of material to produce the joint. The heating is accomplished by friction between the tool and the workpiece and plastic deformation of workpiece. The localized heating softens the material around the pin and combination of tool rotation and translation leads to movement of material from the front of the pin to the back of the pin. As a result of this process, a joint is produced in "solid state." During the FSW process, the material undergoes intense plastic deformation at elevated temperature, resulting in generation of fine and equiaxed

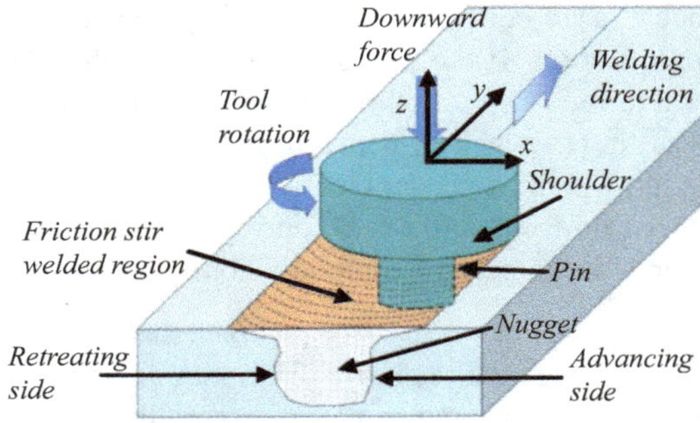

Figure 6.36 Schematic drawing of the principle of FSW [446]

recrystallized grains. The fine microstructure in friction stir welds produces good mechanical properties [446,447].

FSW is emerging as a very effective solid-state joining technique. In a relatively short duration after invention, several successful applications of FSW have been demonstrated. FSW is primarily used on wrought or extruded aluminum and for structures which need very high weld strength. Therefore, FSW is particularly found in modern shipbuilding, trains, and aerospace applications. In addition to aluminum alloys, FSW has been used successfully to join other metallic materials, such as copper, titanium, steel, magnesium, and composites. Because of high melting point and/or low ductility, successful joining of high melting temperature materials by means of FSW was usually limited to a narrow range of FSW parameters. Preheating is beneficial for improving the weld quality as well as increase in the traverse rate for high melting materials such as steel [448]. Despite considerable interests in the FSW technology in past decade, the basic physical understanding of the process is adverse. Some important aspects, including material flow, tool geometry design, wear of welding tool, microstructural stability, welding of dissimilar alloys and metals, still require deeper understanding. However, it can be observed that new technologies are often commercialized before a fundamental science emphasizing the underlying physics can be developed. This seems to be the case for the FSW technology.

6.4.1.1 Imperfections/defects caused by FSW

The solid-state nature of FSW leads to several advantages over fusion welding methods as problems associated with cooling from the liquid phase are avoided. Nevertheless, FSW can be associated with a number of imperfections or defects, if it is not performed properly. Insufficient weld temperature, due to low rotational speed or high traverse speed, for example, mean that the weld material is unable to accommodate the extensive deformation during welding. This may result in long, tunnel-like defects (so-called wormholes) running along the weld, which usually occur inside the welding seam. Low temperatures may also limit the forging action of the tool and so reduce the continuity of the bond between the materials from each side of the weld. The light contact between the materials has given rise to the name *kissing-bond*. This defect is particularly worrying since it is very difficult to detect using nondestructive methods such as X-ray or ultrasonic testing. If the pin is not long enough or the tool rises out of the plate then the interface at the bottom of the weld may not be disrupted and forged by the tool, resulting in a lack-of-penetration (LoP) defect. This is essentially a notch in the material, which can be a potential source of fatigue cracks. It is widely accepted that material flow within the weld during FSW is very complex and still poorly understood. Three different microstructural zones have been identified in friction stir weld:

1. The nugget region experiencing intense plastic deformation and high-temperature exposure and characterized by fine and equiaxed recrystallized grains.
2. The thermo-mechanically affected region experiencing medium temperature and deformation and characterized by deformed and un-recrystallized grains.
3. The heat-affected region experiencing only temperature and characterized by precipitate coarsening.

In each region the physical behavior of the material as well as the selected welding parameters during the welding process will cause different potential defects. The tool geometry is very an important factor for producing sound welds. However, at the present stage, tool designs are generally proprietary to individual researchers and only limited information is available in open literature. From the open literature, it is known that a cylindrical threaded pin and concave shoulder are widely used welding tool features. The welding parameters, including tool rotation rate, traverse speed, spindle tilt angle, and target depth, are crucial to produce sound and defect-free weld. It is well accepted that FSW can be generally described as a local extrusion process and the stirring and mixing of material occurred only at the surface layer of the weld adjacent to the rotating shoulder. Thus, the design of the tool shoulder together with the applied force pushing the tool on the specimen surface, the rotational speed of the tool, and the speed of welding along the specimen has an important impact on the degree of changes of the material plasticity and on the potential appearance of defects.

6.4.1.2 Typical weld seam defects

Wormholes: The so-called wormholes appear if process parameters are not correctly chosen and the heat input is insufficient. This happens, if either the force the tool is pushing with on the material is too low or the rotational speed of the tool is too low. The wormhole channels often extend over the whole seam. The appearance of wormhole defects can be reduced by increasing the rotational speed. On the other hand, if the generation of wormhole channels is a desired effect, this can be achieved by means of tuning the welding speed (Figure 6.37). In this way, e.g. cooling channels could be realized inside the welding seam.

LoP and surface lack of fill: The LoP occurs at the weld root and is caused by insufficient weld penetration caused by insufficient heat input or probe length. It can become later-on the reason for fatigue cracks because the welding zone is not

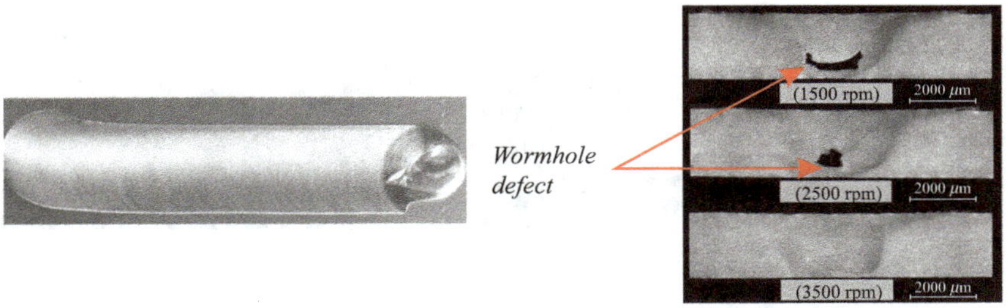

Figure 6.37 Wormholes appearing in the welding zone if the force on the tool or its rotational speed is too low [449]

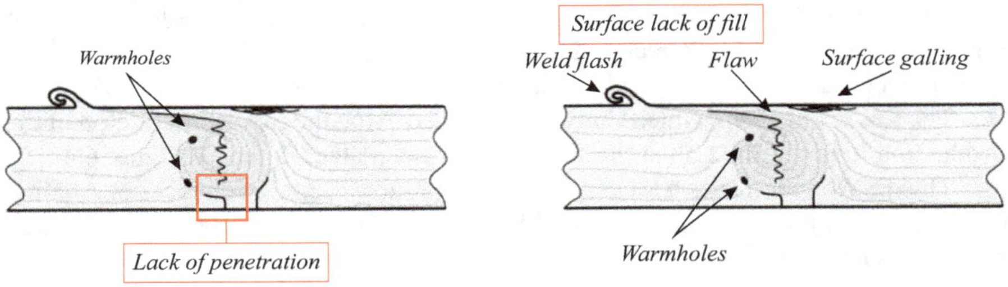

Figure 6.38 Nugget zone constriction and surface lack of fill—welding defects due to the transport of the bad conducting thin aluminum oxide sheet into the welding zone [449]

completely going through the specimen's thickness. This is a defect within the weld which has to be detected from the opposite surface of the specimen (Figure 6.38). This kind of defect is probably favor crack initiation (reduce in bending strength), notch effects, and decrease in tensile strength as well as crosshead travel.

If aluminum specimens are welded, it often happens that the very thin, but bad conducting aluminum oxide surface layer is moved inside the welding zone. The surface lack of fill is caused by oxide layer on material surface and results in a vertical Al_2O_3 layer across the weld seam. Its influence on tensile strength is still not fully understood. It can be reduced by an enhanced material flow or by removing the oxide layer.

Lack-of-Fusion: Lack-of-fusion (LoF) is a welding defect due to different cooling down speeds in the seam causing a LoF between both phase interfaces. This interface is marking a boundary separating regions whose electrical conductivity differs by several orders of magnitude (Figure 6.39).

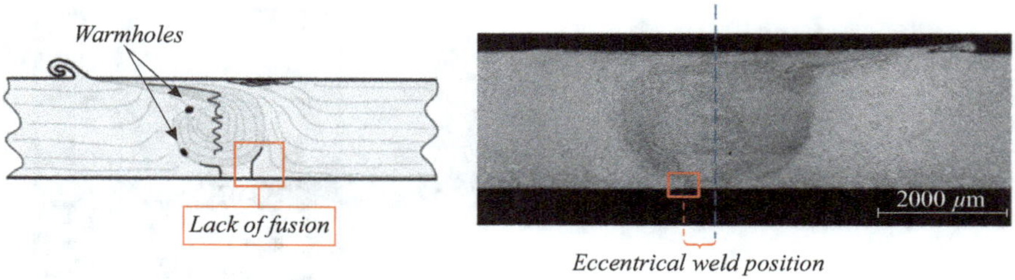

Figure 6.39 LoF is a welding defect due to different cooling down speeds in the seam [449]

6.4.2 FSW experiments

In the literature, there have been defined numerous imperfections or defects, some of them have been described already in the previous chapter [449–451]. Unfortunately, these defects cannot be identified in a nondestructive way so far.

The following Al-specimens have been welded using the robot-based setup of the Production Engineering Lab at the Technische Universitaet Ilmenau (Figure 6.40).

This FSW station uses a KUKA six-axis joint-arm robot. Two pieces of Al alloy (thickness = 8 mm) were jointed on this FSW system (Figure 6.41). Following process parameters have been chosen:

Figure 6.40 Robot-based FSW station, with KUKA joint-arm robot

(a) (b)

Figure 6.41 (a) Robot-based FSW of aluminum specimen in the production engineering lab. (b) FSW on Al-specimen

- Material: Al-alloy, EN AW 6060 T66
- Dimension: 300 mm × 100 mm × 8 mm
- Feed rate: 400 mm/min
- Tool rotation speed: 2,000–3,000 rpm
- Tool force: 8 kN
- Surface treatment: milled

The main goal of this study was to analyze, whether there is a lack of penetration in the welding zone, how deep is it and can be found other imperfections/defects in or close to the welding seam. This has to be shown in an exhaustive nondestructive way. The analysis of the welding seam should be done by means of MIECT measurements which have been described already earlier in the book.

6.4.3 NDT of friction stir welds

FSW is well known for its reproducibility and freedom from traditional fusion welding imperfections such as shrinkage cavities or slag inclusions. But, if FSW should be widely accepted as a joining method, reliable but also cost-effective process-specific quality assurance activities have to be developed. So far there is neither a common standard defect catalogue for FSW, which summarizes all relevant irregularities and describes their allowable sizes for different applications nor a standardized test specification for FSW welds. Moreover, it is even not fully understood, how different imperfections of the weld are affecting its mechanical properties during static and dynamic load.

Off-line methods include nondestructive tests, commonly applied in welding engineering and in testing the quality of FSW joints. Most popular NDT methods include visual testing, penetrant inspection, ultrasonic examination, X-ray testing, and ECT. Other useful techniques include modern methods such as synchrotron radiation and computer tomography [451]. Standards (EN ISO 25239-5 and AWS D17.3) recommend the following NDT methods for testing FSW welding imperfections [452,453]:

- Visual testing (VT),
- Liquid penetrant inspection (LPI),
- X-ray testing (CT)
- Ultrasonic testing (UT)

In addition, the standard AWS D17.3 enables carrying out other tests such as acoustic emission, ECT, neuron radiography, leak tests, etc. Individual types of tests have various intended uses and various levels of detectability of specific welding imperfections.

ECT of FSW joints is frequently used, even if it is not one of the techniques that are explicitly recommended by the standard for the inspection of FSW joints [452]. It should be further noted that the depth of eddy currents induced in the specimen, depends on the frequency, conductivity, and permeability of the material under test. A decrease of the frequency is accompanied by an increasing penetration depth of the electromagnetic field. In the case of thin joints, imperfections could be expected in the whole cross-section. ECT of FSW joints often requires individually designed transducers, which optimizes the technique of the NDT method [454–456]. As mentioned earlier, ECT can be used for a variety of inspection tasks [457]. The idea is the inspection of welded joints using pencil probes as a replacement for the more conventional magnetic particle or liquid penetrant inspection techniques. A major advantage is that the process may be used under water and can be used to scan welds through paint and other coatings. With respect to detection of linear defects such as cracks and LoF the defect should break the lines of the eddy currents ideally at right angles—as with magnetic particle inspection defects parallel to the eddy currents are likely to remain undetected. It is important therefore that the weld is scanned in the correct direction. Cracks as small as 0.5 mm deep and 5mm in length are capable of being detected. A more general overview is given in Table 6.9 where the features, advantages and disadvantages of the eddy current method are presented [451].

Like for established welding techniques it is necessary to provide NDT for assuring the integrity of structural welds by identifying cracking, porosity and other "well-known" flaws that can compromise weld strength. The NDT methods proposed for FSW so far, such as ultrasound and eddy current arrays and x-ray techniques, are suitable to only a limited extent for application in SME production, because they are expensive, time-consuming, and not flexible enough. In addition, however, there are some FSW-specific irregularities affecting fatigue behavior and corrosion resistance, such as flat wormholes, oxides lines and blisters (small pores near to surface), which are difficult to identify with conventional NDT. Therefore, there is still exists the need for further development of NDT techniques. This is especially true for in-process NDT methods, which allow the automatic joint quality inspection during welding. Such devices offer the possibility for a 100% monitoring of welds. Integrated into a feedback control algorithm, they will help to reduce weld imperfections. This is important especially for oxide lines—sometimes also referred as

Table 6.9 ECT in testing FSW joints

Features	Advantages	Disadvantages
Inclusions, cracks, and gas pores	Quick method	Time-consuming manual testing
Incomplete root penetration	Tests are carried out during one run	Possibly difficult interpretation of test results
Overlapping metal	Tests cover the whole joint	Limited depth of imperfection detection
Oxide layer in a weld	C-type scanning facilitates the interpretation of results	Location of an imperfection affects its detectability
Material coatings and the measurement of their thickness	Possibility of process automation	Material must be a conductor
Surface and subsurface imperfection	Continuous archiving of data	Method is sensitive to the physical properties of a material
Measurement of conductivity and permeability	Contact between the test piece surface and the transducer is not necessary	Surface porosity may cause improper indications
Grain size measurement		
Possibility of determining the size of an imperfection		

kissing bonds or joint line remnants—which are difficult to inspect in a post-process inspection. Process optimization is required in order to prevent such imperfections. The detection of conventional flaws, like LoP and LoF, is an important NDT task but they can be reduced or even avoided by a suitable clamping, tool pin length, and machine setup. Similar simple preventive measures do not exist for process-specific flaws, such as collapsed nugget, wormholes, or oxide lines. For FSW, wormholes, are process-specific too, because, unlike to fusion welding, they are not formed by gases. Here, their formation is dependent on the material flow around the tool pin and the generated frictional heat. Wormholes have a significant impact on the fatigue properties. Fatigue tests show, that they act as an initiator for a crack network and result in a forced fracture of the weld [458]. To improve the reliability in FSW nondestructive inspection using ECT, alternative methods or new probes have been investigated. Mandache *et al.* [459] applied the PECT technique to study lacks of penetration as defects. A new NDT EC probe, the so-called IOnic probe, was developed and tested in different FSW defect conditions [460]. This new EC probe allows a 3D-induced eddy currents in the material, deeper field penetration, independence of the deviation between the probe and the material surface, and easy interpretation of the output signal based on a comprehensible qualitative change. This probe can be applied advantageously, in particular if micro-defects with size of 50–200 μm have to be detected. The usage of conventional EC probes demonstrates that there is no distinctive signal feature that can allow to distinct between each defect condition. Indeed, the absolute planar spiral probe can only reproduce the global spread increase of current field due to the FSW bead. Such probes are not able to distinguish small suddenly variations of conductivity, caused by a local root defect with small size. These results illustrate the difficulty of NDT of FSW when using conventional EC probes. Conventional axisymmetric EC probes such as planar circular spiral probes are not able to distinguish FSW micro defects with depth below 200 μm. The experimental results show that the IOnic probe is able to identify different levels of FSW root defects, by a qualitative perturbation of the output signal.

6.4.4 MIECT measurements of friction stir welds

The current situation for NDT of friction stir welds is characterized by the application of only few conventional techniques (visual inspection, ultrasound, liquid penetration inspection, and computer tomography), as mentioned in the standards [452]. Electromagnetic testing methods like ECT is not widely spread. But if metals such as aluminum have to be investigated, it is worth to take into account the utilization of the MIECT technique to identify imperfections or defects in the welding region. It is important to know what happens below the surface and deeper inside the welds or in the welding zones. In the previous chapter, it has been described which aluminum alloy specimen has been jointed by means of the FSW system available in the Production Engineering Lab. This sample has been used for MIECT measurements on the MMP-15 platform, where the main goal was to identify different depths of LoP or perhaps other imperfections/defects in or near the welding zone (Figure 6.42). The parameters used for the MIECT measurements are:

- Magnet: cylindrical Halbach structure (with DiLET coil)
- Orientation of specimen: welded side facing the magnet
- Velocity: 200 mm/s
- Lift-off distance: 1 mm
- Scan width: −30 mm (0.5 mm step) to 30 mm
- Sampling frequency: 1000 Hz
- Low-pass cut-off frequency: 50 Hz
- Scanline repetitions: 25

An aluminum specimen has been analyzed using the MIECT technique applied on the multipurpose measuring platform MMP-15 (Figure 6.42(b)). During the movement of the FSW sample, the profiles of the Lorentz forces exerting on the permanent magnet (lift-off distance = 1 mm) have been measured. Figure 6.43(a) shows the drag force along the length of the specimen and has been cut at the central line ($y = 0$), i.e. just above the welding seam.

If there is no disturbance due to imperfections or defects, the drag force profile has a plateau over the entire length of the specimen. In this case, it can be clearly seen

Figure 6.42 (a) FSW of the Al-specimen. (b) MIECT measurement of the FSW specimen on MMP-15

that the welding changes the conductivity distribution, compared to the region outside the welding zone. The starting point and the endpoint of welding can be identified as well as an additional defect close to the endpoint.

In the zoomed plot (Figure 6.43(b)) it can be observed that at the defect position the force signal is significantly smaller than in the neighborhood. This means, that there is a remarkable difference in the electrical conductivity, e.g. due to a hole in the seam. This has been confirmed by an afterward visual inspection.

Figure 6.44 presents the lift force recorded along a band with a width of 60 mm at both sides of the centerline ($y = 0$) where the welding seam is marked by the two red-dashed lines. It is obvious that the lift force is more sensitive to the conductivity changes in the welding zone. Again, the edge effect caused by the ends of the sample gives the largest signal. In addition, it should be noted that the lift force signals contain obviously more information about the internal defect than the drag force signals due to more details in the force profile in the vicinity of the defect.

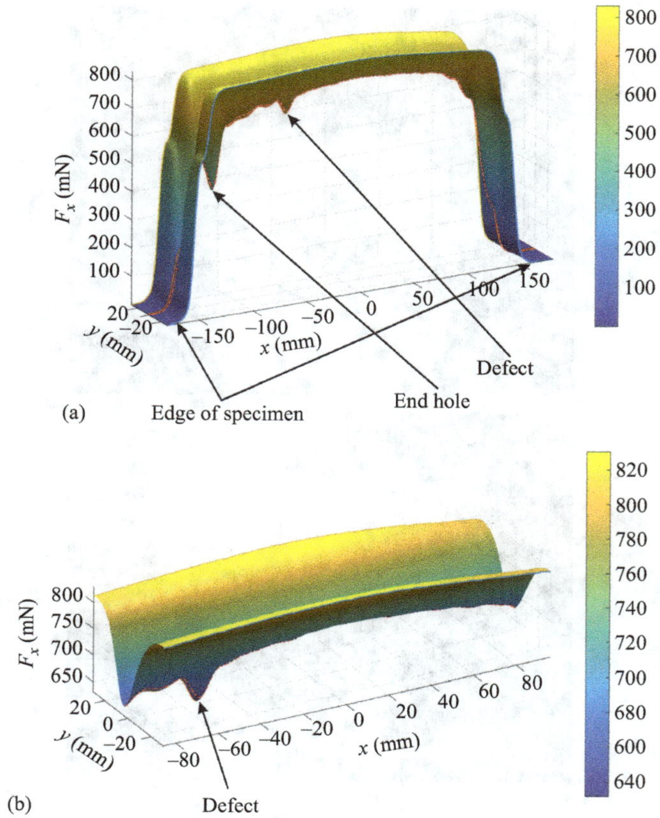

Figure 6.43 Drag force F_x measured along the friction stir weld (red dashed line). (a) Overall specimen. (b) Defect region

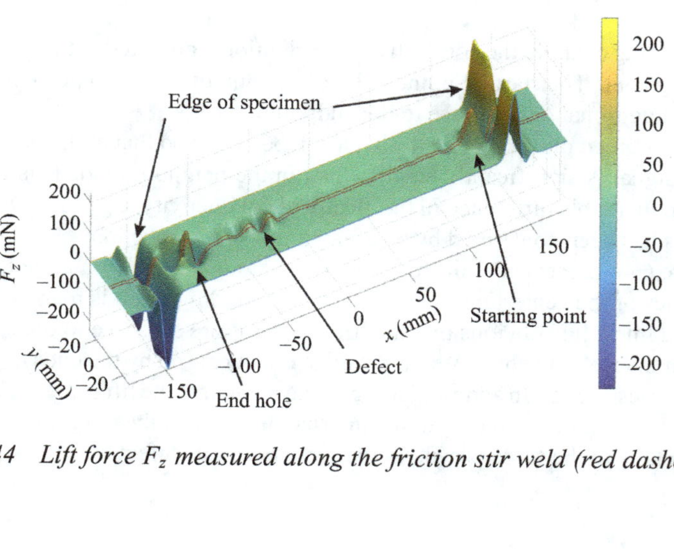

Figure 6.44 Lift force F_z measured along the friction stir weld (red dashed lines)

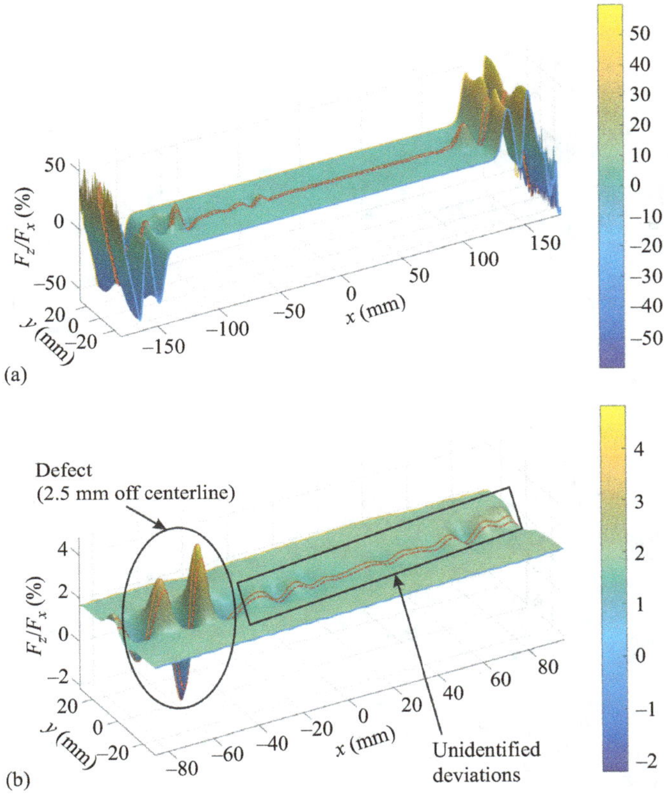

Figure 6.45 Lift-to-drag-ratio F_z/F_x along the friction stir weld (red dashed lines). (a) Overall specimen. (b) Defect region

The signals shown in Figure 6.45 are the result of some postprocessing of drag force and lift force. It turned out that the *lift-to-drag-ratio*, i.e. the relation of lift force to drag force, reduces the sensitivity to the lift-off distance and conductivity deviations. This results in data which show higher contrasts than the signals separately. The position of the defect can be identified in this figure at about 2.5 mm beside the centerline (red-dashed lines). Furthermore, there could be localized a new, so far unidentified deviation close to the starting point. This unexpected finding has been identified afterward as a micro-defect, not visible at the surface.

In a second measurement run a DiLET sensor has been used. This is a differential sensor consisting of a permanent magnet configuration combined with a coil containing a large number of windings. In this sensor system, the motion of the specimen in the static magnetic field will induce a voltage in the coil that can be measured and then used for the defect detection. If the magnet system has been optimized with respect to the desired kind of measurement task, a much higher magnetic flux density can be achieved in the region of interest. The magnitude of the measured voltage is

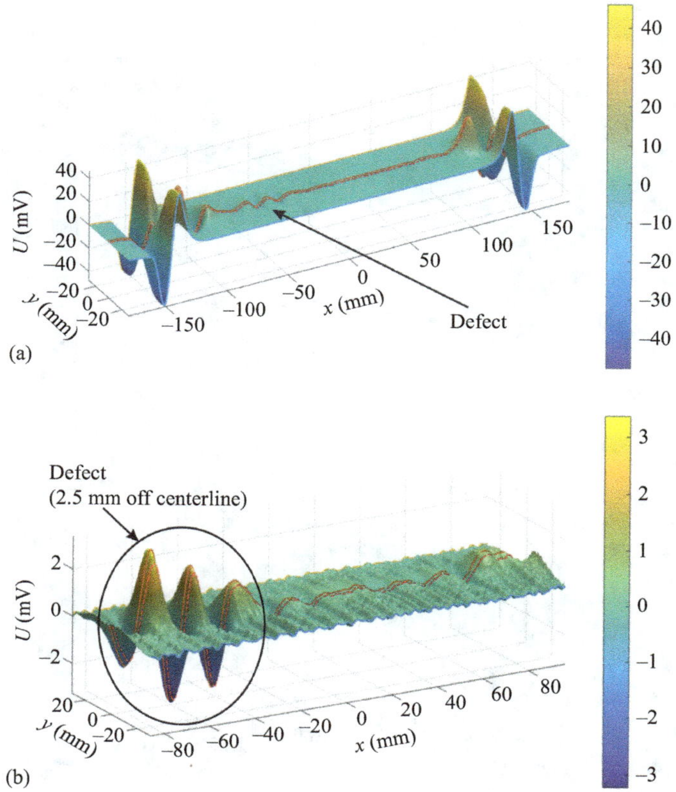

Figure 6.46 DiLET signal measured the full length along the friction stir weld (red-dashed lines). (a) Overall specimen. (b) Defect region

proportional to the velocity, i.e. increasing the measurement speed will lead to higher voltage signals, presupposed the sampling rate is high enough.

The DiLET sensor generally yields to a higher sensitivity of the entire sensor system (Figure 6.46(a)), but it leads also to rather noisy data which can be observed as the ripples in Figure 6.46(b). On the other hand, the voltage profile confirms the position of the defect at 2.5 mm beside the centerline (Figure 6.45).

6.4.5 *Potential applications of MIECT*

The encouraging results of the investigation of FSW specimen give reason to continue in this direction. The following three regions of interest will be taken into consideration for the next steps:

6.4.5.1 Nondestructive defect detection of FSW

The general aim of NDT techniques applied to FSW is the detection of imperfections and/or defects. This includes the determination of the electrical properties (e.g. electrical conductivity) as an alternative for the peak-to-peak measurement, the estimation of the oxide band (so-called kissing bonds), the identification of nugget collapses, or the investigation at both the advancing and the retreating side of the welding joint. Further, it would be very helpful to provide a technique to identify any LoP from the welding surface in a nondestructive way, especially for thick material components, i.e. much thicker than the penetration depth of the electromagnetic field.

6.4.5.2 Material science

There is a high potential for the development of the MIECT technique if the structure transformation in different materials during welding can be evaluated. The investigation of grain size–affected properties in similar and dissimilar aluminum alloys is a promising field of application, with a high impact on lightweight constructions or aircraft engineering. Furthermore, the interaction between process parameter and grain structure is highly interesting. This include the improvement of the understanding of the necessary structural adjustment for components with high static and dynamic stresses, the entire friction stirring process or the electrical characterization of the interfaces between different grain sizes.

6.4.5.3 Process control and monitoring

An important improvement would be the integration of the MIECT into the welding process itself, either for surface-based inspection or for weld root-based inspection. This would generally qualify the whole process monitoring, both with a tracking system for the process control and with an independent, nondestructive weld seam inspection for the quality control of the welding process.

6.5 Application to ferromagnetic materials

The content of the book is focused on the application of MIECT methods and in particular LET in the framework of non-ferromagnetic materials this far. A lot of parallels to the traditional eddy current method could be identified. However, the moderate rate of change of the magnetic field with respect to the specimen together with the application of NdFeB magnets obeying high-energy densities lead to considerable advantages in terms of penetration depth, which is the physical limit of traditional ECT.

This section is devoted to provide an outlook and to reveal the potential of the MIECT/LET method regarding its usage in combination with ferromagnetic materials. Typical areas of application are for example railway testing, steel casting, or pipeline inspection, where safe operation is paramount. In most application scenarios, motion is inherently present and already part of the operation, which privileges MIECT in general especially when it comes to online and in process inspection.

The general approach of MIECT in case of ferromagnetic specimen resembles the principle of magnetic flux leakage (MFL) [461], which is one of the most popular methods of pipeline inspection. In MFL, the specimen is subjected to an external magnetic field. Defects are detected on the basis of the leakage magnetic field, i.e. $\mathbf{B}^{(p,s)}$, which is sensed by for example Hall effect sensors or GMRs [461]. However, MIECT extends the principle to moving parts.

In the following, areas which are subject to review compared with the testing of nonferromagnetic material are enlightened and discussed. Because of distinct differences in the underlying physics, it cannot make claims to completeness since it would require a comprehensive retreatment of the content presented this far. However, the major issues are emphasized.

The nonlinear relationship between the magnetic field and the magnetic flux requires a reformulation of the governing equations. The primary magnetic field $\mathbf{B}^{(p)}$ is now altered in the presence of the ferromagnetic specimen. Therefore, it can be further divided into a part containing only the magnetic flux generated from the permanent magnet itself as in case of nonferromagnetic specimen $\mathbf{B}^{(p,m)}$ and into a distorted part, resulting from the ferromagnetic specimen $\mathbf{B}^{(p,s)}$ such that $\mathbf{B}^{(p)} = \mathbf{B}^{(p,m)} + \mathbf{B}^{(p,s)}$. The backreaction of the motional eddy currents, i.e. their magnetic flux is the secondary part $\mathbf{B}^{(s)}$. One fundamental difference compared with nonferromagnetic specimen is that the distorted part of the primary field $\mathbf{B}^{(p,s)}$ leads to considerable Kelvin forces attracting the magnet to the specimen. In this way, the Kelvin force acts against the Lorentz force and a complex interaction between both can be observed. Another elementary difference arises when considering the presence of a defect and the time derivatives of the mentioned field components. Now, additionally to a varying secondary magnetic field from the induced eddy currents ($\partial\mathbf{B}^{(s)}/\partial t \neq 0$) the distorted part of the primary magnetic field changes over time as well ($\partial\mathbf{B}^{(p,s)}/\partial t \neq 0$). The latter plays a key role and privileges the application of differential magnetic field sensors, e.g. coils, to sense field variations in addition to the force as it is presented in Chapter 3.

This requires changes of the numerical simulation environments and additional effort to propose efficient approaches since nonlinear magnetic materials involve higher computational cost because of the usage of iterative solvers. Note that some of the semianalytical approaches presented so far in Section 2.2 already include the possibility to model linear ferromagnetic materials. It is also incorporated in the analysis of oscillatory motion in Section 2.2.4, which could serve as a reference solution for first numerical simulations and implementations thereof.

The availability of efficient numerical approaches would enable the revision of optimal magnet designs by replacing the proposed forward model and a possible redefinition of the quantity of interest depending on the applied sensor system. The remaining procedures of the proposed optimization strategy remains unaltered because of its general attempt.

In the same course, the uncertainty and sensitivity analysis should be retreated in order to identify most influencing parameters. The uncertainty analysis of nonlinear magnetic materials in the framework of the gPC is more intricate but information about its treatment can be found in [462]. Note that because of the attraction and magnification of the magnetic flux into the specimen, the useful signal will be positively influenced by being more sensitive with respect to the material properties of the specimen.

The major differences also necessitates a revision of the experimental setup and the applied sensors. Because strong Kelvin forces are present, the lift-off distance between the magnet and the specimen plays a major role during the design process of new systems. Besides of the Lorentz force, magnetic field sensors such as coils, Hall effect sensors or GMRs are becoming of particular interest depending on the final application.

The maybe greatest influence of introducing ferromagnetic materials can be observed in terms of defect evaluation and in solving the inverse problem. The difficulty, as pointed out in the preceding discussion, lies in the estimation of the primary magnetic field distribution $\mathbf{B}^{(p)} = \mathbf{B}^{(p,m)} + \mathbf{B}^{(p,s)}$ and the associated eddy current distribution. Efficient forward models would again provide an indispensable starting point. A promising solution approach lies in the use of magnetic dipoles in addition to presently used current dipoles to determine the total magnetic flux and the resulting eddy current distribution.

Notwithstanding the above, the introduction of ferromagnetic materials engenders a series of new challenges. The calculation of the electromagnetic fields is more complex and costly and so are all related tasks depending on those. However, applications like in-service high-speed railway inspection give reason enough to pursue the effort to further elaborate the MIECT method in this direction.

References

[1] Shull PJ. Nondestructive Evaluation. Theory, Techniques, and Applications. New York, NY: Marcel Dekker Inc.; 2002.

[2] Garcia-Martin J, Gomez-Gil J. Comparative evaluation of coil and Hall probes in hole detection and thickness measurement on aluminum plates using eddy current testing. Russian Journal of Nondestructive Testing. 2013;49(8): 482–91.

[3] Aastroem T. From fifteen to two hundred NDT-methods in fifty years. In: 17th World Conference on Nondestructive Testing, Shanghai, China; 25–28 Oct. 2008. p. 283.

[4] Wu X, Zhang C, Goldberg P, *et al.* Early pottery at 20000 years ago in Xianrendong cave. Science. 2012;336(6089):1696–1700.

[5] Krautkrämer J, Krautkrämer H. Werkstoffprüfung mit Ultraschall. Berlin: Springer-Verlag; 1986.

[6] Hellier CJ. Handbook of Nondestructive Evaluation. 2nd ed. New York, NY: McGraw-Hill Education LLC; 2013.

[7] Foerster F. Theoretische und experimentelle Grundlagen der zerstörungsfreien Werkstoffprüfung mit Wirbelstromverfahren. I. Das Tastspulverfahren. Zeitschrift für Metallkunde. 1952;43:163–71.

[8] Morgner W. In Memoriam Friedrich Foerster. In: Invited Lecture at 100th Birthday of Friedrich Förster (in German), German Society of Nondestructive Testing (DGZfP), Hundisburg, Germany; Feb. 2008. p. 1–61.

[9] Mix PE. Introduction to Nondestructive Testing: A Training Guide. 2nd ed. Hoboken, NJ: Wiley; 2005.

[10] Non-Destructive Testing and Inspection Market by Technique (Visual Testing, Magnetic Particle, Liquid Penetrant, Eddy Current, Ultrasonic, Radiographic, Acoustic Emission), Method, Service, Vertical, and Geography—Global Forecast to 2023 [Market Report]. MARKETSandMARKETS; 2017.

[11] Eddy Current Testing Market by Type (conventional Eddy Current, ACFM, RFT, Eddy Current Array, Pulsed Eddy Current, Near-Field Testing, Near-Field Array, Partial Saturation Eddy Current), Service, Vertical, and Geography—Global Forecast to 2022 [Market Report]. MARKETSandMARKETS; 2016.

[12] Jiles DC. Review of magnetic methods for nondestructive evaluation (Part 2). NDT International. 1990;23(2):83–92.

[13] García-Martín J, Gómez-Gil J, Vázquez-Sánchez E. Non-destructive techniques based on eddy current testing. Sensors. 2011;11(3):2525–65.

[14] Sophian A, Tian GY, Fan M. Pulsed eddy current non-destructive testing and evaluation. Journal of Mechanical Engineering. 2017;30:500–14.
[15] Huang S, Wang S. New Technologies in Electromagnetic Non-Destructive Testing. Singapore, China: Springer; 2016.
[16] Omar M. Nondestructive Testing Methods and New Applications. Rijeka, Croatia: InTech; 2012.
[17] Dodd CV, Deeds WE. Analytical solutions to eddy-current probe-coil problems. Journal of Applied Physics. 1968;39(6):2829–38.
[18] Nair NV, Melapudi VR, Jimenez HR, *et al.* A GMR-based eddy current system for NDE of aircraft structures. IEEE Transactions on Magnetics. 2006;42(10):3312–14.
[19] Wincheski B. Deep flaw detection with giant magnetoresistive (GMR) based self-nulling probe. 26th Annual Review of Progress in Quantitative Nondestructive Evaluation. 2010; p. 465–72.
[20] Chomsuwan K, Yamada S, Iwahara M, *et al.* Application of eddy-current testing technique for high-density double-layer printed circuit board inspection. IEEE Transactions on Magnetics. 2005;41(10):3619–21.
[21] Cheng J, Ji H, Qiu J, *et al.* Role of interlaminar interface on bulk conductivity and electrical anisotropy of CFRP laminates measured by eddy current method. NDT & E International. 2014;68(1):1–12.
[22] Mizukami K, Mizutani Y, Todoroki A, *et al.* Detection of delamination in thermoplastic CFRP welded zones using induction heating assisted eddy current testing. NDT & E International. 2015;74(1):106–11.
[23] Kreutzbruck M, Allweins K, Heiden C. Fluxgate-magnetometer for the detection of deep lying defects. In: Proceedings of the 15th World Conference on Non-Destructive Testing, Rome, Italy; 15–21 Oct. 2000. p. 1–5.
[24] Ramos HG, Ribeiro AL. Present and future impact of magnetic sensors in NDE. Procedia Engineering. 2014;86(1):406–19.
[25] Wincheski B, Yu F, Simpon J, *et al.* Development of SDT sensor based eddy current probe for detection of deep fatigue cracks in multi-layer structure. NDT & E International. 2010;43(8):718–25.
[26] Krause HJ, Kreutzbruck M. Recent developments in SQUID NDE. Physica C: Superconductivity. 2002;368(1-4):70–79.
[27] Mueck M, von Kreutzbruck M, Baby U, *et al.* Eddy current nondestructive material evaluation based on HTS SQUIDs. Physica C: Superconductivity. 1997;282(1):407–10.
[28] Weinstock H. A review of SQUID magnetometry applied to nondestructive evaluation. IEEE Transactions on Magnetics. 1991;27(2):3231–36.
[29] Lobera-Serrano JA, Claycomb JR, Miller JH, *et al.* Hybrid double-D sheet-inducer for SQUID-based NDT. IEEE Transactions on Appiled Superconductivity. 2001;11(1):1283–86.
[30] Tavrin Y, Krause HJ, Wolf W, *et al.* Eddy current technique with high temperature SQUID for non-destructive evaluation of non-magnetic metallic structures. Cryogenics. 1996;36(2):83–86.

[31] Jenks WG, Sadeghi, SSH, Wikswo JP. SQUIDs for nondestructive evaluation. Journal of Physics D: Applied Physics. 1997;30(3):293–323.

[32] Braginski AI, Krause HJ. Nondestructive evaluation using high-temperature SQUIDs. Physica C: Superconductivity. 2000;335(1–4):179–83.

[33] Chady T, Enokizono M. Multi-frequency exciting and spectrogram-based ECT method. Journal of Magnetism and Magnetic Materials. 2000;215–16:700–03.

[34] Chady T, Sikora R. Optimization of eddy-current sensor for multi-frequency systems. IEEE Transactions on Magnetics. 2003;39(3):1313–16.

[35] Egorov AV, Polyakov VV, Salita DS, *et al.* Inspection of aluminum alloys by a multi-frequency eddy current method. Defence Technology. 2015;11(2): 99–103.

[36] Mook G, Michel F, Simonin J. Electromagnetic imaging using probe arrays. SV - JME (Strojniški vestnik - Journal of Mechanical Engineering). 2011;57(3):227–36.

[37] Postolache O, Ribeiro AL, Ramos HG. GMR array uniform eddy current probe for defect detection in conductive specimens. Measurement. 2013;46(10):4369–78.

[38] Jun J, Hwang J, Lee J. Quantitative nondestructive evaluation of the crack on the austenite stainless steel using the induced eddy current and the Hall sensor array. In: Proceedings of the IEEE Instrumentation & Measurement Technology Conference IMTC, Warsaw, Poland; 1–3 May 2007. p. 1–6.

[39] He Y, Pan M, Luo F, *et al.* Pulsed eddy current imaging and frequency spectrum analysis for hidden defect nondestructive testing and evaluation. NDT & E International. 2011;44(4):344–52.

[40] Atherton DL. Remote field eddy current inspection. IEEE Transactions on Magnetics. 1995;31(6):4142–47.

[41] Auld BA, Moulder JC. Review of advances in quantitative eddy current nondestructive evaluation. Journal of Nondestructive Evaluation. 1999;18(1):3–36.

[42] Mook G, Hesse O, Uchanin V. Deep penetrating eddy currents and probes. In: Proceeding of 9th European Conference on NDT: ECNDT; 25–29 Sept. 2006. p. Tu.3.6.2(1–14).

[43] Almeida G, Gonzalez J, Rosado L, *et al.* Advances in NDT and materials characterization by eddy currents. Procedia CIRP. 2013;7(1):359–64.

[44] Carlstedt M, Porzig K, Uhlig RP, *et al.* Application of Lorentz force eddy current testing and eddy current testing on moving nonmagnetic conductors. International Journal of Applied Electromagnetics and Mechanics. 2014;45(1):519–26.

[45] Gasparics A, Daroczi CS, Vertesy G, *et al.* Improvement of ECT probes based on fluxset type magnetic field sensor. Studies in Applied Electromagnetics and Mechanics, Electromagnetic Non-Destructive Evaluation (II). 1998;14(1):146–51.

[46] Dogaru T, Smith ST. Giant magnetoresistance-based eddy current sensor. IEEE Transactions on Magnetics. 2001;37(5):3831–38.

[47] Sikora R, Chady T, Gratkowski S, *et al.* Eddy current testing of thick aluminum plates with hidden cracks. Review of Quantitative Nondestructive Evaluation. 2003;22(1):427–34.

[48] Tsukada K, Kiwa T, Kawata T, *et al.* Low-frequency eddy current imaging using MR sensor detecting tangential magnetic field components for nondestructive evaluation. IEEE Transactions on Magnetics. 2006;42(10): 3315–17.

[49] Yamada H, Hasegawa T, Ishihara Y, *et al.* Difference in the detection limits of flaws in the depths of multi-layered and continuous aluminum plates using low-frequency eddy current testing. NDT & E International. 2008;41(2): 108–11.

[50] Hamia R, Cordier C, Saez S, *et al.* Eddy-current nondestructive testing using an improved GMR magnetometer and a single wire as inducer: a FEM performance analysis. IEEE Transactions on Magnetics. 2010;46(10):3731–37.

[51] Cacciola M, Megali G, Pellicano D, *et al.* Modelling and validating ferrite-core probes for GMR-eddy current testing in metallic plates. PIERS Online. 2010;6(3):237–41.

[52] Horng HE, Jeng JT, Yang HC, *et al.* Evaluation of the flaw depth using high-Tc SQUID. Physica C: Superconductivity. 2002;367(1–4):303–07.

[53] Jeng JT, Horng HE, Yang HC. Crack detection for the graphite slab using the high-Tc SQUID in unshielded environment. Physica C: Superconductivity. 2002;372–376(1):174–77.

[54] Allweins K, Gierelt G, Krause HJ, *et al.* Defect detection in thick aircraft samples based on HTS SQUID-magnetometry and pattern recognition. IEEE Transactions on Appiled Superconductivity. 2003;13(2):250–53.

[55] Fardmanesh M, Sarreshtedari F, Pourhashemi A, *et al.* Optimization of NDE characterization parameters for a RF-SQUID based system using FEM analysis. IEEE Transactions on Applied Superconductivity. 2009;19(3):791–95.

[56] van Bladel J. Motion of a conducting loop in a magnetic field. IEE Proceedings Pt A. 1988;135(4):217–22.

[57] van Bladel J. Foucault currents in a conducting sphere moving with constant velocity. IEE Proceedings Pt A. 1988;135(7):463–69.

[58] Saslow WM. Maxwell's theory of eddy currents in thin conducting sheets, and applications to electromagnetic shielding and MAGLEV. American Journal of Physics. 1992;60(8):693–711.

[59] Bachelet E. Foucault and eddy currents put to service. The Engineer. Oct 1912;114:420–21.

[60] Reitz JR. Forces on moving magnets due to eddy currents. Journal of Applied Physics. 1970;41(5):2067–71.

[61] Reitz JR, Davis LC. Force on a rectangular coil moving above a conducting slab. Journal of Applied Physics. 1972;43(4):1547–53.

[62] Richards PL. Magnetic suspension and propulsion systems for high-speed transportation. Journal of Applied Physics. 1972;43(6):2680–91.

[63] Borcherts RH. Force on a coil moving over a conducting surface including edge and channel effects. Journal of Applied Physics. 1972;43(5):2418–27.

[64] Borcherts RH, Davis LC, Reitz JR, *et al.* Baseline specifications for a magnetically suspended high-speed vehicle. Proceedings of the IEEE. 1973;61(5):569–78.

[65] Davis LC. Drag force on a magnet moving near a thin conductor. Journal of Applied Physics. 1972;43(10):4256–57.

[66] Shung-Wu L, Menendez RC. Force on current coils moving over a conducting sheet with application to magnetic levitation. In: Proceedings of the IEEE. 1974;62(5):567–77.

[67] van Bladel J, de Zutter D. Magnetic Levitation: The track currents. Applied Physics B: Lasers and Optics. 1984;34(4):193–201.

[68] de Zutter D. Levitation force acting on a three-dimensional static current source moving over a stratified medium. Journal of Applied Physics. 1985;58(7):2751–58.

[69] Panas S, Kriezis EE. Eddy current distribution due to a rectangular current frame moving above a conducting slab. Archiv für Elektrotechnik. 1986;69(3):185–91.

[70] Chady T, Spychalski I. Eddy current transducer with rotating permanent magnets. In: 22nd International Workshop on Electromagnetic Nondestructive Evaluation (ENDE 2017), Saclay, France; 2017. p. 2.

[71] Brauer H, Ziolkowski M. Eddy current testing of metallic sheets with defects using force measurements. Serbian Journal of Electrical Engineering. 2008;5(1):11–20.

[72] Ziolkowski M, Brauer H. Fast computation technique of forces acting on moving permanent magnet. IEEE Transactions on Magnetics. 2010;46(8): 2927–30.

[73] Ramos HMG, Rocha T, Pasadas D, *et al.* Velocity induced eddy currents technique to inspect cracks in moving conducting media. In: IEEE International Instrumentation and Measurement Technology Conference (I2MTC) Proceedings, The Depot, Minneapolis, MN, USA; 6–9 May 2013. p. 931–34.

[74] Ramos HG, Rocha T, Pasadas D, *et al.* Faraday induction effect applied to the detection of defects in a moving plate. Review of Progress in Quantitative Nondestructive Evaluation. 2013;32(1):1490–97.

[75] Rocha TJ. Velocity Induced Eddy Current Testing [Dissertation]. Instituto Superior Técnico Lisboa. Lisboa, Portugal; 2017.

[76] Rocha TJ, Ramos HG, Lopes Ribeiro A, *et al.* Studies to optimize the probe response for velocity induced eddy current testing in aluminium. Measurement. 2015;67(1):108–15.

[77] Ramos HG, Rocha T, Ribeiro AL, *et al.* GMR versus differential coils in velocity induced eddy current testing. In: IEEE International Instrumentation and Measurement Technology Conference (I2MTC) Proceedings, Montevideo, Uruguay; 12–15 May 2014. p. 915–18.

[78] Rocha TJ, Ramos HG, Lopes Ribeiro A, *et al.* Magnetic sensors assessment in velocity induced eddy current testing. Sensors and Actuators A: Physical. 2015;228(1):55–61.

[79] Tan Y, Wang X, Moreau R. An innovative contactless method for detecting defects in electrical conductors by measuring a change in electromagnetic torque. Measurement Science and Technology. 2015;26: Art. ID 035602.

[80] Brauer H, Porzig K, Mengelkamp J, *et al.* Lorentz force eddy current testing: a novel NDE - technique. COMPEL. 2014;33(6):1965–77.

[81] Pasadas DJ, Rocha TJ, Ramos HG, *et al.* Remote field eddy current inspection of metallic tubes using GMR sensors. In: IEEE International Instrumentation and Measurement Technology Conference (I2MTC) Proceedings, The Depot, Minneapolis, MN, USA; 6–9 May 2013. p. 1–4.

[82] Pasadas DJ, Ribeiro AL, Ramos HG, *et al.* Comparative evaluation of coil and hall probes in hole detection and thickness measurement on aluminum plates using eddy current testing. ACTA IMEKO. 2015;4(2): 62–67.

[83] Tajima N, Yusa N, Hashizume H. Application of low-frequency eddy current testing to the inspection of a double-walled tank in a reprocessing plant. Nondestructive Testing and Evaluation. 2018;33(2):189–97.

[84] Gowatski S, Miner G. The use of the low frequency electromagnetic technique to detect and quantify the amount of magnetite deposits in stainless steel superheater tubes due to exfoliation. Power Plant Chemistry. 2011;13(8): 484–91.

[85] Faraday M. The Bakerian lecture. Experimental researches in electricity. Philosophical Transactions of the Royal Society of London. 1832;122: 163–94.

[86] Shercliff JA. The Theory of Electromagnetic Flow-Measurement. Cambridge: Cambridge University Press; 1962.

[87] Feng CC, Deeds WE, Dodd CV. Analysis of eddy current flow meters. Journal of Applied Physics. 1975;46:2935–40.

[88] Baumgartl J, Hubert A, Müller G. The use of magnetohydrodynamic effects to investigate fluid flow in electrically conducting melts. Physics of Fluids A. 1993;5:3280–89.

[89] Thess A, Votyakov E, Kolesnikov Y. Lorentz force velocimetry. Physical Review Letters. 2006;96(16):Art. ID 164501.

[90] Thess A, Votyakov E, Knaepen B, *et al.* Theory of the Lorentz force flowmeter. New Journal of Physics. 2007;9(8):299.

[91] Uhlig RP. An experimental validation of Lorentz force eddy current testing. Ilmenau, Germany: Universitätsverlag Ilmenau; 2014.

[92] Zec M, Uhlig RP, Ziolkowski M, *et al.* Finite element analysis of nondestructive testing eddy current problems with moving parts. IEEE Transactions on Magnetics. 2013;49(8):4785–94.

[93] Uhlig RP, Zec M, Brauer H, *et al.* Lorentz force eddy current testing: a prototype model. Journal of Nondestructive Evaluation. 2012;31(4): 357–72.

[94] Donoso G, Ladera CL, Martín P. Damped fall of magnets inside a conducting pipe. American Journal of Physics. 2011;79(2):193–200.

[95] Uhlig RP, Zec M, Ziolkowski M, *et al.* Lorentz force sigmometry: a contactless method for electrical conductivity measurements. Journal of Applied Physics. 2012;111(9):Art. ID 094914.

[96] Uhlig RP, Zec M, Ziolkowski M, *et al.* Lorentz force eddy current testing: validation of numerical results. In: Proceedings of the Electrotechnical Institute. 2011;251:135–45.

[97] Zec M, Uhlig RP, Ziolkowski M, *et al.* Three-dimensional numerical investigations of Lorentz force eddy current testing. Studies in Applied Electromagnetics and Mechanics, Electromagnetic Nondestructive Evaluation XVI. 2014;38(1):83–93.

[98] Petković B, Haueisen J, Zec M, *et al.* Lorentz force evaluation: a new approximation method for defect reconstruction. NDT & E International. 2013;59:57–67.

[99] Mengelkamp J, Carlstedt M, Weise K, *et al.* Current density reconstructions for Lorentz force evaluation. Research in Nondestructive Evaluation. 2017;28(2):76–100.

[100] Mengelkamp J, Ziolkowski M, Weise K, *et al.* Permanent magnet modeling for Lorentz force evaluation. IEEE Transactions on Magnetics. 2015;51(7): Art. ID 6301211.

[101] Mengelkamp J, Lattner D, Haueisen J, *et al.* Lorentz force evaluation with differential evolution. IEEE Transactions on Magnetics. 2016;52(5): Art. ID 6001310.

[102] Zec M. Theory and Numerical Modelling of Lorentz Force Eddy Current Testing [Dissertation]. Technische Universität Ilmenau; 2013.

[103] Carlstedt M. A Contribution to the Experimental Validation in Lorentz Force Eddy Current Testing [Dissertation]. Technische Universität Ilmenau. Ilmenau, Germany; 2016.

[104] Data Sheet - K3D40. ME-Meßsysteme GmbH. Available from: http://www.me-systeme.de/de/datasheets/k3d40.pdf.

[105] Carlstedt M, Porzig K, Ziolkowski M, *et al.* Comparison of Lorentz force eddy current testing and common eddy current testing – measurements and simulations. Studies in Applied Electromagnetics and Mechanics. 2013;39(1): 218–25.

[106] Porzig K, Carlstedt M, Ziolkowski M, *et al.* Reverse engineering of ECT probes for nondestructive evaluation of moving conductors. In: AIP Conference Proceedings. vol. 1581; 2014. p. 1519–25.

[107] Jackson JD. Classical Electrodynamics. 3rd ed. New York, NY: John Wiley & Sons; 1999.

[108] Woodson HH, Melcher JR. Electromechanical Dynamics (Part I: Discrete Systems). New York, NY: Wiley; 1968.

[109] Knoepfel HE. Magnetic Fields. A Comprehensive Theoretical Treatise for Practical Use. New York, NY: Wiley; 2000.

[110] Furlani EP. Permanent Magnet and Electromechanical Devices: Materials, Analysis, and Applications. San Diego, CA: Academic Press; 2001.

[111] Zec M, Uhlig RP, Ziolkowski M, *et al.* Fast technique for Lorentz force calculations in non-destructive testing applications. IEEE Transactions on Magnetics. 2014;50(2):133–36.

[112] Ziółkowski M. Modern methods for selected electromagnetic field problems. Szczecin, Poland: Wydawnictwo Uczelniane Zachodniopomorskiego Uniwersytetu Technologicznego; 2015.

[113] Bracewell RN. The Fourier Transform & Its Applications. 3rd ed. Boston, MA: McGraw-Hill; 2000.

[114] Abramowitz M, Stegun IA. Handbook of Mathematical Functions: with Formulas, Graphs, and Mathematical Tables. New York, USA: Dover Books on Mathematics; 1965.

[115] Langerholc J. Torques and forces on a moving coil due to eddy currents. Journal of Applied Physics. 1973;44(4):1587–94.

[116] Votyakov EV, Thess A. Interaction of a magnetic dipole with a slowly moving electrically conducting plate. Journal of Engineering Mathematics. 2012;77(1):147–61.

[117] Derby N, Olbert S. Cylindrical magnets and ideal solenoids. American Journal of Physics. 2010;78(3):229–35.

[118] Singh A. Theory of eddy-current brakes with thick rotating disc. In: Proceedings of the Institution of Electrical Engineers. 1977;124(4):373–76.

[119] Sharif S, Faiz J, Sharif K. Performance analysis of a cylindrical eddy current brake. IET Electric Power Applications. 2012;6(9):661–68.

[120] Lee HW, Kim KC, Lee J. Review of Maglev train technologies. IEEE Transactions on Magnetics. 2006;42(7):1917–25.

[121] Kanamori M, Ishihara Y. Finite-element analysis of an electromagnetic damper taking into account the reaction of the magnetic field. JSME International Journal Series III - Vibration Control Engineering, Engineering for Industry. 1989;32(1):36–43.

[122] Zuo L, Chen X, Nayfeh S. Design and analysis of a new type of electromagnetic damper with increased energy density. Journal of Vibration and Acoustics. 2011;133(4):Art. ID 041006.

[123] Ebrahimi B, Khamesee MB, Golnaraghi F. Permanent magnet configuration in design of an eddy current damper. Microsystem Technologies. 2010;16(1):19–24.

[124] Palomera-Arias R, Connor JJ, Ochsendorf JA. Feasibility study of passive electromagnetic damping systems. Journal of Structural Engineering. 2008;134(1):164–70.

[125] Jansen JW, Lomonova EA, Rovers JMM. Effects of eddy currents due to a vacuum chamber wall in the airgap of a moving-magnet linear actuator. Journal of Applied Physics. 2009;105(7):Art. ID 07F111.

[126] Zhang Y, Huang K, Yu F, *et al.* Experimental verification of energy-regenerative feasibility for an automotive electrical suspension system. In: Proceedings of the International Conference on Vehicular Electronics and Safety. 2007; p. 1–5.

[127] Jansen JW, Gysen B, H Paulides JJ, *et al.* Advanced electromagnetic modeling applied to anti-vibration systems for high precision and automotive applications. International Compumag Society Newsletter. 2012;1(19): 3–16.

[128] Collins L. Harvest for the world: energy harvesting techniques. Power Engineer. 2006;20(1):34–37.

[129] Mierické J, Urankar L. Theory of electrodynamic levitation with a continuous sheet track—Part I. Applied Physics. 1973;2(4):201–11.

[130] Urankar L, Miericke J. Theory of electrodynamic levitation with a continuous sheet track—Part II. Applied Physics. 1974;3(1):67–76.

[131] Xypteras JE, Kriezis EE. Force on a vertical circular current loop moving above a conducting slab and eddy current distribution. Archiv für Elektrotechnik. 1981;63(6):309–15.

[132] Zuo L, Scully B, Shestani J, *et al.* Design and characterization of an electromagnetic energy harvester for vehicle suspensions. Smart Materials and Structures. 2010;19(4):Art. ID 04500310.

[133] Tang X, Lin T, Zuo L. Design and optimization of a tubular linear electromagnetic vibration energy harvester. IEEE/ASME Transactions on Mechatronics. 2014;19(2):615–22.

[134] Pluk KJW, van Beek TA, Jansen JW, *et al.* Modeling and measurements on a finite rectangular conducting plate in an eddy current damper. IEEE Transactions on Industrial Electronics. 2014;61(8):4061–72.

[135] Amati N, Tonoli A, Canova A, *et al.* Dynamic behavior of torsional eddy-current dampers: sensitivity of the design parameters. IEEE Transactions on Magnetics. 2007;43(7):3266–77.

[136] Ooi BT, Eastham AR. Transverse edge effects of sheet guideways in magnetic levitation. IEEE Transactions on Power Apparatus and Systems. 1975;94(1):72–80.

[137] Ooi BT. Electromechanical stiffness and damping coefficients in the repulsive magnetic levitation system. IEEE Transactions on Power Apparatus and Systems. 1976;95(3):936–43.

[138] Ooi BT. A dynamic circuit theory of the repulsive magnetic levitation system. IEEE Transactions on Power Apparatus and Systems. 1977;96(4):1094–100.

[139] Ooi BT, Jain OP. Force transients at guideway butt joints in repulsive magnetic levitation system. IEEE Transactions on Power Apparatus and Systems. 1979;PAS-98(1):323–30.

[140] Ooi BT, Jain OP. Moments and force densities of the electrodynamic levitation system. IEEE Transactions on Magnetics. 1979;15(3):1102–08.

[141] Weidermann C, Sokolov I, Thess A. Lorentz force and Joule heat induced in an electrically conducting plate moving with time-dependent velocity under the influence of a homogeneous magnetic field. IEEE Transactions on Magnetics. 2014;50(8):Art. ID 7027209.

[142] van Bladel J. Electromagnetic Fields. 2nd ed. Hoboken, NJ: Wiley; 2007.

[143] Weise K, Ziolkowski M, Carlstedt M, *et al.* Oscillatory motion of permanent magnets above a conducting slab. IEEE Transactions on Magnetics. 2015;51(10):7209113.

[144] Shiozawa T. Phenomenological and electron-theoretical study of the electrodynamics of rotating systems. Proceedings of the IEEE. 1973;61(12):1694–1702.

[145] van Bladel J. Relativistic theory of rotating disks. Proceedings of the IEEE. 1973;61(3):260–68.

[146] Mo TC. Theory of electrodynamics in media in noninertial frames and applications. Journal of Mathematical Physics. 1970;11(8):2589–610.

[147] Haykin SS. Communication Systems. 4th ed. New York, NY: Wiley; 2001.

[148] Tolstov GP, Silverman RA. Fourier Series. New York, NY, and London: Dover Publications and Constable; 1976.

[149] Shampine LF. Vectorized adaptive quadrature in MATLAB. Journal of Computational and Applied Mathematics. 2008;211(2):131–40.

[150] Shampine LF. Weighted quadrature by change of variable. Neural, Parallel & Scientific Computations. 2010;18(2):195–206.

[151] COMSOL Inc, Burlington, MA, USA. COMSOL Multiphysics. Available from: http://www.comsol.de.

[152] Zuo L, Tang X. Large-scale vibration energy harvesting. Journal of Intelligent Material Systems and Structures. 2013;24(11):1405–30.

[153] Canova A, Vusini B. Analytical modeling of rotating eddy-current couplers. IEEE Transactions on Magnetics. 2005;41(1):24–35.

[154] Moon P, Spencer DE. Field theory for engineers. Toronto, London, New York, NY: D. Van Nostrand Company, Inc.; 1961.

[155] Kraiger M, Schnizer B. Potential and field of a homogeneous magnetic spheroid of arbitrary direction in a homogeneous magnetic field in Cartesian coordinates. COMPEL. 2013;32(3):936–60.

[156] Mengelkamp J. Forward and Inverse Calculation Methods for Lorentz Force Evaluation Applied to Laminated Composites [Dissertation]. Technische Universität Ilmenau. Ilmenau, Germany; 2016.

[157] Weise K, Carlstedt M, Ziolkowski M, *et al.* Optimal magnet design for Lorentz force eddy current testing. IEEE Transactions on Magnetics. 2015;51(9):Art. ID 6201415.

[158] Zhou Pb. Numerical Analysis of Electromagnetic Fields. Berlin, Heidelberg, Germany: Springer; 1993.

[159] Malik NH. A Review of the charge simulation method and its applications. IEEE Transactions on Electrical Insulation. 1989;24(1):3–20.

[160] Singer H, Steinbigler H, Weiss P. A charge simulation method for the calculation of high voltage fields. IEEE Transactions on Power Apparatus and Systems. 1974;93(5):1660–68.

[161] Canova A, Gruosso G. 3D source simulation method for static fields in inhomogeneous media. International Journal for Numerical Methods in Engineering. 2007;70(9):1096–111.

[162] Sato S, Zaengl WS. Effective 3-dimensional electric field calculation by surface charge simulation method. IEE Proceedings A (Physical Science, Measurement and Instrumentation, Management and Education, Reviews). 1986;133(2):77–86.

[163] Fujishima Y, Wakao S. Surface charge analysis in eddy current problems. IEEE Transactions on Magnetics. 2003;39(3):1123–26.

[164] Petković B, Weise K, Haueisen J. Computation of Lorentz force and 3-D eddy current distribution in translatory moving conductors in the field of a permanent magnet. IEEE Transactions on Magnetics. 2017;53(2):Art. ID 7000109.

[165] Ziolkowski M, Schmidt R, Petković B, *et al.* SCSM for calculation of motion-induced eddy currents in isotropic and anisotropic conductive objects. IEEE Transactions on Magnetics. 2018;54(3):Art. ID 7204804.

[166] Carlstedt M, Weise K, Ziolkowski M, *et al.* Estimation of Lorentz force from dimensional analysis: similarity solutions and scaling laws. IEEE Transactions on Magnetics. 2016;52(8):Art. ID 7004813.

[167] Krawczyk A, Tegopoulos JA. Numerical Modelling of Eddy Currents. 1st ed. New York, NY: Oxford University Press; 1993.

[168] Ida N. Numerical Modeling for Non-Destructive Evaluation. 1st ed. London, UK: Chapman & Hall; 1995.

[169] Sykulski JK. Computational Magnetics. 1st ed. London, UK: Chapman & Hall; 1995.

[170] Ida N, Bastos JPA. Electromagnetics and Calculation of Fields. 2nd ed. New York, NY: Springer; 1997.

[171] Popović Z, Popović BD. Introductory Electromagnetics. 1st ed. Upper Saddle River, NJ: Prentice-Hall; 2000.

[172] Zienkiewicz OC, Taylor RL, Zhu JZ. The Finite Element Method: Its Basis and Fundamentals. 6th ed. Amsterdam: Elsevier Butterworth-Heinemann; 2005.

[173] Bastos JP, Sadowski N. Electromagnetic Modeling by Finite Element Methods. vol. 117 of Electrical and Computer Engineering. New York, NY: Marcel Dekker; 2003.

[174] Humphries S. Field Solutions on Computers. Boca Raton, FL: CRC Press; 1998.

[175] Sadiku MNO. Numerical Techniques in Electromagnetics. 2nd ed. Boca Raton, FL: CRC Press; 2001.

[176] Zimmerman WBJ. Process Modelling and Simulation with Finite Element Methods. vol. 15 of Series on Stability, Vibration, and Control of Systems. Hackensack, NJ: World Scientific; 2004.

[177] Ying P, Jiangjun R, Yu Z, *et al.* A composite grid method for moving conductor eddy-current problem. IEEE Transactions on Magnetics. 2007;43(7): 3259–65.

[178] Demenko A. Movement simulation in finite element analysis of electric machine dynamics. IEEE Transactions on Magnetics. 1996;32(3): 1553–56.

[179] Emson CRI, Riley CP, Walsh Da, *et al.* Modelling eddy currents induced by rotating systems. IEEE Transactions on Magnetics. 1998;34(5):2593–6.

[180] Biddlecombe C, Simkin J, Jay AP, *et al.* Transient electromagnetic analysis coupled to electric circuits and motion. IEEE Transactions on Magnetics. 1998;34(5):3182–85.

[181] Rodger D, Eastham J. Characteristics of a linear induction tachometer – a 3D moving conductor eddy current problem. IEEE Transactions on Magnetics. 1985;21(6):2412–15.

[182] Rodger D, Lai HC, Leonard PJ. Coupled elements for problems involving movement (switched reluctance motor). IEEE Transactions on Magnetics. 1990;26(2):548–50.

[183] Rodger D, Leonard PJ, Eastham JF. Modelling electromagnetic rail launchers at speed using 3D finite elements. IEEE Transactions on Magnetics. 1991;27(1):314–17.

[184] Binns KJ, Lawrenson PJ, Trowbridge CW. The analytical and numerical solution of electric and magnetic fields. Chichester: Wiley; 1992.

[185] Bird J, Lipo T. Modeling the 3-D rotational and translational motion of a halbach rotor above a split-sheet guideway. IEEE Transactions on Magnetics. 2009;45(9):3233–42.

[186] Kirpo M, Tympel S, Boeck T, *et al.* Electromagnetic drag on a magnetic dipole near a translating conducting bar. Journal of Applied Physics. 2011; 109(11).

[187] Chan EKC, Williamson S. Factors influencing the need for upwinding in two-dimensional field calculation. IEEE Transactions on Magnetics. 1992;28(2):1611–14.

[188] Enokizono M, Nagata S. Convection-diffusion analysis at high peclet number by the boundary element method. IEEE Transactions on Magnetics. 1992;28(2):1651–54.

[189] Kurz S, Fetzer J, Lehner G, *et al.* A novel formulation for 3D eddy current problems with moving bodies using a Lagrangian description and BEM-FEM coupling. IEEE Transactions on Magnetics. 1998;34(5):3068–73.

[190] Heinrich JC, Huyakorn PS, Zienkiewicz OC. An 'upwind' finite element scheme for two-dimensional convective transport equation. International Journal for Numerical Methods in Engineering. 1977;11:131–43.

[191] Hughes TJR. A simple scheme for developing 'upwind' finite elements. International Journal for Numerical Methods in Engineering. 1978;12:1359–65.

[192] Trowbridge CW, Sykulski JK. Some key developments in computational electromagnetics and their attribution. IEEE Transactions on Magnetics. 2006;42(4):503–08.

[193] Davat B, Ren Z, Lajoie-Mazenc M. The movement in field modeling. IEEE Transactions on Magnetics. 1985;21(6):2296–98.

[194] Salon SJ, Palma R, Hwang CC. Dynamic modeling of an induction motor connected to an adjustable speed drive. IEEE Transactions on Magnetics. 1989;25(4):3061–63.

[195] Lee K, DeBortoli M, Lee MJ, *et al.* Coupling finite elements and analytical solution in the airgap of electric machines. IEEE Transactions on Magnetics. 1991;27(5):3955–57.

[196] Rajanathan CB, Hu G. Electromechanical transient characteristics of an induction actuator by finite element analysis. IEEE Transactions on Magnetics. 1993;29(2):2001–05.

[197] Allegre JM, Marchand C, Ren Z. On the influence of mesh deformation to the force calculation when accounting for the movement. In: 3rd International Conference on Computation in Electromagnetics (CEM 96); 1996. p. 253–58.

[198] Perrin-Bit R, Coulomb JL. A three dimensional finite element mesh connection for problems involving movement. IEEE Transactions on Magnetics. 1995;31(3):1920–23.

[199] Hannalla AY, MacDonald DC. Numerical analysis of transient field problems in electrical machines. Proceedings of the Institution of Electrical Engineers. 1976;123(9):893.

[200] Ratnajeevan S, Hoole H. Rotor motion in the dynamic finite element analysis of rotating electrical machinery. IEEE Transactions on Magnetics. 1985;21(6):2292–95.

[201] Vassent E, Meunier G, Foggia A, *et al.* Simulation of induction machine operation using a step-by-step finite-element method. Journal of Applied Physics. 1990;67(9):5809–11.

[202] Vassent E, Meunier G, Foggia A, *et al.* Simulation of induction machine operation using a step-by-step finite element method coupled with circuits and mechanical equations. IEEE Transactions on Magnetics. 1991;27(6): 5232–34.

[203] Abdel-Razek A, Coulomb J, Feliachi M, *et al.* Conception of an air-gap element for the dynamic analysis of the electromagnetic field in electric machines. IEEE Transactions on Magnetics. 1982;18(2):655–59.

[204] Feliachi M, Coulomb J, Mansir H. Second order air-gap element for the dynamic finite-element analysis of the electromagnetic field in electric machines. IEEE Transactions on Magnetics. 1983;19(6):2300–03.

[205] Mohellebi H, Latreche ME, Feliachi M. Coupled axisymmetric analytical and finite element analysis for induction devices having moving parts. IEEE Transactions on Magnetics. 1998;34(5):3308–10.

[206] De Gersem H, Weiland T. A computationally efficient air-gap element for 2-D FE machine models. IEEE Transactions on Magnetics. 2005;41(5):1844–47.

[207] Gaussens B, Hoang E, de la Barriere O, *et al.* Analytical approach for air-gap modeling of field-excited flux-switching machine: no-load operation. IEEE Transactions on Magnetics. 2012;48(9):2505–17.

[208] Salon S, Schneider J. A hybrid finite element-boundary integral formulation of the Eddy current problem. IEEE Transactions on Magnetics. 1982;18(2):461–46.

[209] Pichon L, Razek A. Force calculation in axisymmetric induction devices using a hybrid FEM-BEM Technique. IEEE Transactions on Magnetics. 1990;26(2):1050–53.

[210] Onuki T, Wakao S, Yoshizawa T. Eddy current computations in moving conductors by the hybrid FE-BE method. IEEE Transactions on Magnetics. 1995;31(3):1436–39.

[211] Fetzer J, Kurz S, Lehner G, *et al.* Analysis of an actuator with eddy currents and iron saturation: comparison between a FEM and a BEM-FEM coupling approach. IEEE Transactions on Magnetics. 1999;35(3):1793–96.

[212] Rucker WM, Kurz S, Haas M, *et al.* 3D transient analysis of electromechanical devices using parallel BEM coupled to FEM. IEEE Transactions on Magnetics. 2000;36(4):1360–63.

[213] Rischmüller V. Eine Parallelisierung der Kopplung der Methode der finiten Elemente und der Randelementmethode. vol. 366 of Fortschritt-Berichte VDI Reihe 21, Elektrotechnik. Düsseldorf: VDI-Verlag; 2005.

[214] Buffa A, Maday Y, Rapetti F. Calculation of eddy currents in moving structures by a sliding mesh-finite element method. IEEE Transactions on Magnetics. 2000;36(4):1356–59.

[215] Preston TW, Reece ABJ, Sangha PS. Induction motor analysis by time-stepping techniques. IEEE Transactions on Magnetics. 1988;24(1): 471–74.

[216] Muramatsu K, Nakata T, Takahashi N, *et al.* Linear AC steady-state eddy current analysis of high speed conductor using moving coordinate system. IEEE Transactions on Magnetics. 1996;32(3):749–52.

[217] Yamazaki K. Generalization of 3D eddy current analysis for moving conductors due to coordinate systems and gauge conditions. IEEE Transactions on Magnetics. 1997;33(2):1259–62.

[218] Muramatsu K, Takahashi N, Hashio T, *et al.* 3-D eddy current analysis in moving conductor of permanent magnet type of retarder using moving coordinate system. IEEE Transactions on Energy Conversion. 1999;14(4): 1312–17.

[219] Yamazaki K. 3D eddy current formulation for moving conductors with variable velocity of coordinate system using edge finite elements. IEEE Transactions on Magnetics. 1999;35(3):1594–97.

[220] Lai H, Rodger D, Leonard P. A finite element method for problems with moving parts. In: Proceedings of 8th International Conference on Computation in Electromagnetics (CEM91); 1991. p. 211–13.

[221] Marechal Y, Meunier G, Coulomb JL, *et al.* A general purpose tool for restoring inter-element continuity. IEEE Transactions on Magnetics. 1992;28(2):1728–31.

[222] Leonard PJ, Lai HC, Hainsworth G, *et al.* Analysis of the performance of tubular pulsed coil induction launchers. IEEE Transactions on Magnetics. 1993;29(1):686–90.

[223] Golovanov C, Coulomb J, Marechal Y, *et al.* 3D mesh connection techniques applied to movement simulation. IEEE Transactions on Magnetics. 1998;34(5):3359–62.

[224] Antunes OJ, Bastos JPa, Sadowski N, *et al.* Torque calculation with conforming and non-conforming movement interface. IEEE Transactions on Magnetics. 2006;42(4):983–86.

[225] Antunes OJ, Bastos JPA, Sadowski N, *et al.* Comparison between non-conforming movement methods. IEEE Transactions on Magnetics. 2006 Apr;42(4):599–602.

[226] Rapetti F, Bouillault E, Santandrea L, *et al.* Calculation of eddy currents with edge elements on non-matching grids in moving structures. IEEE Transactions on Magnetics. 2000;36(4):1351–55.

[227] Belgacem F, Buffa A, Maday Y. The mortar finite element method for 3D maxwell equations: first results. SIAM Journal on Numerical Analysis. 2001;39(3):880–901.

[228] Rapetti F, Maday Y, Bouillault F, *et al.* Eddy-current calculations in three-dimensional moving structures. IEEE Transactions on Magnetics. 2002;38(2):613–16.

[229] Antunes OJ, Bastos JPa, Sadowski N. Using high-order finite elements in problems with movement. IEEE Transactions on Magnetics. 2004;40(2): 529–32.

[230] Shi X, Menach YL, Ducreux JP, *et al.* Comparison of slip surface and moving band techniques for modelling movement in 3D with FEM. COMPEL. 2006;25(1):17–30.

[231] Tsukerman Ia. Overlapping finite elements for problems with movement. IEEE Transactions on Magnetics. 1992;28(5):2247–49.

[232] Lai HC, Coles PC, Rodger D, *et al.* Transient analysis of an electromagnetic actuator using an overlapping finite element scheme. IEEE Transactions on Magnetics. 2000;36(4):1462–67.

[233] Lai HC, Rodger D, Coles PC. A 3-D overlapping finite-element scheme for modeling movement. IEEE Transactions on Magnetics. 2004;40(2): 533–36.

[234] Watanabe T, Todaka T, Enokizono M. Analysis of a new induction heating device by using permanent magnets. IEEE Transactions on Magnetics. 2005;41(5):1884–87.

[235] Yu Z, Jiangjun R, Yan G, *et al.* Application of a composite grid method in the analysis of 3-D eddy current field involving movement. IEEE Transactions on Magnetics. 2008;44(6):1298–301.

[236] Davidson PA. An introduction to magnetohydrodynamics. In: Cambridge Texts in Applied Mathematics. Cambridge: Cambridge University Press; 2001.

[237] Hoburg JF. Principles of quasistatic magnetic shielding with cylindrical and spherical shields. IEEE Transactions on Electromagnetic Compatibility. 1995;37(4):574–79.

[238] Jankoski R, Porzig K, Brauer H. Extended version of the weak reaction approach in Lorentz force eddy current testing. In: Proceedings of the 9th International Workshop of Electromagnetic Compatibility (CEM). 2014; p. 1–4.

[239] Davis PJ, Rabinowitz P. Methods of Numerical Integration. 2nd ed. Boston, MA: Academic Press; 1984.

[240] Borwein JM, Borwein PB. Pi and the AGM: A Study in Analytic Number Theory and Computational Complexity. New York, NY: John Wiley & Sons; 1987.

[241] Zec M, Uhlig RP, Ziolkowski M, *et al.* Fast technique for Lorentz force calculations in nondestructive testing applications. IEEE Transactions on Magnetics. 2014;50(2):Art. ID 70031044.

[242] Chari MVK, Konrad A, Palmo MA, *et al.* Simulation analysis of magnetic sensor for nondestructive testing by boundary element method. IEEE Transactions on Magnetics. 1990;26(2):877–80.

[243] Zec M, Uhlig RP, Ziolkowski M, *et al.*, inventors; Institut Dr. Foerster Gmbh & Co. Kg, assignee. Differentieller Sensor, Prüfsystem und Verfahren zur Detektion von Anomalien in elektrisch leitfähigen Materialien. EP2893336 A1; 2015.

[244] Haus H, Melcher J. Electromagnetic Fields and Energy. Englewood Cliffs, NJ: Prentice-Hall; 1989.

[245] Weise K, Schmidt R, Carlstedt M, *et al.* Optimal magnet design for Lorentz force eddy current testing. IEEE Transactions on Magnetics. 2015;51(9):Art. ID 6201415.

[246] Mallinson JC. One-sided fluxes – a magnetic curiosity? IEEE Transactions on Magnetics. 1973;9(4):678–82.

[247] Halbach K. Design of permanent multipole magnets with oriented rare-earth cobalt material. Nuclear Instrumentation & Methods. 1980;169(1): 1–10.

[248] Zhu ZQ, Howe D. Halbach permanent magnet machines and applications: a review. IEE Proceedings – Electric Power Applications. 2001;148(4): 299–308.

[249] Lemarquand G. Ironless loudspeakers. IEEE Transactions on Magnetics. 2007;43(8):3371–74.

[250] Ravaud R, Lemarquand G, Lemarquand V, *et al.* Discussion about the analytical calculation of the magnetic field created by permanent magnets. Progress In Electromagnetics Research B. 2009;11(1):281–97.

[251] Vučković AN, Raičević NB, Ilić SS, *et al.* Axial force calculation of passive magnetic bearing. Serbian Journal of Electrical Engineering. 2014;11(4): 649–60.

[252] Dogan N, Topkaya R, Subaşi H, *et al.* Development of Halbach magnet for portable NMR device. Journal of Physics: Conference Series. 2009;153: Art. ID 012047.

[253] Werner M, Halbedel B. Optimization of NdFeB magnet arrays for improvement of Lorentz force velocimetry. IEEE Transactions on Magnetics. 2012;48(11):2925–28.

[254] Quanling Peng, McMurry SM, Coey JMD. Cylindrical permanent-magnet structures using images in an iron shield. IEEE Transactions on Magnetics. 2003;39(4):1983–89.

[255] Kumada M, Fujisawa T, Hirao Y, *et al.* Development of 4 Tesla permanent magnet. Proceedings of the Particle Accelerator Conference. 2001; p. 3221–23.

[256] Weichmagnetische Kobalt-Eisen-Legierungen: Vacoflux 48, Vacoflux 50 Vacodur 49, Vacodur 50, Vacodur S Plus, Vacoflux 17, Vacoflux 18 HR, Vacoflux 9 CR. Hanau: VACUUMSCHMELZE GmbH & Co. KG, Hanau. Available from: http://www.vacuumschmelze.de/fileadmin/docroot/medialib/documents/broschueren/htbrosch/Pht-004_d.pdf.

[257] VACUUMSCHMELZE GmbH & Co KG, Hanau. Selten-Erd-Dauermagnete: Vacodym, Vacomax. Hanau. Available from: http://www.vacuumschmelze.de/fileadmin/Medienbiliothek_2010/Downloads/DM/Vdym_Vmax_de.pdf.

[258] VACUUMSCHMELZE GmbH & Co KG, Hanau. Hanau. Available from: http://www.vacuumschmelze.de/.

[259] Mihara T, Iwashita Y, Kumada M, *et al.* Super strong permanent magnet quadrupole for a linear collider. IEEE Transactions on Applied Superconductivity. 2004;14(2):469–72.

[260] Iwashita Y, Mihara T, Antokhin E, *et al.* Permanent magnet quadrupole for final focus for linear collider. Proceedings of the Particle Accelerator Conference. 2003; p. 2198–200.

[261] Gill PE, Murray W, Wright MH. Practical Optimization. London: Academic Press; 1981.

[262] Davidon WC. Variable Metric Method for Minimization. SIAM Journal on Optimization. 1991;1(1):1–17.

[263] Biggs MC. Constrained minimization using recursive quadratic programming. In: Proceedings of a Workshop Towards Global Optimisation. 1975; p. 341–49.

[264] Han SP. A globally convergent method for nonlinear programming. Journal of Optimization Theory and Applications. 1977;22(3):297–309.

[265] Powell MJD. A fast algorithm for nonlinearly constrained optimization calculations. In: Proceedings of the Biennial Conference on Numerical Analysis. 1978; p. 144–57.

[266] Broyden CG. The convergence of a class of double-rank minimization algorithms 1. General considerations. Journal of the Institute of Mathematics and Its Applications. 1970;6(3):76–90.

[267] Broyden CG. The convergence of a class of double-rank minimization algorithms 2. The New Algorithm. Journal of the Institute of Mathematics and Its Applications. 1970;6(3):222–31.

[268] Fletcher R. A new approach to variable metric algorithms. Computer Journal. 1970;13(3):317–22.

[269] Goldfarb D. A family of variable-metric methods derived by variational means. Mathematics of Computation. 1970;24(109):23–26.

[270] Shanno DF. Conditioning of quasi-Newton methods for function minimization. Mathematics of Computation. 1970;24(111):647–56.

[271] Karush W. Minima of Functions of Several Variables with Inequalities as Side Conditions [Dissertation]. Department of Mathematics, University of Chicago. Chicago, IL, USA; 1939.

[272] Alt W. Nichtlineare Optimierung: Eine Einfuehrung in Theorie, Verfahren und Anwendungen. 2nd ed. Wiesbaden: Vieweg und Teubner; 2011.

[273] Nocedal J, Wright SJ. Numerical Optimization. 2nd ed. New York, NY: Springer; 2006.

[274] Han SP. Superlinearly convergent variable metric algorithms for general nonlinear programming problems. Mathematical Programming. 1976;11(1): 263–82.

[275] Powell MJD. Some global convergence properties of a variable metric algorithm for minimization without exact line searches. SIAM-AMS-Proceedings: Nonlinear Programming. 1976;9(1):53–72.

[276] Powell MJD. The convergence of variable metric methods for nonlinearly constrained optimization calculations. In: Nonlinear Programming 3. Academic Press; 1978. p. 27–63.

[277] Boggs PT, Tolle JW. Sequential quadratic programming. Acta Numerica. 1995;4(1):1–52.

[278] Gill PE, Wong E. Sequential Quadratic Programming Methods: Technical Report NA-10-03. University of California, San Diego, La Jolla; 2010.

[279] Schittkowski K. The Nonlinear Programming Method of Wilson, Han, and Powell with an Augmented Lagrangian Type Line Search Function Part 1: Convergence Analysis. Numerische Mathematik. 1981;38(1):83–114.

[280] Schittkowski K. The nonlinear programming method of Wilson, Han, and Powel with an augmented Lagrangian type line search function Part 2: an efficient implementation with linear least squares subproblems. Numerische Mathematik. 1981;38(1):115–27.

[281] Schittkowski K. On the convergence of a sequential quadratic programming method with an augmented Lagrangian line search function. Mathematische Operationsforschung und Statistik Series Optimization: A Journal of Mathematical Programming and Operations Research. 1983;14(2):197–216.

[282] The MathWorks, Inc , Natick, Massachusetts, USA. MATLAB. Available from: https://www.mathworks.com.

[283] Gill PE, Murray W, Saunders MA, *et al.* User's Guide for NPSOL 5.0: A Fortran Package for Nonlinear Programming. Technical Report SOL 86-6. 2001; p. 1–43.

[284] Gill PE, Murray W, Saunders MA. SNOPT: An SQP algorithm for large-scale constrained optimization. SIAM Review. 2005;47(1):99–131.

[285] Gill PE, Murray W, Saunders MA. User's Guide for SNOPT Version 7: Software for Large-Scale Nonlinear Programming; 2008. Available from: http://web.stanford.edu/group/SOL/guides/sndoc7.pdf.

[286] Schittkowski K. NLPQLP: A Fortran Implementation of a Sequential Quadratic Programming Algorithm with Distributed and Non-Monotone Line Search - User's Guide, Version 3.1; 2010. Available from: http://www.ai7.uni-bayreuth.de/NLPQLP.pdf.

[287] Weise K, Carlstedt M, Schmidt R. Defect Depth Study Considering a Long Slit and a Cylindrical Magnet. TU Ilmenau Technical Report TET-2015-05. 2015;(5).

[288] Gautschi G. Piezoelectric Sensorics: Force, Strain, Pressure, Acceleration and Acoustic Emission Sensors, Materials and Amplifiers. Berlin: Springer; 2002.

[289] Schmidt R, Weise K, Ziolkowski M, *et al.* Optimized Lorentz force eddy current testing system for small metal injection molding specimens. In: ENDE'2015, 20th International Workshop on Electromagnetic Nondestructive Evaluation, Sendai, Japan; 2015. p. 35–36.

[290] International vocabulary of metrology – Basic and general concepts and associated terms (VIM). Joint Committee for Guides in Metrology; 2007. Available from: http://www.iso.org/iso/catalogue_detail.htm?csnumber=45324.

[291] Inoue H. Force feedback in precise assembly tasks. Bulletin of the Electrotechnical Laboratory. 1974;38(12):775–89.

[292] Rahneberg I, Dontsov D, Pöschel W, *et al.* Charakterisierung von Linearführungen mit Mehrstrahlinterferometern. In: Beiträge der 17. ITG/

GMA-Fachtagung vom 3. bis 4. Juni 2014; 2014. Available from: https://www.vde-verlag.de/proceedings-de/453622027.html.

[293] Weise K, Carlstedt M, Ziolkowski M, *et al.* Uncertainty analysis in Lorentz force eddy current testing. IEEE Transactions on Magnetics. 2016;52(3):6200104.

[294] Data Sheet – NLS4 Series. Newmark Systems Inc. Available from: http://www.newmarksystems.com/linear-positioners/nls4-series.

[295] Data Sheet – NSC-G Series. Newmark Systems Inc. Available from: http://www.newmarksystems.com/motion-controllers/nsc-g-series.

[296] Schleichert J, Carlstedt M, Marangoni R, *et al.* Dynamic characterization of a multi-component force transducer using a Lorentz force load changer. In: 58th IWK, Ilmenau Scientific Colloquium: Proceedings; 2014. p. 1–11. Available from: https://www.db-thueringen.de/servlets/MCRFileNodeServlet/dbt_derivate_00030737/ilm1-2014iwk-181.pdf.

[297] Data Sheet – GSV-1A4. ME-Meßsysteme GmbH. Available from: http://www.me-systeme.de/docs/de/datasheets/gsv-1a4.pdf.

[298] Data Sheet – GSV-1M. ME-Meßsysteme GmbH. Available from: http://www.me-systeme.de/de/datasheets/gsv1l-gsv1m.pdf.

[299] Data Sheet – ASC 5511LN-002. ASC GmbH. Available from: http://www.asc-sensors.de/uploads/tx_ascproducts/ASC_5511LN_5515LN_Dez15.pdf.

[300] Data Sheet – AD624. Analog Devices Inc. Available from: http://www.analog.com/media/en/technical-documentation/data-sheets/AD624.pdf.

[301] Data Sheet – T1000-50A. Renishaw plc. Available from: http://www.renishaw.com/media/pdf/en/ffda12c3199d44ea8d74f7c625fee290.pdf.

[302] Data Sheet – RGSZ20 scale. Renishaw plc. Available from: http://www.renishaw.com/media/pdf/en/ccc958b233d1407ca03dab2bec32139a.pdf.

[303] Data Sheet – T0100 A 40 A. Renishaw plc. Available from: http://www.renishaw.com/media/pdf/en/dcc4b48f350d487981abc7d1ce6b4806.pdf.

[304] Data Sheet – NI 4472 Series. National Instruments Corporation. Available from: http://www.ni.com/pdf/products/us/3sv414-416.pdf.

[305] Data Sheet – NI PXI-1036. National Instruments Corporation. Available from: http://www.ni.com/datasheet/pdf/en/ds-447.

[306] Dougherty ER. Random Processes for Image and Signal Processing. SPIE/IEEE Series on Imaging Science & Engineering. Bellingham, WA: SPIE Optical Engineering Press and Institute of Electrical and Electronics Engineers; 1999.

[307] Chatfield C. The Analysis of Time Series: An Introduction. 4th ed. London and New York, NY: Chapman and Hall; 1989.

[308] Oppenheim AV, Schafer RW, Buck JR. Discrete-Time Signal Processing. 2nd ed. Prentice Hall Signal Processing Series. Upper Saddle River, NJ: Prentice Hall; 1999.

[309] Weise K. Advanced Modeling in Lorentz Force Eddy Current Testing [Dissertation]. Technische Universität Ilmenau. Ilmenau, Germany; 2016.

[310] Evaluation of measurement data – Guide to the Expression of Uncertainty in Measurement (GUM): GUM 1995 with minor corrections. Genève, Switzerland: International Organization for Standardization; 2008.

[311] Feller W. An Introduction to Probability Theory and Its Applications. 3rd ed. Wiley Series in Probability and Mathematical Statistics. New York, NY: John Wiley & Sons; 1968.

[312] Friedman C. The frequency interpretation in probability. Advances in Applied Mathematics. 1999;23(3):234–54.

[313] Kacker R, Jones A. On use of Bayesian statistics to make the guide to the expression of uncertainty in measurement consistent. Metrologia. 2003; 40(5):235–48.

[314] Data Sheet – Elotest N300. Rohmann GmbH; 2006. Available from: http://www.rohmann.de/datasheets/instruments/N300_eng_04.pdf.

[315] Moreau O, Beddek K, Clenet S, *et al.* Stochastic nondestructive testing simulation: sensitivity analysis applied to material properties in clogging of nuclear powerplant steam generators. IEEE Transactions on Magnetics. 2013;49(5):1873–76.

[316] Beddek K, Clenet S, Moreau O, *et al.* Adaptive method for non-intrusive spectral projection—application on a stochastic eddy current NDT problem. IEEE Transactions on Magnetics. 2012;48(2):759–62.

[317] Fishman GS. Monte Carlo: Concepts, Algorithms, and Applications. Springer Series in Operations Research. New York, NY: Springer; 1996.

[318] Mckay MD, Beckman RJ, Conover WJ. A comparison of three methods for selecting values of input variables in the analysis of output from a computer code. Technometrics. 2000;42(1):55–61.

[319] Morokoff WJ, Caflisch RE. Quasi-Monte Carlo integration. Journal of Computational Physics. 1995;122(2):218–30.

[320] Plaskota L, Woźniakowski H. Monte Carlo and Quasi-Monte Carlo methods 2010. vol. 23. Berlin, Heidelberg: Springer; 2012.

[321] Xiu D. Fast numerical methods for stochastic computations: a review. Communications in Computational Physics. 2009;5(2–4):242–72.

[322] Ghanem R, Higdon D, Owhadi H, editors. Handbook of Uncertainty Quantification. Cham, Switzerland: Springer International Publishing; 2016.

[323] Le Maitre, OP, Knio OM. Spectral Methods for Uncertainty Quantification. Dordrecht, Netherland: Springer; 2010.

[324] Knio OM, Le Maitre, OP. Uncertainty propagation in CFD using polynomial chaos decomposition. Fluid Dynamics Research. 2006;38(9):616–40.

[325] Xiu D, Karniadakis GE. Modeling uncertainty in flow simulations via generalized polynomial chaos. Journal of Computational Physics. 2003;187(1): 137–67.

[326] Hosder S, Perez R, Walters R. A non-intrusive polynomial chaos method for uncertainty propagation in CFD simulations. In: 44th AIAA Aerospace Sciences Meeting and Exhibit, Aerospace Sciences Meetings; 2006. p. 1–19. Available from: https://arc.aiaa.org/doi/pdf/10.2514/6.2006-891.

[327] Wan X, Xiu D, Karniadakis GE. Modeling uncertainty in three-dimensional heat transfer problems. In: Advanced Computational Methods in Heat Transfer (VII). vol. 46 of WIT Transactions on Engineering Sciences. WITpress; 2004. p. 1–10.

[328] Xiu D, Karniadakis GE. A new stochastic approach to transient heat conduction modeling with uncertainty. International Journal of Heat and Mass Transfer. 2003;46(24):4681–93.

[329] Sandu A, Sandu C, Ahmadian M. Modeling multibody systems with uncertainties. Part I: theoretical and computational aspects. Multibody System Dynamics. 2006;15(4):369–91.

[330] Sandu C, Sandu A, Ahmadian M. Modeling multibody systems with uncertainties. Part II: numerical applications. Multibody System Dynamics. 2006;15(3):241–62.

[331] Zein S. A polynomial chaos expansion trust region method for robust optimization. Communications in Computational Physics. 2013;14(2): 412–24.

[332] Codecasa L, Di Rienzo L, Weise K, *et al.* Fast MOR-based Approach to uncertainty quantification in transcranial magnetic stimulation. IEEE Transactions on Magnetics. 2016;52(3):Art. ID 7200904.

[333] Askey R, Wilson JW. Some basic hypergeometric orthogonal polynomials that generalize Jacobi-polynomials. Memoirs of the American Mathematical Society. 1985;54(319):1–55.

[334] Eldred M, Burkardt J. Comparison of non-intrusive polynomial chaos and stochastic collocation methods for uncertainty quantification. In: 47th AIAA Aerospace Sciences Meeting including The New Horizons Forum and Aerospace Exposition, Aerospace Sciences Meetings; 2009. p. 1–20. Available from: https://arc.aiaa.org/doi/pdf/10.2514/6.2009-976.

[335] Babuška I, Tempone R, Zouraris GE. Galerkin finite element approximations of stochastic elliptic partial differential equations. SIAM Journal on Numerical Analysis. 2004;42(2):800–25.

[336] Sobol IM. Global sensitivity indices for nonlinear mathematical models and their Monte Carlo estimates. Mathematics and Computers in Simulation. 2001;55(1–3):271–80.

[337] Sudret B. Global sensitivity analysis using polynomial chaos expansions. Reliability Engineering & System Safety. 2008;93(7):964–79.

[338] Delignette-Muller ML, Dutang C. fitdistrplus: An R package for fitting distributions. Journal of Statistical Software. 2015;64(4):1–34.

[339] Achenbach JD. Quantitative nondestructive evaluation. International Journal of Solids and Structures. 2000;37(1–2):13–27.

[340] Jiles DC. Review of magnetic methods for nondestructive evaluation. NDT International. 1988;21(5):311–19.

[341] Ma X. Electromagnetic NDT and condition monitoring – a personal view. In: Proceedings of the 17th International Conference on Automation and Computing. 2011; p. 266–71.

[342] Grimberg R. Electromagnetic nondestructive evaluation: present and future. Strojniški vestnik – Journal of Mechanical Engineering. 2011;2011(03): 204–17.

[343] Lim MK, Cao H. Combining multiple NDT methods to improve testing effectiveness. Construction and Building Materials. 2013;38:1310–15.

[344] Heckel T, Thomas HM, Kreutzbruck M, *et al.* High speed non-destructive rail testing with advanced ultrasound and eddy-current testing techniques. In: NDE 2009 (Proceedings); 2009. p. 261–15.

[345] Petković B. Assessment of Linear Inverse Problems in Magnetocardiography and Lorentz Force Eddy Current Testing [Dissertation]. Technische Universität Ilmenau. Ilmenau, Germany; 2013.

[346] Fuchs M, Wagner M, Köhler T, *et al.* Linear and nonlinear current density reconstructions. Journal of Clinical Neurophysiology. 1999;16(3): 267–95.

[347] Eichardt R, Haueisen J. Influence of sensor variations on the condition of the magnetostatic linear inverse problem. IEEE Transactions on Magnetics. 2010;46(8):3449–52.

[348] Natterer F. The mathematics of computerized tomography. Society for Industrial and Applied Mathematics; 2001.

[349] Courant R, Hilbert D. Methods of Mathematical Physics. Volume 1, Weinheim, Germany: Wiley-VCH; 2009.

[350] Colton D, Kress R. Inverse Acoustic and Electromagnetic Scattering Theory. vol. 93 of Applied Mathematical Sciences. 3rd ed. New York, NY: Springer; 2013.

[351] Kirsch A. An Introduction to the Mathematical Theory of Inverse Problems. vol. 120. New York, NY: Springer; 2011.

[352] Hadamard J. Sur les problèmes aux dérivés partielles et leur signification physique. Princeton University Bulletin. 1902;13(28):49–52.

[353] Hadamard J. Lectures on Cauchy's Problem in Linear Partial Differential Equations. Mineola, NY: Dover Publications, 2003

[354] Bertero M, Poggio TA, Torre V. Ill-posed problems in early vision. Proceedings of the IEEE. 1988;76(8):869–89.

[355] Hämäläinen MS, Ilmoniemi RJ. Interpreting magnetic fields of the brain: minimum norm estimates. Medical & Biological Engineering & Computing. 1994;32(1):35–42.

[356] Leder U, Haueisen J, Huck M, *et al.* Non-invasive imaging of arrhythmogenic left-ventricular myocardium after infarction. The Lancet. 1998;352(9143): 1825.

[357] Wagner M, Wischmann HA, Fuchs M, *et al.* Current density reconstructions using the L1 norm. In: Aine CJ, Stroink G, Wood CC, *et al.*, editors. Proceedings of the 10th International Conference on Biomagnetics. New York, NY: Springer New York; 1996. p. 393–96.

[358] Baumgarten D, Liehr M, Wiekhorst F, *et al.* Magnetic nanoparticle imaging by means of minimum norm estimates from remanence measurements. Medical & Biological Engineering & Computing. 2008;46(12):1177–85.

[359] Eichardt R, Baumgarten D, Di Rienzo L, *et al.* Localisation of buried ferromagnetic objects based on minimum–norm–estimations. COMPEL. 2009;28(5):1327–37.
[360] Haueisen J, Unger R, Beuker T, *et al.* Evaluation of inverse algorithms in the analysis of magnetic flux leakage data. IEEE Transactions on Magnetics. 2002;38(3):1481–88.
[361] Ma L, Soleimani M. Hidden defect identification in carbon fibre reinforced polymer plates using magnetic induction tomography. Measurement Science and Technology. 2014;25(5):Art. ID 055404.
[362] Pirani A, Ricci M. 3D reconstruction of flaws in metallic materials by eddy current inspections. Studies in Applied Electromagnetics and Mechanics. 2008;31(1):109–16.
[363] Nelder JA, Mead R. A simplex method for function minimization. The Computer Journal. 1965;7(4):308–13.
[364] Press WH. Numerical Recipes in C: The Art of Scientific Computing. 2nd ed. Cambridge: Cambridge University Press; 2002.
[365] Levenberg K. A method for the solution of certain non-linear problems in least squares. Quarterly of Applied Mathematics. 1944;2(2):164–68.
[366] Marquardt DW. An algorithm for least-squares estimation of nonlinear parameters. Journal of the Society for Industrial and Applied Mathematics. 1963;11(2):431–41.
[367] Kirkpatrick S, Gelatt CD, Vecchi MP. Optimization by simulated annealing. Science (New York, NY). 1983;220(4598):671–80.
[368] Kennedy J, Eberhart R. Particle swarm optimization. In: Proceedings of the IEEE International Conference on Neural Networks; 1995. p. 1942–48.
[369] Dorigo M, Stützle T. Ant Colony Optimization. Cambridge, MA: MIT Press; 2004.
[370] Rechenberg I. Optimierung technischer Systeme nach Prinzipien der biologischen Evolution. Stuttgart: Friedrich Frommann Verlag; 1973.
[371] Schwefel HP. Evolution and Optimum Seeking. Sixth-Generation Computer Technology Series. New York, NY: John Wiley & Sons; 1995.
[372] Goldberg DE. Genetic Algorithms in Search, Optimization, and Machine Learning. 30th ed. Boston, MA: Addison-Wesley; 2012.
[373] Holland JH. Outline for a logical theory of adaptive systems. Journal of the ACM. 1962;9(3):297–314.
[374] Zitzler E, Deb K, Thiele L. Comparison of multiobjective evolutionary algorithms: empirical results. Evolutionary Computation. 2000;8(2):173–95.
[375] Cacciola M, Calcagno S, Morabito FC, *et al.* Swarm optimization for imaging of corrosion by impedance measurements in Eddy current test. IEEE Transactions on Magnetics. 2007;43(4):1853–56.
[376] Duca A, Rebican M, Janousek L, *et al.* PSO based techniques for NDT-ECT inverse problems. Unpublished. Available from: https://doi.org/10.13140/2.1.3886.7203.
[377] Kojima F, Kubota N, Kobayashi F, *et al.* Shape recovery of natural crack using evolutionary programming related to eddy current testing. International

Journal of Applied Electromagnetics and Mechanics. 2001;15(1–4): 243–47.
[378] Li Y, Udpa L, Udpa SS. Three-dimensional defect reconstruction from eddy-current NDE signals using a genetic local search algorithm. IEEE Transactions on Magnetics. 2004;40(2):410–17.
[379] Massa A, Pastorino M, Randazzo A. Reconstruction of two-dimensional buried objects by a differential evolution method. Inverse Problems. 2004;20(6):135–50.
[380] Rocca P, Oliveri G, Massa A. Differential evolution as applied to electromagnetics. IEEE Antennas and Propagation Magazine. 2011;53(1):38–49.
[381] Qing A, Xu X, Gan YB. Anisotropy of composite materials with inclusion with orientation preference. IEEE Transactions on Antennas and Propagation. 2005;53(2):737–44.
[382] Eichardt R, Baumgarten D, Petković B, *et al.* Adapting source grid parameters to improve the condition of the magnetostatic linear inverse problem of estimating nanoparticle distributions. Medical & Biological Engineering & Computing. 2012;50(10):1081–89.
[383] Mosher JC, Lewis PS, Leahy RM. Multiple dipole modeling and localization from spatio-temporal MEG data. IEEE Transactions on Biomedical Engineering. 1992;39(6):541–57.
[384] Brandstein M, Ward D, editors. Microphone arrays: signal processing, techniques and applications. Engineering online library. Berlin: Springer; 2001.
[385] Yang S, Ho SL, Yao Y, *et al.* Studies on numerical methodologies for inverse problems and optimizations in China. COMPEL. 2013;33(1–2):56–64.
[386] Hansen PC. The truncated SVD as a method for regularization. BIT. 1987;27(4):534–53.
[387] Gebhardt S, Scheinert G. Position determination based on electrical capacitance tomography. COMPEL. 2011;30(4):1224–36.
[388] Gebhardt S. Entwurf und Realisierung eines kapazitiven Tomographie-Systems nach dem Dreielektroden-Messprinzip [Dissertation]. Technische Universität Ilmenau. Ilmenau, Germany; 2014.
[389] Phillips DL. A technique for the numerical solution of certain integral equations of the first kind. Journal of the ACM. 1962;9(1):84–97.
[390] Tikhonov AN. Solution of incorrectly formulated problems and the regularization method. Soviet Mathematics Doklady. 1963;4(4):1035–38.
[391] Hansen PC, O'Leary DP. The use of the L-curve in the regularization of discrete Ill-posed problems. SIAM Journal on Scientific Computing. 1993;14(6):1487–503.
[392] Blitz J, Simpson G. Ultrasonic methods of non-destructive testing. 1st ed. Non-destructive evaluation series. London: Chapman & Hall; 1996.
[393] Mouritz AP, Townsend C, Shah Khan MZ. Non-destructive detection of fatigue damage in thick composites by pulse-echo ultrasonics. Composites Science and Technology. 2000;60(1):23–32.

[394] Alaknanda, Anand RS, Kumar P. Flaw detection in radiographic weld images using morphological approach. NDT & E International. 2006;39(1):29–33.

[395] D'Orazio T, Guaragnella C, Leo M, *et al.* Defect detection in aircraft composites by using a neural approach in the analysis of thermographic images. NDT & E International. 2005;38(8):665–73.

[396] Meola C, Carlomagno GM, Squillace A, *et al.* Non-destructive evaluation of aerospace materials with lock-in thermography. Engineering Failure Analysis. 2006;13(3):380–88.

[397] Mandal K, Atherton DL. A study of magnetic flux-leakage signals. Journal of Physics D: Applied Physics. 1998;31(22):3211–17.

[398] Heuer H, Schulze M, Meyendorf N. High resolution inspection of carbon fiber materials by eddy current techniques. In: Proceedings of the 2nd International Symposium on NDT in Aerospace. 2010;2(3):1–13.

[399] Schulze MH, Heuer H, Küttner M, *et al.* High-resolution eddy current sensor system for quality assessment of carbon fiber materials. Microsystem Technologies. 2010;16(5):791–97.

[400] Sinke J. Some inspection methods for quality control and in-service inspection of GLARE. Applied Composite Materials. 2003;10(4–5):277–91.

[401] Lenz E. Ueber die Bestimmung der Richtung der durch elektrodynamische Verteilung erregten galvanischen Ströme. Annalen der Physik. 1834;107(31):483–94.

[402] Lorentz HA. La Théorie Électromagnétique de Maxwell et son Application aux Corps Mouvants. Archives Néerlandaises des Sciences Exactes et Naturelles. 1892;25:363–552.

[403] Uhlig RP, Zec M, Brauer H, *et al.* Lorentz force eddy current testing: a prototype model. Journal of Nondestructive Evaluation. 2012; p. 1–40.

[404] Hahn KD. Eddy current damping of a magnet moving through a pipe. American Journal of Physics. 1998;66(12):1066.

[405] Weise K, Carlstedt M, Ziolkowski M, *et al.* Lorentz force on permanent magnet rings by moving electrical conductors. IEEE Transactions on Magnetics. 2015;51(12):6202211.

[406] Sun YS, Lord W, Katragadda G, *et al.* Motion induced remote field eddy current effect in a magnetostatic non-destructive testing tool: a finite element prediction. IEEE Transactions on Magnetics. 1994;30(5):3304–07.

[407] Park GS, Park SH. Analysis of the velocity-induced eddy current in MFL type NDT. IEEE Transactions on Magnetics. 2004;40(2):663–66.

[408] Nestleroth JB, Davis RJ. The effects of magnetizer velocity on magnetic flux leakage signals. In: Thompson DO, Chimenti DE, editors. Review of Progress in Quantitative Nondestructive Evaluation. Boston, MA: Springer; 1993. p. 1891–98.

[409] Zec M, Uhlig RP, Ziolkowski M, *et al.*, inventors; Technische Universität Ilmenau, assignee. Verfahren und Anordnung zur Bestimmung der elektrischen Leitfähigkeit eines Werkstoffes. DE 10 2011 056 650 (B4); 2014.

[410] van der Pauw LJ. A method of measuring specific resistivity and hall effect of discs of arbitrary shape. Philips Research Reports. 1958;13(1):1–9.

[411] Delaney JA, Pippard AB. Electrodeless methods for conductivity measurement in metals. Reports on Progress in Physics. 1972;35(2):677–715.

[412] Suvarna RP, Rao KR, Subbarangaiah K. A simple technique for A.C. conductivity measurements. Bulletin of Materials Science. 2002;25(7):647–51.

[413] Rietveld G, Koijmans CV, Henderson LCA, *et al.* DC conductivity measurements in the Van Der Pauw geometry. IEEE Transactions on Instrumentation and Measurement. 2003;52:449–53.

[414] Mallick PK. Fiber-reinforced composites: materials, manufacturing, and design. 3rd ed. Boca Raton, FL: CRC Press; 2008.

[415] Storn R, Price K. Differential Evolution – a simple and efficient heuristic for global optimization over continuous spaces. Journal of Global Optimization. 1997;11(4):341–59.

[416] Das S, Suganthan P. Differential evolution: a survey of the state of the art. IEEE Xplore: IEEE Transactions on Evolutionary Computation. 2011;15(1):4–31.

[417] Gorges S, Brauer H, Ziolkowski M, *et al.* Motion-induced eddy current testing of composite materials. In: Proceedings of the 19th World Conference on Non-Destructive Testing (WCNDT). 2016; p. Fr.1.F.

[418] Brauer H, Gorges S, Ziolkowski M. Bewegungsinduzierte Wirbelstromprüfung von Verbundmaterialien. In: Proceedings of the DGZfP-Jahrestagung. 2017; p. 1–8.

[419] Thomas Beumler. Flying GLARE: A contribution to aircraft certification issues on strength properties in non-damaged and fatigue damaged GLARE structures [Dissertation]. Technische Universiteit Delft. Delft; 2004.

[420] Wu G, Yang JM. The mechanical behavior of GLARE laminates for aircraft structures. JOM. 2005;57(1):72–79.

[421] Haase W, Maurer A. Latest developments on industrial ultrasonic testing of aircraft components. In: Proceedings of the World Conference on Non-Destructive Testing. 2004; p. 225–36.

[422] Datenblatt: 3D-Magnetfeldsonden AS-N3DM und AS-L3DM. Projekt Elektronik Mess- und Regelungstechnik GmbH. Available from: http://www.projekt-elektronik.de/wp-content/uploads/AS-N3DM-L3DM-3D-AS-Aktivsonde.pdf.

[423] TETRA Gesellschaft für Sensorik, Robotik und Automation mbH. Betriebsanleitung: BASALT®-C MMP-15.

[424] Karbhari VM, editor. Non-Destructive Evaluation of Polymer Matrix Composites: Techniques and Applications. vol. 43 of Woodhead Publishing Series in Composites Science and Engineering. Cambridge and Philadelphia, PA: Woodhead Pub; 2013.

[425] EADS. Taking the lead: A350XWB presentation; 2006.

[426] Hale J. Boeing 787 from the Ground Up. AERO Quarterly. 2006;6(4):17–23.

[427] Gardiner G. Thermoplastic composites gain leading edge on the A380. High-Performance Composites. 2006;14(2):50–55.

[428] Pora J. Composite Materials in the Airbus A380 – From History to Future. Airbus; 2018. Available from: http://www.iccm-central.org/Proceedings/ICCM13proceedings/SITE/PAPERS/paper-1695.pdf.

[429] Solares ERG. A Novel Structural Health Monitoring Method for Full-Scale CFRP Structures [Ph.D. Dissertation]. École Polytechnique Fédérale de Lausanne. Lausanne, Switzerland; 2015.

[430] Li X. Eddy Current Techniques for Non-destructiveTesting of Carbon Fibre Reinforced Plastic (CFRP) [Ph.D. Dissertation]. University of Manchester. Manchester, United Kingdom; 2012.

[431] Cheng J, Ji H, Qiu J, *et al.* Novel electromagnetic modeling approach of carbon fiber-reinforced polymer laminate for calculation of eddy currents and eddy current testing signals. Journal of Composite Materials. 2015;49(5): 617–31.

[432] Mook G. Zerstörungsfreie Charakterisierung von carbonfaserverstärkten Kompositen mit Hilfe des Wirbelstromverfahrens. ZfP in Anwendung, Entwicklung und Forschung, Berichtsband 75-CD. 2001.

[433] Mook G, Pohl J, Michel F. Non-destructive characterization of smart CFRP structures. Smart Material Structures. 2003;12:997–1004.

[434] Heuer H, Schulze MH. Eddy current testing of carbon fiber materials by high resolution directional sensors. In: Proceedings of the Conference NDT in Canada. 2011; p. 1–10.

[435] Menana H, Feliachi M. 3-D Eddy current computation in carbon-fiber reinforced composites. IEEE Transactions on Magnetics. 2009;45(3):1008–11.

[436] Menana H, Feliachi M. An integro-differential model for 3-D eddy current computation in carbon fiber reinforced polymer composites. IEEE Transactions on Magnetics. 2011;47(4):756–63.

[437] Cacciola M, Calcagno S, Megali G, *et al.* Eddy current modeling in composite materials. PIERS Online. 2009;5(6):591–55.

[438] Barbato L, Petrarca C, Rubinacci G, *et al.* Numerical models for composite materials in E-NDT. In: Electromagnetic Nondestructive Evaluation (XVII). Studies in Applied Electromagnetics and Mechanics. IOS Press; 2014. p. 47–54.

[439] Barbato L, Minucci S, Rubinacci G, *et al.* Numerical behavior of models of composite materials in E'NDT at "low" frequencies. In: Electromagnetic Nondestructive Evaluation (XVII). Studies in Applied Electromagnetics and Mechanics. IOS Press; 2014. p. 77–84.

[440] He Y, Tian G, Pan M, *et al.* Non-destructive testing of low-energy impact in CFRP laminates and interior defects in honeycomb sandwich using scanning pulsed eddy current. Composites Part B: Engineering. 2014;59:196–203.

[441] Li X, Yin W, Liu Z, *et al.* Characterization of carbon fibre reinforced composite by means of non-destructive Eddy current testing and FEM modeling. In: Proceedings of the 17th World Conference on Nondestructive Testing. 2008; p. 1–6.

[442] Bui HK, Wasselynck G, Trichet D, *et al.* Thermo inductive nondestructive testing method applied to CFRP. COMPEL. 2013;33(1–2):167–80.

[443] Bui HK, Wasselynck G, Trichet D, *et al.* Application of degenerated hexahedral Whitney elements in the modeling of NDT induction thermography of laminated CFRP composite. IEEE Transactions on Magnetics. 2016;52(3):1–4.

[444] Thomas WM, Nicholas ED, Needham JC, *et al.*, inventors; W. M. Thomas, assignee. Friction stir butt welding (International Patent Application). PCT/GB92/02203; 1991.

[445] Dawes CJ, Thomas WM. Friction stir process welds aluminum alloys. Welding Journal. 1996;75(3):41–45.

[446] Mishra RS, Ma ZY. Friction stir welding and processing. Materials Science and Engineering: R: Reports. 2005;50(1–2):1–78.

[447] Mishra RS, De PS, Kumar N. Friction Stir Welding and Processing: Science and Engineering. Cham, Switzerland: Springer International Publishing; 2014.

[448] Reddy NR, Reddy GM. Friction stir welding of aluminium alloys – A review. International Journal of Mechanical Engineering and Technology. 2016;7(2):73–80.

[449] Voellner G. Rührreibschweißen mit Schwerlast-Industrierobotern. Forschungsberichte IWB. München: Herbert Utz Verlag; 2010.

[450] Kim YG, Fujii H, Tsumura T, *et al.* Three defect types in friction stir welding of aluminum die casting alloy. Materials Science and Engineering: A. 2006;415(1–2):250–54.

[451] Pietras A, Węglowski MS. Imperfections in FSW joints and NDT methods of their detection. Biuletyn Instytutu Spawalnictwa w Gliwicach. 2014;58(2):23–32.

[452] Ruehrreibschweißen - Aluminium - Teil 5: Qualitaets- und Pruefungsanforderungen. German Institute for Standardization; 2015.

[453] American Welding Society. Specification of Friction Stir Welding of Aluminum Alloys for Aerospace.

[454] dos Santos TG, Ramos PM, dos Santos Vilaca P. Non destructive testing of friction stir welding: comparison of planar eddy current probes. Proceedings of the 16th IMEKO – TC4 Symposium. 2008; p. 507–12.

[455] Rosado LS, Santos TG, Piedade M, *et al.* Advanced technique for non-destructive testing of friction stir welding of metals. Measurement. 2010;43(8):1021–30.

[456] Pitkänen J, Haapalainen J, Lipponen A, *et al.* NDT of friction stir welding PLFW1 to PLFW5 (FSWL98, FSWL100, FSWL101, FSWL102, FSWL103) NDT data report. Working Report POSIVA OY. 2014.

[457] Yashan A. Über die Wirbelstromprüfung und magnetische Streuflussprüfung mittels GMR-Sensoren: Dissertation [Dissertation]. Universität des Saarlandes. Saarbrücken; 2008.

[458] Wolter B, Jene T, Conrad C, *et al.* New NDT – approaches to monitor the friction stir welding process and to inspect the welding quality. In: Proceedings of the 4th International Conference – Innovative Technologies for Joining Advanced Materials. 2010; p. 130–35.

[459] Mandache C, Levesque D, Dubourg L, *et al.* Non-destructive detection of lack of penetration defects in friction stir welds. Science and Technology of Welding and Joining. 2013;17(4):295–303.

[460] Santos TG, Vilaca P, Rosado L, *et al.* Developments in NDT of friction stir welding using Eddy currents. In: Proceedings European Conference on Nondestructive Testing (ECNDT). 2010; p. 88–89.

[461] Shi Y, Zhang C, Li R, *et al.* Theory and application of magnetic flux leakage pipeline detection. Sensors. 2015;15(12):31036–55.

[462] Roemer U, Schoeps S, Weiland T. Approximation of moments for the nonlinear magnetoquasistatic problem with material uncertainties. IEEE Transactions on Magnetics. 2014;50(2):417–20.

Index

www.ingramcontent.com/pod-product-compliance
Lightning Source LLC
LaVergne TN
LVHW011842200926
848484LV00002B/388